Lecture Notes in Computer Science 8497

Commenced Publication in 1973
Founding and Former Series Editors:
Gerhard Goos, Juris Hartmanis, and Jan van Leeuwen

Editorial Board

Jianer Chen   John E. Hopcroft
Jianxin Wang (Eds.)

Frontiers in Algorithmics

8th International Workshop, FAW 2014
Zhangjiajie, China, June 28-30, 2014
Proceedings

 Springer

Volume Editors

Jianer Chen
Texas A&M University
Department of Computer Science and Engineering
College Station, TX 77843-3112, USA
E-mail: chen@cs.tamu.edu

John E. Hopcroft
Cornell University
Computer Science Department
5144 Upson Hall, Ithaca, NY 14853, USA
E-mail: jeh@cs.cornell.edu

Jianxin Wang
Central South University
School of Information Science and Engineering
Changsha 410083, China
E-mail: jxwang@mail.csu.edu.cn

ISSN 0302-9743   e-ISSN 1611-3349
ISBN 978-3-319-08015-4   e-ISBN 978-3-319-08016-1
DOI 10.1007/978-3-319-08016-1
Springer Cham Heidelberg New York Dordrecht London

Library of Congress Control Number: Applied for

LNCS Sublibrary: SL 1 – Theoretical Computer Science and General Issues

Typesetting: Camera-ready by author, data conversion by Scientific Publishing Services, Chennai, India

Printed on acid-free paper

Springer is part of Springer Science+Business Media (www.springer.com)

Preface

The 8th International Frontiers of Algorithmics Workshop (FAW 2014) was held during June 28–30, 2014, at Zhangjiajie, China. The workshop brings together researchers working on all aspects of computer science for the exchange of ideas and results.

FAW 2014 was the eighth conference in the series. The previous seven meetings were held during August 1–3, 2007, in Lanzhou, June 19–21, 2008, in Changsha, June 20–23, 2009, in Hefei, August 11–13, 2010, in Wuhan, May 28–31, 2011, in Jinhua, May 14–16, 2012, in Beijing, and June 26–28, 2013, in Dalian. FAW is already playing an important regional and international role, promising to become a focused forum on current research trends on algorithms seen throughout China and other parts of Asia.

In total, 65 submissions were received from more than 12 countries and regions. The FAW 2014 Program Committee selected 30 papers for presentation at the conference. In addition, we had three plenary speakers, John Hopcroft (Cornell University, USA), Ming Li (University of Waterloo, Canada), and Ying Xu (University of Georgia, USA). We thank them for their contributions to the conference and proceedings.

We would like to thank the Program Committee members and external reviewers for their hard work in reviewing and selecting papers. We also are very grateful to all authors who submitted their work to FAW 2014.

We wish to thank the editors at Springer and the local organizing chairs for their hard work and cooperation through the preparation of this conference.

Finally, we would like to thank the conference sponsors, Central South University, China, and the National Science Foundation of China.

June 2014

Jianer Chen
John Hopcroft
Jianxin Wang

Organization

Program Committee

Marat Arslanov	Kazan State University, Russia
Hans Bodlaender	Utrecht University, The Netherlands
Yixin Cao	Hungarian Academy of Sciences, Hungary
Xi Chen	Columbia University, USA
Marek Chrobak	University of California, Riverside, USA
Barry Cooper	University of Leeds, UK
Qilong Feng	Central South University, China
Henning Fernau	University of Trier, Germany
Mordecai Golin	Hong Kong University of Science and Technology China
Gregory Gutin	University of London, UK
Tomio Hirata	Nagoya University, Japan
Hiro Ito	University of Electro-Communications, Japan
Klaus Jansen	University of Kiel, Germany
Iyad Kanj	DePaul University, USA
Ming-Yang Kao	Northwestern University, USA
Naoki Katoh	Kyoto University, Japan
Michael Langston	University of Tennessee, USA
Guohui Lin	University of Alberta, Canada
Tian Liu	Peking University, China
Daniel Lokshtanov	University of California San Diego, USA
Dániel Marx	Hungarian Academy of Sciences, Hungary
Venkatesh Raman	Institute of Mathematical Sciences, India
Peter Rossmanith	RWTH Aachen University, Germany
Ulrike Stege	University of Victoria, Canada
Xiaoming Sun	Institute of Computing Technology, CAS, China
Gerhard Woeginger	Eindhoven University of Tech., The Netherlands
Ge Xia	Lafayette College, USA
Ke Xu	Beihang University, China
Boting Yang	University of Regina, Canada
Binhai Zhu	Montana State University, USA

Program Chairs

Jianer Chen	Texas A&M University, USA and Central South University, China
John Hopcroft	Cornell University, USA
Jianxin Wang	Central South University, China

Organizing Chairs

Jianxin Wang	Central South University, China
Qingping Zhou	Jishou University, China

Organizing Committee

Qilong Feng	Central South University, China
Mingxing Zeng	Jishou University, China
Yu Sheng	Central South University, China
Guihua Duan	Central South University, China
Li Wang	Jishou University, China
Yanping Yang	Jishou University, China

Sponsoring Institutions

Central South University, China
The National Natural Science Foundation of China

External Reviewers

René Van Bevern	Charles Phillips
Hans-Joachim Boeckenhauer	Pawel Pralat
Kevin Buchin	Sasanka Roy
Tatiana Gutiérrez Bunster	Saket Saurabh
Viktor Engelmann	Nitin Saurabh
Takuro Fukunaga	Kazuhisa Seto
Jiong Guo	Hadas Shachnai
Ronald Hagan	Bin Sheng
Leo van Iersel	Jack Snoeyink
Mark Jones	Takeyuki Tamura
M Kaluza	Junichi Teruyama
Kim-Manuel Klein	Weitian Tong
Stefan Kraft	Akihiro Uejima

Felix Land
Xingwu Liu
Marten Maack
Mehrdad Mansouri
Catherine Mccartin
Benjamin Niedermann
Rolf Niedermeier
Sang-Il Oum
Vinayak Pathak

Fernando Sanchez Villaamil
Kai Wang
Mathias Weller
Sue Whitesides
Meirav Zehavi
Jialin Zhang
Huili Zhang
Ruben van der Zwaan

Invited Talks

Expanding Algorithmic Research at FAW

John Hopcroft

Cornell University

Abstract. This talk explores a wide range of topics in the algorithmic area to expand the scope of FAW. Some of the topics are why the 1-norm often works as a proxy for the 0-norm, how a quadratic constraint can be formulated as a matrix being semi definite, dilation equations where a mapping changes all distances by a fixed scale factor. It is hoped that the talk will lead to some exciting results at next years FAW conference.

Applying Spaced Seeds Beyond Homology Search

Ming Li

University of Waterloo

Abstract. Optimal spaced seeds were invented in bioinformatics and are playing a key role in modern homology search systems. They helped to improve sensitivity of homology search in DNA sequences while not decreasing the search speed. Can this beautiful idea be applied to other fields or problems? We look at several situations (including financial market prediction, image search, and internet search) where the spaced seeds sometimes apply but some other times do not apply.

Table of Contents

Contributed Papers

Broadcast Problem in Hypercube of Trees

Puspal Bhabak and Hovhannes A. Harutyunyan

Department of Computer Science and Software Engineering
Concordia University
Montreal, QC, H3G 1M8, Canada

Abstract. *Broadcasting* is an information dissemination problem in a
connected network in which one node, called the *originator*, must distribute a message to all other nodes by placing a series of calls along the
communication lines of the network. Every time the informed nodes aid
the originator in distributing the message. Finding the broadcast time of
any vertex in an arbitrary graph is NP-complete. The polynomial time
solvability is shown only for certain graphs like trees, unicyclic graphs,
tree of cycles, necklace graphs, fully connected trees and tree of cliques.
In this paper we study the broadcast problem in a hypercube of trees
for which we present a 2-approximation algorithm for any originator. We
also provide a linear algorithm to find the broadcast time in hypercube
of trees with one tree.

1   Introduction

In today's world, due to massive parallel processing, processors have become
faster and more efficient. In recent years, a lot of work has been dedicated to
studying properties of interconnection networks in order to find the best communication structures for parallel and distributed computing. One of the main
problems of information dissemination investigated in this research area is broadcasting. The broadcast problem is one in which the knowledge of one processor
must spread to all other processors in the network. For this problem we can view
any interconnection network as a connected undirected graph $G = (V, E)$, where
V is the set of vertices (or processors) and E is the set of edges (or communication lines) of the network.

Formally, *broadcasting* is the message dissemination problem in a connected
network in which one informed node, called the *originator*, must distribute a
message to all other nodes by placing a series of calls along the communication
lines of the network. Every time the informed nodes aid the originator in distributing the message. This is assumed to take place in discrete time units. The
broadcasting is to be completed as quickly as possible subject to the following
constraints: (1) Each call requires one unit of time. (2) A vertex can participate
in only one call per unit of time. (3) Each call involves only two adjacent vertices,
a sender and a receiver.

Given a connected graph G and a message originator, vertex u, the natural
question is to find the minimum number of time units required to complete

J. Chen, J.E. Hopcroft, and J. Wang (Eds.): FAW 2014, LNCS 8497, pp. 1–12, 2014.
© Springer International Publishing Switzerland 2014

broadcasting in graph G from vertex u. We define this number as the *broadcast time* of vertex u, denoted $b(u, G)$ or $b(u)$. The broadcast time $b(G)$ of the graph G is defined as $\max\{b(u)|u \in V\}$. It is easy to see that for any vertex u in a connected graph G with n vertices, $b(u) \geq \lceil \log n \rceil$ (all log's in the paper are base 2), since during each time unit the number of informed vertices can at most double. Determining $b(u)$ for an arbitrary originator u in an arbitrary graph G has been proved to be NP-complete in [19]. The problem remains NP-Complete even for 3-regular planar graphs [15]. The best theoretical upper bound is obtained by the approximation algorithm in [3] which produces a broadcast scheme with $O(\frac{\log(|V|)}{\log\log(|V|)} b(G))$ rounds. Research in [18] has showed that the broadcast time cannot be approximated within a factor $\frac{57}{56} - \epsilon$. However this result has been improved within a factor of $3 - \epsilon$ in [3]. As a result research has been made in the direction of finding approximation or heuristic algorithms to determine the broadcast time in arbitrary graphs (see [13], [1], [2], [3], [4], [5], [7], [14], [16], [17], [6], [12]).

Since the broadcast problem in general is very difficult, another direction is to design polynomial algorithms for some classes of graphs. The first result in this direction was a linear algorithm to determine the broadcast time of any tree [19]. Recent research shows that there are polynomial time algorithms for the broadcast problem in tree-like graphs where two cycles intersect at one vertex - tree of cycles, necklace graphs, or in graphs containing cliques, however with no intersecting cliques - fully connected trees and tree of cliques ([8], [9], [10], [11]). However, the problem remains NP-Hard for restricted classes of graphs. No other results are known in this area. The main reason is that the broadcasting problem becomes very difficult when two cycles intersect at more than one vertex.

In this paper we continue the work in [11] and consider broadcasting in a hypercube graph where each vertex of the hypercube is the root of a tree, called hypercube of trees. Although there is a simple minimum time broadcast scheme for hypercube but the problem is much more difficult for hypercube of trees because in hypercube any pair of vertices are not neighbors as in clique. However we were able to design a non-trivial algorithm to find the broadcast time of any originator for the hypercube of trees containing one tree. For the general case we present a linear time 2-approximation algorithm.

2   Hypercube of Trees

Assume that we have a hypercube graph where every vertex is the root of a tree. We will call the resulting graph hypercube of trees.

Definition 1. *The hypercube of dimension k, denoted by H_k, is a simple graph with vertices representing 2^k binary strings of length k, $k \geq 1$ such that adjacent vertices have binary strings differing in exactly one bit position.*

Definition 2. *Consider 2^k trees $T_i = (V_i, E_i)$ rooted at r_i where $1 \leq i \leq 2^k$. We define the hypercube of trees, $HT_{k,n} = (V, E)$, to be a graph where $V = V_1 \cup V_2 \cup ... \cup V_{2^k}$ and $E = E_1 \cup E_2 \cup ... \cup E_{2^k} \cup E_{H_k}$ where $E_{H_k} = \{(r_i, r_j)|\ r_i,$*

r_j are vertices of H_k}. The roots of the trees, r_i, will be called root vertices and the rest of the vertices will be called tree vertices (see Fig. 1).
$|V| = n \geq 2^k$ and $|E| = |V| - 2^k + k2^{k-1} = |V| + 2^{k-1}(k-2)$.

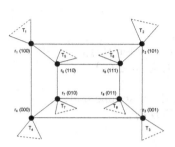

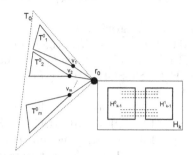

Fig. 1. $HT_{3,n}$ with 8 trees T_i rooted at r_i, $1 \leq i \leq 2^3$. The roots r_i include the hypercube H_3 as subgraph.

Fig. 2. Hypercube of Trees $HT_{k,n}$ with only tree T_0 rooted at r_0

3   Broadcasting in Hypercube of Trees Containing One Tree

As mentioned above to find the broadcast time in hypercube of trees is difficult in general. In this section we design a linear algorithm to determine the broadcast time of $HT_{k,n}$ containing one tree.

Let G_1 be a $HT_{k,n}$ graph where r_0 is a root vertex and r_0 is the root of a tree T_0. The remaining $2^k - 1$ root vertices do not contain any tree. Let us also assume that r_0 has m neighbors in T_0, vertices $v_1, v_2, ..., v_m$. v_i is the root of the subtree T_i^0, $1 \leq i \leq m$. Let us consider $b(v_i, T_i^0) = t_i$ and without loss of generality we assume that $t_1 \geq t_2 \geq ... \geq t_m$. Then it follows from [19] that $b(r_0, T_0) = \max\{i + t_i\}$, where $1 \leq i \leq m$. Let $b(r_0, T_0) = \tau$ and $\tau \geq 1$ (see Fig. 2).

3.1   Broadcast Algorithm When Originator Is r_0

Consider two cases depending on the relationship between τ and the dimension of the hypercube in G_1. Let all the root vertices will be informed by $b(r)$ time units. The algorithm A calls another algorithm Broadcast-Hypercube which returns $b(r)$. When a tree vertex is informed there is not much it can do other than following the well known broadcast algorithm in trees [19], called A_T.

Tree Broadcast Algorithm A_T:
INPUT: originator r_i and tree rooted at r_i: T_i
OUTPUT: Broadcast time $b_{A_T}(r_i, T_i)$
TREE-BROADCAST(r_i, T_i)

1. r_i informs a child vertex in T_i that has the maximum broadcast time in the subtree rooted at it.
2. Let $\alpha_1, ..., \alpha_f$ be the broadcast times of the f subtrees rooted at r_i and $\alpha_1 \geq ... \geq \alpha_f$. Then, $b_{A_T}(r_i, T_i) = \max\{j + \alpha_j\}$ for $1 \leq j \leq f$.

Broadcast Algorithm A:

INPUT: $HT_{k,n} = (V, E)$, originator r_0, $b(r_0, T_0) = \tau$, m, $t_1 \geq t_2 \geq ... \geq t_m$
OUTPUT: Broadcast time $b_A(r_0)$ and broadcast scheme for $HT_{k,n}$
BROADCAST-SCHEME-A($HT_{k,n}$, r_0, τ, m, $t_1 \geq t_2 \geq ... \geq t_m$)

1. If $\tau \leq k$
    1.1. r_0 informs another root vertex r_1 in the first time unit.
    1.2. $b(r) = $ BROADCAST-HYPERCUBE($HT_{k,n}$, r_1, 1).
    1.3. For each time unit $i = 2$ to $m + 1$
        1.3.1. r_0 informs tree vertex v_{i-1}.
2. If $\tau > k$
    2.1. If $\tau \geq k + m$
        2.1.1. For each time unit $i = 1$ to m
            2.1.1.1. r_0 informs tree vertex v_i.
        2.1.2. For each time unit $i = m + 1$ to $m + k$
            2.1.2.1. an informed root vertex informs another uninformed root vertex using any shortest path.
    2.2. If $k + m > \tau \geq k + m_1$, where $1 \leq m_1 < m$
        2.2.1. For each time unit $i = 1$ to $m_1 - 1$
            2.2.1.1. r_0 informs tree vertex v_i.
        2.2.2. At time unit m_1, r_0 informs another root vertex r_1.
        2.2.3. $b(r) = $ BROADCAST-HYPERCUBE($HT_{k,n}$, r_1, m_1).
        2.2.4. For each time unit $i = m_1 + 1$ to $m + 1$
            2.2.4.1. r_0 informs tree vertex v_{i-1}.
3. TREE-BROADCAST(v_i, T_i^0) for $1 \leq i \leq m$.

Broadcast-Hypercube:

INPUT: $HT_{k,n} = (V, E)$, originator r_1, time at which r_1 is informed: t_{r_1}
OUTPUT: $b(r)$
BROADCAST-HYPERCUBE($HT_{k,n}$, r_1, t_{r_1})

1. Assume r_1 is 10...0 (last $k - 1$ bits consist of zeroes)
2. For each time unit $i = t_{r_1} + 1$ to $t_{r_1} + k - 1$
    2.1. For all $a_1, ..., a_{i-t_{r_1}-1} \in \{0, 1\}$ do in parallel
        2.1.1. $1a_1...a_{i-t_{r_1}-1}00...0$ sends to $1a_1...a_{i-t_{r_1}-1}10...0$
3. For all $a_1, ..., a_{k-1} \in \{0, 1\}$ except r_1 do in parallel
    3.1. $1a_1...a_{k-1}$ sends to $0a_1...a_{k-1}$
4. Return $t_{r_1} + k$

Complexity Analysis

Broadcast-Hypercube takes $O(\log 2^k) = O(k)$ time to inform the root vertices. Algorithm A: Steps 1.1 and 1.3 take constant time to run. Step 2.1.2 can be completed in $O(k)$ time. Also steps 2.1.1 and 2.2 run in constant time. Again, the tree broadcast algorithm in step 3 takes $O(|V| - 2^k) = O(|V_T|)$ time to run, where $|V_T|$ is the number of tree vertices in G_1. Thus, complexity of algorithm is $O(|V_T| + k)$.

Theorem 1. *Algorithm A always generates the minimum broadcast time $b(r_0)$.*

Proof. Case 1: $m \leq \tau \leq k$
At least k time units are necessary to inform all the root vertices of G_1. Since r_0 is the root of the tree T_0, at least one more time unit is required to broadcast a tree vertex in T_0. So, $b(r_0) \geq k+1$. Under algorithm A, the subroutine Broadcast-Hypercube informs the root vertices by time $k + 1$. Since starting at time two onwards, r_0 informs the adjacent tree vertices in T_0, hence $b_A(r_0, T_0) = \tau + 1 \leq k + 1$ (as $\tau \leq k$). So, $b_A(r_0) = b(r_0) = k + 1$.

Case 2: $\tau > k$

SubCase 2.1: $\tau \geq k + m$
At least τ time units are necessary to inform all the tree vertices of G_1. So, $b(r_0) \geq \tau$. Under algorithm A, r_0 first informs all the adjacent tree vertices. So all the vertices in T_0 will receive the message by time τ. Starting at time $m + 1$ onwards, r_0 informs the root vertices. Since it takes exactly k time units to inform all the root vertices, hence the root vertices will be informed by $m + k$ time units and $m + k \leq \tau$. So, $b_A(r_0) = b(r_0) = \tau$

SubCase 2.2: $k + m > \tau \geq k + m_1$, where $1 \leq m_1 < m$
Note that initially r_0 is the only informed vertex from which all other root vertices can receive the message. Let us assume under any broadcast scheme, r_0 informs another root vertex r_1 at time l_r, where $1 \leq l_r \leq m + 1$. We know $\tau = \max\{i + t_i\}$ for $1 \leq i \leq m$ where t_i is the broadcast time of the m subtrees rooted at r_0. Let us consider j is the largest index such that $\tau = t_j + j$ for $1 \leq j \leq m$ and $b(r_0, H_k) \leq \tau$.

(i) if $l_r > j$: Under any broadcast scheme, r_0 informs $v_1,..., v_j,..., v_{l_r-1}, r_1, v_{l_r},..., v_m$ at time units $1, ..., j, ..., l_r - 1, l_r, l_r + 1, ..., m + 1$ respectively. Thus, $b(r_0, T_0) = \max\{t_1 + 1, ..., t_{l_r-1} + l_r - 1, t_{l_r} + l_r + 1, ..., t_m + m + 1\}$. It is clear that all the subtrees $T_1^0, ..., T_{l_r-1}^0$ will be informed by time τ. Since, j is the largest index such that $\tau = t_j + j$ and $l_r > j$, hence $t_i + i < \tau \Rightarrow t_i + i + 1 \leq \tau$ for $l_r \leq i \leq m$. So, $b(r_0) = \tau$ as $b(r_0, H_k) \leq \tau$.
(ii) if $1 \leq l_r \leq j$: Similar to the proof of (i), the subtrees $T_1^0, ..., T_{l_r-1}^0$ will be informed by time τ. Since $1 \leq l_r \leq j$, under any broadcast scheme, r_0 informs $v_{l_r},..., v_j,..., v_m$ at time units $l_r + 1, ..., j + 1, ..., m + 1$ respectively. Also, j is the largest index such that $\tau = t_j + j$. Thus, $t_i + i \leq \tau \Rightarrow t_i + i + 1 \leq \tau + 1$ for $l_r \leq i \leq j - 1$ and $t_j + j + 1 = \tau + 1$. Similarly, $t_i + i + 1 \leq \tau$ for $j + 1 \leq i \leq m$. So, $b(r_0, T_0) = b(r_0) = \tau + 1$ as $b(r_0, H_k) \leq \tau$.

Under algorithm A, r_1 receives the message at time m_1. The subroutine Broadcast-Hypercube informs the root vertices by time $k + m_1 \leq \tau$. r_0 informs its adjacent tree vertices $v_1, v_2, ..., v_{m_1-1}, v_{m_1}, ..., v_m$ at time units $1, 2, ..., m_1 -$

$1, m_1 + 1, ..., m + 1$ respectively. As a result, $b_A(r_0, T_0) = \max\{t_1 + 1, t_2 + 2, ..., t_{m_1-1} + m_1 - 1, t_{m_1} + m_1 + 1, ..., t_m + m + 1\} \leq \tau + 1$ as $\tau = \max\{t_i + i\}$ for $1 \leq i \leq m$.

So, $\tau \leq b_A(r_0) \leq \tau + 1$.   □

In the next section we will develop a broadcast algorithm for any originator in an arbitrary hypercube of trees, G_1. First we assume that the originator is any root vertex other than r_0. Finally we will discuss the broadcast algorithm in G_1 when the originator is any tree vertex.

3.2   Broadcasting from a Root Vertex Other Than r_0

In this section we present the broadcast algorithm A_r for graph G_1 when the originator is any root vertex (say r_j) other than r_0. Let us assume that r_j is at a distance d_r from vertex r_0, where $k \geq d_r \geq 1$. The algorithm A_r in G_1 starts by informing along the path $\overline{r_j r_0}$ (the shortest among all paths between r_j and r_0). r_0 receives the message at time d_r, and then it sends the message to the tree attached to it.

Broadcast Algorithm A_r:

INPUT: $HT_{k,n} = (V, E)$, originator r_j, $b(r_0, T_0) = \tau$
OUTPUT: Broadcast time $b_{A_r}(r_j)$ and broadcast scheme for $HT_{k,n}$
BROADCAST-SCHEME-$A_r(HT_{k,n}, r_j, \tau)$

1. r_j informs along the path $\overline{r_j r_0}$ (the shortest among all paths between r_j and r_0) in the first time unit.
2. r_j continues to inform the other root vertices using any shortest path.
   r_0 receives the message at time d_r.
3. TREE-BROADCAST(r_0, T_0).

Complexity Analysis

Steps 1 and 2 can be completed in $O(k)$ time. The tree broadcast algorithm in step 3 takes $O(|V_T|)$time to run. Complexity of algorithm is $O(|V_T| + k)$.

Theorem 2. *Algorithm A_r always generates the minimum broadcast time $b(r_j)$.*

Proof. Under algorithm A_r, r_0 receives the message at time d_r. Starting at time $d_r + 1$ onwards, r_0 informs the adjacent tree vertices. As a result all the vertices of T_0 will be informed by time $\tau + d_r$. Since r_0 does not play any role in informing a root vertex, it will take at most $k + 1$ time units for all the root vertices in G_1 to receive the message.

Case 1: $\tau + d_r \leq k + 1$
Algorithm A_r in this case generates $b_{A_r}(r_j) = k + 1$.
Under any broadcast scheme, at least k time units are necessary to inform all the root vertices of G_1. Since r_0 is the root of the tree T_0, at least one more time unit is required to broadcast a tree vertex in T_0. So, $b(r_j) \geq k + 1$.

Case 2: $\tau + d_r > k + 1$
Algorithm A_r in this case generates $b_{A_r}(r_j) = \tau + d_r$.

Under any broadcast scheme, r_0 is informed no earlier than d_r time units. It takes another τ time units to inform all the tree vertices in T_0. So, $b(r_j) \geq \tau + d_r$. □

3.3   Broadcasting from a Tree Vertex

In this section we will develop a broadcast algorithm from any tree vertex in an arbitrary hypercube of trees G_1. Assume we are given a graph G_1 such that the originator v is in the subtree T_i^0 rooted at the root vertex r_0. There is a unique path P in T_i^0 connecting r_0 to the originator v. The vertex on the path P adjacent to r_0 is denoted by v_i. Let $u_1, u_2, ..., u_z$ be the z neighbors of v in the subtree. One of these vertices falls on the path P, call this vertex u_i. As shown in Fig. 3, the graph G_1 can be restructured by drawing the tree T_i^0 rooted at the originator v and vertex v_i as one of its nodes. It can be observed that the remaining subgraph of G_1, denoted by G_1' is attached to T_i^0 by a bridge (v_i, r_0). Since the graph G_1' is connected to tree T_i^0 by a bridge, the broadcast algorithm

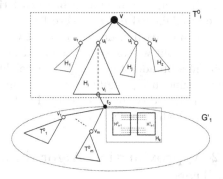

Fig. 3. Hypercube of Trees G_1 with originator v. The subtree T_i^0 is separated from rest of the graph G_1'.

in G_1' is independent of the broadcast algorithm in T_i^0. Once vertex r_0 is informed from v_i, it cannot inform any other vertex in T_i^0. It can only inform the vertices in G_1' in the minimum possible time unit. However, since r_0 is a root vertex and G_1' contains the hypercube H_k as its subgraph, broadcast in G_1' from r_0 can be considered as the broadcast problem in hypercube of trees with one tree rooted at r_0 where the originator is the root vertex r_0. We have a broadcast algorithm to solve this problem in G_1' from r_0. Let $T_{G_1'}$ be the broadcast tree of G_1' from r_0 obtained from algorithm A and τ_{m-1} be the broadcast time of r_0 in the remaining $m-1$ subtrees.

Broadcast Algorithm A_v:
INPUT: G_1', originator v in subtree T_i^0, r_0, τ_{m-1}, $m-1$, $t_1 \geq t_2 \geq ... \geq t_{m-1}$
OUTPUT: Broadcast time $b_{A_v}(v)$ and broadcast scheme for G_1
BROADCAST-SCHEME-$A_v(G_1', v, T_i^0, \tau_{m-1}, m-1, r_0, t_1 \geq t_2 \geq ... \geq t_{m-1})$

1. $T_{G'_1}$ = BROADCAST-SCHEME-A$(G'_1, r_0, \tau_{m-1}, m-1, t_1 \geq t_2 \geq ... \geq t_{m-1})$

2. Attach $T_{G'_1}$ with T_i^0 by the bridge (r_0, v_i) and let the resulting tree be labelled as T_v.

3. TREE-BROADCAST(v, T_v).

Complexity Analysis: Finding the broadcast time of a tree vertex in an arbitrary hypercube of trees with one tree is equivalent to solving two problems: (1) Finding the broadcast time of a root vertex in a hypercube of trees with one tree. As discussed before the complexity of this algorithm is linear. (2) Finding the broadcast time of a tree vertex in a tree. The complexity of this algorithm is also linear. Hence, the complexity is $O(|V|)$.

Proof of Correctness: We can use the optimal broadcast tree of G'_1 obtained from algorithm A and attach it to the tree T_i^0 and solve the broadcast problem in the resulting tree. According to the broadcast algorithm in trees [19], v informs a child vertex that has the maximum broadcast time in the subtree rooted at it. The subtrees are labelled by H_j, where $1 \leq j \leq z$ (see Fig. 3). The broadcast times $b(u_j, H_j)$ can be easily calculated except when $u_j = u_i$, since in this case G'_1 is attached to v_i. But we can solve the broadcast problem in G'_1 for the originator r_0 and obtain a broadcast tree $T_{G'_1}$. The weight of r_0 will then be initialized as the broadcast time in $T_{G'_1}$, call this $\tau_{G'}$. The optimal time required to inform all the vertices of H_i and G'_1 from u_i is equal to the broadcast time in the tree $H_i + (v_i, r_0)$ from u_i where weight$(r_0) = \tau_{G'}$.

4 Linear Time 2-Approximation Algorithm in General Hypercube of Trees

In this section we will study the broadcast problem in general hypercube of trees.

4.1 Lower Bound on the Broadcast Time

First we assume the originator is any root vertex.

Lemma 1. *Let G be a $HT_{k,n}$ where the originator r_0 is a root vertex. If $b(r_i, T_i)$ is the broadcast time of the root vertex r_i in the tree T_i where $0 \leq i \leq 2^k - 1$ then,*
i) $b(r_0) \geq max\{b(r_i, T_i)\}$ ii) $b(r_0) \geq k$.

Proof. (i): Under any broadcast scheme, it takes at least $max\{b(r_i, T_i)\}$ time units to inform all the vertices of G. Hence, $b(r_0) \geq max\{b(r_i, T_i)\}$.
(ii) goes as follows: At least k time units are necessary to inform all the root vertices of G. So, $b(r_0) \geq k$. □

Observations: 1. $b(r_0) = k$ when no trees are attached in $HT_{k,n}$. 2. Consider a graph $HT_{k,n}$ where only one tree is being attached at the originator r_0. The tree is a path P of length $l \geq k + 1$. r_0 first informs along P and then informs the other root vertices. It is easy to see that the root vertices will be informed by $k + 1$ time units. Similarly all the vertices in P will receive the message by $l \geq k+1$ time units. Thus, $b(r_0) = max\{b(r_i, T_i)\}$. Therefore, both lower bounds from Lemma 1 are achievable.

Let us now consider the originator is any tree vertex w in a tree T_i, where $0 \leq i \leq 2^k - 1$. Let us assume that w is at a distance d from the nearest root vertex r_0, where $d \geq 1$ (see Fig. 4).

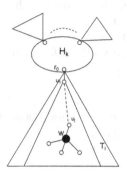

Fig. 4. Hypercube of Trees G where the originator is a tree vertex w

Lemma 2. *Let G be a $HT_{k,n}$ where the originator w is any tree vertex in a tree T_i and the length of the path $\overline{wr_0}$ is d, where $d \geq 1$. If $b(w, T_i)$ is the broadcast time of the tree T_i from w for $0 \leq i \leq 2^k - 1$ then,*
i) $b(w) \geq b(w, T_i)$ ii) $b(w) \geq max\{b(r_j, T_j)\}$ for all $j \neq i$ iii) $b(w) \geq k + d$.

Proof. (i): w is any tree vertex in a tree T_i for $0 \leq i \leq 2^k - 1$ and it takes at least $b(w, T_i)$ time units to inform all the vertices of G. Hence, $b(w) \geq b(w, T_i)$.
(ii): For the remaining $2^k - 1$ trees T_j, where $0 \leq j \leq 2^k - 1$ & $j \neq i$, under any broadcast scheme, initially r_j is the only informed vertex from which the tree vertices can receive the message. Therefore at least $max\{b(r_j, T_j)\}$ time units are necessary to broadcast in all the vertices. Hence, $b(w) \geq max\{b(r_j, T_j)\}$ for all $j \neq i$.
Proof of (iii): Under any broadcast scheme, r_0 is informed no earlier than d time units. It takes at least another k time units to inform all the root vertices of G. So, $b(w) \geq k + d$.   □

4.2   Approximation Algorithm

In this section we present the broadcast algorithm S for graph G. We consider any vertex x to be the originator. When the originator is r_0 then the algorithm S in G starts by informing all the vertices of the hypercube. When all the root vertices are informed, each vertex informs the tree attached to it.

When the originator is w then the algorithm S in G starts by informing along the path $\overline{wr_0}$. r_0 receives the message at time d. During the next k time units all the vertices of the hypercube are being informed. Each root vertex will now send the message to the tree attached to it.

Approximation Algorithm S:

INPUT: $HT_{k,n} = (V, E)$ and any originator x

OUTPUT: Broadcast time $b_S(x)$ and broadcast scheme for $HT_{k,n}$

BROADCAST-SCHEME-$S(HT_{k,n}, x)$

1. If $x = w$
    1.1. w broadcasts along the shortest path $\overline{wr_0}$ in the first time unit. r_0 gets informed at time d.
    1.2. For each time unit $i = d + 1$ to $d + k$
        1.2.1. an informed root vertex informs another uninformed root vertex using any shortest path.
2. If $x = r_0$
    2.1. For each time unit $i = 1$ to k
        2.1.1. an informed root vertex informs another uninformed root vertex using any shortest path.
3. TREE-BROADCAST(r_i, T_i) for $0 \leq i \leq 2^k - 1$.

Complexity Analysis

Steps 1.2 and 2.1 take $O(k)$ time to inform the root vertices. In steps 1.1 and 3, the tree broadcast algorithm takes $O(|V_T|)$ time to run. Complexity of algorithm S is $O(|V_T| + k)$.

Theorem 3. *Algorithm S is a 2-approximation for any originator in the graph $HT_{k,n}$*

Proof. **When originator is r_0:** Considering algorithm S, an upper bound on broadcast time can be obtained when the tree T_i with broadcast time max $\{b(r_i, T_i)\}$ is being attached at the root vertex r_i which is at distance $\log 2^k = k$ from originator r_0, where $0 \leq i \leq 2^k - 1$. By time k all the root vertices will be informed. Each root vertex r_i will take $b(r_i, T_i)$ time units to broadcast in T_i. As a result all the tree vertices will be informed by time $\max\{b(r_i, T_i)\}$. Thus, $b_S(r_0) \leq k + \max\{b(r_i, T_i)\}$, where $0 \leq i \leq 2^k - 1$. Combining Lemma 1(i) and Lemma 1(ii) we can write $b(r_0) \geq \frac{1}{2}(\max\{b(r_i, T_i)\} + k)$. Hence, $\frac{b_S(r_0)}{b(r_0)}$ $\leq 2\frac{max\{b(r_i, T_i)\} + k}{max\{b(r_i, T_i)\} + k} = 2$.

When originator is w: w is any tree vertex in T_i and it is at a distance d from the nearest root vertex r_0. Considering algorithm S, an upper bound on broadcast time can be obtained when the tree T_j with broadcast time max $\{b(r_j, T_j)\}$ is being attached at the root vertex r_j which is at distance k from r_0, where $0 \leq j \leq 2^k - 1$ & $i \neq j$. r_0 receives the message at time d. By time $d + k$ all the root vertices will be informed. Each root vertex r_j will take $b(r_j, T_j)$ time units to broadcast in T_j.

Let $u_1, u_2, .., u_q$ be the q neighbors of w in T_i. One of these vertices falls on the path $\overline{wr_0}$, call this vertex u_r. u_j is the root of the subtree T_i^j, $1 \leq j \leq q$ and $b(u_j, T_i^j) = b_j$. If $b_1 \geq b_2 \geq ... \geq b_q$, then it follows from [19] that $b(w, T_i) = \max\{j + b_j\}$, where $1 \leq i \leq q$. Under algorithm S, w informs $u_r, u_1, ..., u_{r-1}, u_{r+1}, u_q$ at time units $1, 2, ..., r, r + 1, q$ respectively. If $b(w, T_i) = j + b_j$ for any $1 \leq j \leq r - 1$, then $b_S(w, T_i) \leq b(w, T_i) + 1$.

Let $u_{r_1}, u_{r_2}, .., u_{r_p}$ be the p neighbors of u_r in T_i^r and let u_{r_r} be the vertex that falls on the path $\overline{wr_0}$. Similarly if b_{r_j}, where $1 \leq j \leq p$ be the broadcast times of p subtrees rooted at u_{r_j} and $b_{r_1} \geq ... \geq b_{r_p}$, then $b(u_r, T_i^r) = \max\{j + b_{r_j}\}$. Under algoithm S, u_r informs $u_{r_r}, u_{r_1}, ..., u_{r_{r-1}}, u_{r_{r+1}}, ..., u_{r_p}$ at time units $1, 2, ..., r, r+1, ..., p$ respectively (time units are considered after u_r is informed). If $b(u_r, T_i^r) = j + b_{r_j}$ for any $1 \leq j \leq r - 1$, then $b_S(u_r, T_i^r) \leq b(u_r, T_i^r) + 1$. Thus, $b_S(w, T_i^r) \leq 1 + b(u_r, T_i^r) + 1 = b(w, T_i^r) + 1$. Since the path $\overline{wr_0}$ has been given the priority in algorithm S, similarly in the worst case, at every level (upto d levels) the broadcast time of the subtrees will be delayed by one time unit. Therefore, $b_S(w, T_i) \leq b(w, T_i) + d$. As a result all the tree vertices in G will be informed by time $\max\{b(w, T_i) + d, k + d + max\{b(r_j, T_j)\}\}$.

If $b(w, T_i) + d \geq \max\{b(r_j, T_j)\} + k + d$, then $b_S(w) \leq b(w, T_i) + d$. Combining Lemma 2(i) and Lemma 2(iii) we can write $b(w) \geq \frac{1}{2}(b(w, T_i) + k + d)$. Hence, $\frac{b_S(w)}{b(w)} \leq 2\frac{b(w, T_i) + d}{b(w, T_i) + k + d} < 2$ as $k \geq 1$.

If $b(w, T_i) + d < \max\{b(r_j, T_j)\} + k + d$, then $b_S(w) \leq k + d + \max\{b(r_j, T_j)\}$. Combining Lemma 2(ii) and Lemma 2(iii) we can write $b(w) \geq \frac{1}{2}(\max\{b(r_j, T_j)\} + k + d)$. Hence, $\frac{b_S(w)}{b(w)} \leq 2\frac{max\{b(r_j, T_j)\} + k + d}{max\{b(r_j, T_j)\} + k + d} = 2$.   □

5   Conclusion

The only known result for such graphs is the $O(n \log \log n)$ algorithm for fully connected trees and tree of cliques. In this paper we study broadcasting in hypercube of trees. The algorithm for fully connected trees and tree of cycles can not be applied to hypercube of trees because the non-tree vertices in hypercube of trees are not connected. We presented a linear time algorithm to find the broadcast time from any originator for hypercube of trees containing one tree. For the general case we presented a 2-approximation algorithm to find the broadcast time from any originator. The two main directions for future work are proving the NP-completeness or designing a polynomial algorithm for the broadcast problem in hypercube of trees.

References

1. Bar-Noy, A., Guha, S., Naor, J., Schieber, B.: Multicasting in heterogeneous networks. In: Proceedings of the Thirtieth Annual ACM Symposium on Theory of Computing (STOC 1998), pp. 448–453 (1998)

2. Beier, R., Sibeyn, J.F.: A powerful heuristic for telephone gossiping. In: Proceedings of the 7th International Colloquium on Structural Information Communication Complexity (SIROCCO 2000), pp. 17–36 (2000)
3. Elkin, M., Kortsarz, G.: Combinatorial logarithmic approximation algorithm for directed telephone broadcast problem. In: Proceedings of the Thirty-fourth Annual ACM Symposium on Theory of Computing (STOC 2002), pp. 438–447 (2002)
4. Elkin, M., Kortsarz, G.: Sublogarithmic approximation for telephone multicast: path out of jungle (extended abstract). In: Proceedings of the Fourteenth Annual ACM-SIAM Symposium on Discrete Algorithms (SODA 2003), pp. 76–85 (2003)
5. Fraigniaud, P., Vial, S.: Approximation algorithms for broadcasting and gossiping. J. Parallel and Distrib. Comput. 43(1), 47–55 (1997)
6. Fraigniaud, P., Vial., S.: Heuristic algorithms for personalized communication problems in point-to-point networks. In: Proceedings of the 4th Colloquium on Structural Information Communication Complexity (SIROCCO 1997), pp. 240–252 (1997)
7. Fraigniaud, P., Vial, S.: Comparison of heuristics for one-to-all and all-to-all communication in partial meshes. Parallel Processing Letters 9, 9–20 (1999)
8. Harutyunyan, H.A., Laza, G., Maraachlian, E.: Broadcasting in necklace graphs. In: Proceedings of the 2nd Canadian Conference on Computer Science and Software Engineering (C3S2E 2009), pp. 253–256 (2009)
9. Harutyunyan, H., Maraachlian, E.: Linear algorithm for broadcasting in unicyclic graphs. In: Lin, G. (ed.) COCOON 2007. LNCS, vol. 4598, pp. 372–382. Springer, Heidelberg (2007)
10. Harutyunyan, H.A., Maraachlian, E.: On broadcasting in unicyclic graphs. J. Comb. Optim. 16(3), 307–322 (2008)
11. Harutyunyan, H.A., Maraachlian, E.: Broadcasting in fully connected trees. In: Proceedings of the 2009 15th International Conference on Parallel and Distributed Systems (ICPADS 2009), pp. 740–745 (2009)
12. Harutyunyan, H.A., Shao, B.: An efficient heuristic for broadcasting in networks. J. Parallel Distrib. Comput. 66(1), 68–76 (2006)
13. Harutyunyan, H.A., Wang, W.: Broadcasting algorithm via shortest paths. In: Proceedings of the 2010 IEEE 16th International Conference on Parallel and Distributed Systems (ICPADS 2010), pp. 299–305 (2010)
14. Kortsarz, G., Peleg, D.: Approximation algorithms for minimum time broadcast. SIAM J. Discrete Math. 8, 401–427 (1995)
15. Middendorf, M.: Minimum broadcast time is np-complete for 3-regular planar graphs and deadline 2. Inf. Proc. Lett. 46, 281–287 (1993)
16. Ravi, R.: Rapid rumor ramification: approximating the minimum broadcast time. In: Proceedings of the 35th Annual Symposium on Foundations of Computer Science (FOCS 1994), pp. 202–213 (1994)
17. Scheuermann, P., Wu, G.: Heuristic algorithms for broadcasting in point-to-point computer networks. IEEE Trans. Comput. 33(9), 804–811 (1984)
18. Schindelhauer, C.: On the inapproximability of broadcasting time. In: Jansen, K., Khuller, S. (eds.) APPROX 2000. LNCS, vol. 1913, pp. 226–237. Springer, Heidelberg (2000)
19. Slater, P.J., Cockayne, E.J., Hedetniemi, S.T.: Information dissemination in trees. SIAM J. Comput. 10(4), 692–701 (1981)

Direct and Certifying Recognition of Normal Helly Circular-Arc Graphs in Linear Time[*]

Yixin Cao[1,2]

[1] Institute for Computer Science and Control, Hungarian Academy of Sciences
[2] Department of Computing, The Hong Kong Polytechnic University

Abstract. A normal Helly circular-arc graph is the intersection graph of arcs on a circle of which no three or less arcs cover the whole circle. Lin et al. [Discrete Appl. Math. 2013] presented the first recognition algorithm for this graph class by characterizing circular-arc graphs that are not in it. They posed as an open problem to design a direct recognition algorithm, which is resolved by the current paper. When the input is not a normal Helly circular-arc graph, our algorithm finds in linear time a minimal forbidden induced subgraph. Grippo and Safe [arXiv:1402.2641] recently reported the forbidden induced subgraphs characterization of normal Helly circular-arc graphs. The correctness proof of our algorithm provides, as a byproduct, an alternative proof to this characterization.

1 Introduction

This paper will be only concerned with simple undirected graphs. A graph is a *circular-arc graph* if its vertices can be assigned to arcs on a circle such that two vertices are adjacent iff their corresponding arcs intersect. Such a set of arcs is called a *circular-arc model* of this graph. If some point on the circle is not in any arc in the model, then the graph is an *interval graph*, and it can be represented by a set of intervals on the real line, which is called an *interval model*. Circular-arc graphs and interval graphs are two of the most famous intersection graph classes, and both have been studied intensively for decades. However, in contrast to the nice result of Lekkerkerker and Boland [5], characterizing circular-arc graphs by forbidden induced subgraphs remains a notorious open problem in this area.

The complication of circular-arc graphs should be attributed to two special intersection patterns of circular-arc models that are not possible in interval models. The first is two arcs intersecting in both ends, and a circular-arc model is called *normal* if no such pair exists. The second is a set of arcs intersecting pairwise but containing no common point, and a circular-arc model is called *Helly* if no such set exists. Normal and Helly circular-arc models are precisely those with no set of three or less arcs covering the whole circle [10,6]. A graph that admits such a model is called a *normal Helly circular-arc graph*.

One fundamental problem on a graph class is its recognition, i.e., to efficiently decide whether a given graph belongs to this class or not. For intersection graph

---

[*] Supported by ERC under the grant 280152 and OTKA under the grant NK105645.

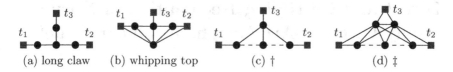

Fig. 1. Chordal minimal forbidden induced graphs

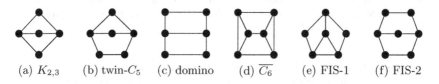

Fig. 2. Non-chordal and finite minimal forbidden induced graphs

classes, all recognition algorithms known to the author provide an intersection model when the membership is asserted. Most of them, on the other hand, simply return "NO" otherwise, while one might also want some verifiable certificate for some reason [9]. A recognition algorithm is *certifying* if it provides both positive and negative certificates. A minimal forbidden (induced) subgraph is arguably the simplest and most preferable among all forms of negative certificates [3].

For example, a graph is an interval graph iff it contains neither hole nor any graph in Fig. 1 [5]. Recall that a graph is *chordal* if it contains no holes. Kratsch et al. [4] reported a certifying recognition algorithm for interval graphs, which in linear time returns either an interval model of an interval graph or a forbidden induced subgraph for a non-interval graph. Although the forbidden induced subgraph returned by [4] is unnecessarily minimal, a minimal one can be easily retrieved from it (see [7] for another approach). Likewise, a minimal forbidden induced subgraph of chordal graphs, i.e., a hole, can be detected from a non-chordal graph in linear time [11]. However, although a circular-arc model of a circular-arc graph can be produced in linear time [8], it remains a challenging open problem to find a negative certificate for a non-circular-arc graph.

Indeed, all efforts attempting to characterize circular-arc graphs by forbidden induced subgraphs have been of no avail. For normal Helly circular-arc graphs, partial results were reported by [6], who listed all Helly circular-arc graphs that are not normal Helly circular-arc graphs. Very recently, Grippo and Safe completed this task by proving the following result. A wheel (resp., C^*) comprises a hole and another vertex completely adjacent (resp., nonadjacent) to it.

Theorem 1 ([2]). *A graph is a normal Helly circular-arc graph iff it contains no C^*, wheel, or any graph depicted in Figs. 1 and 2.*

It is easy to use definition to verify that a normal Helly circular-arc graph is chordal iff it is an interval graph. An interval model is always a normal and Helly circular-arc model, but an interval graph might have circular-arc model that is neither normal nor Helly, e.g., K_4. On the other hand,

Theorem 2 ([10,6]). *If a normal Helly circular-arc graph G is not chordal, then every circular-arc model of G is normal and Helly.*

These observations inspire us to recognize normal Helly circular-arc graphs as follows. If the input graph is chordal, it suffices to check whether it is an interval graph. Otherwise, we try to build a circular-arc model of it, and if success, verify whether the model is normal and Helly. Lin et al. [6] showed that this approach can be implemented in linear time. Moreover, if there exists a set of at most three arcs covering the circle, then their algorithm returns it as a certificate.

This algorithm, albeit conceptually simple, suffers from twofold weakness. First, it needs to call some recognition algorithm for circular-arc graphs, while all known algorithms are extremely complicated. Second, it is very unlikely to deliver a negative certificate in general. Therefore, Lin et al. [6] posed as an open problem to design a direct recognition algorithm for normal Helly circular-arc graphs, which would be desirable for both efficiency and the detection of negative certificates. The main result of this paper is the following algorithm— $n := |V(G)|$ and $m := |E(G)|$ are used throughout:

Theorem 3. *There is an $O(n + m)$-time algorithm that given a graph G, either constructs a normal and Helly circular-arc model of G, or finds a minimal forbidden induced subgraph of G.*

We remark that the proof of Thm. 3 will not rely on Thm. 1. Indeed, since our algorithm always finds a subgraph specified in Thm. 1 when the graph is not a normal Helly circular-arc graph, the correctness proof of our algorithm provides another proof of Thm. 1.

Let us briefly discuss the basic idea behind our disposal of a non-chordal graph G. If G is a normal Helly circular-arc graph, then for any vertex v of G, both $N[v]$ and its complement induce nonempty interval subgraphs. The main technical difficulty is how to combine interval models for them to make a circular-arc model of G. For this purpose we build an auxiliary graph $\mho(G)$ by taking two identical copies of $N[v]$ and appending them to the two ends of $G - N[v]$ respectively. The shape of symbol $\mho$ is a good hint for understanding the structure of the auxiliary graph. We show that $\mho(G)$ is an interval graph and more importantly, a circular-arc model of G can be produced from an interval model of $\mho(G)$. On the other hand, if G is not a normal Helly circular-arc graph, then $\mho(G)$ cannot be an interval graph. In this case we use the following procedure to obtain a minimal forbidden induced subgraph of G.

Theorem 4. *Given a minimal non-interval induced subgraph of $\mho(G)$, we can in $O(n + m)$ time find a minimal forbidden induced subgraph of G.*

The crucial idea behind our certifying algorithm is a novel correlation between normal Helly circular-arc graphs and interval graphs, which can be efficiently used for algorithmic purpose. This was originally proposed in the detection of small forbidden induced subgraph of interval graphs [1], i.e., the opposite direction of the current paper. In particular, in [1] we have used a similar definition of the auxiliary graph and pertinent observations. However, the main structural analyses, i.e., the detection of forbidden induced subgraphs, divert completely. For example, the most common forbidden induced subgraphs in [1] are 4- and 5-holes, which, however, are allowed in normal Helly circular-arc graphs. Their existence makes the interaction between $N[v]$ and $G - N[v]$ far more subtle, and thus the detection of minimal forbidden induced subgraphs in the current paper is significantly more complicated than that of [1].

2   The Recognition Algorithm

All graphs are stored as adjacency lists. We use the customary notation $v \in G$ to mean $v \in V(G)$, and $u \sim v$ to mean $uv \in E(G)$. Exclusively concerned with induced subgraphs, we use F to denote both a subgraph and its vertex set.

Consider a circular-arc model $\mathcal{A}$. If every point of the circle is contained in some arc in $\mathcal{A}$, then we can find an inclusive-wise minimal set X of arcs that cover the entire circle. If $\mathcal{A}$ is normal and Helly, then X consists of at least four arcs and thus corresponds to a hole. Therefore, a normal Helly circular-arc graph G is chordal iff it is an interval graph, for which it suffices to call the algorithms of [4,7]. We are hence focused on graphs that are not chordal. We call the algorithm of Tarjan and Yannakakis [11] to detect a hole H.

Proposition 1. *Let H be a hole of a circular-arc graph G. In any circular-arc model of G, the union of arcs for H covers the whole circle, i.e., $N[H] = V(G)$.*

Indices of vertices in H should be understood as modulo $|H|$, e.g., $h_0 = h_{|H|}$. By Prop. 1, every vertex should have neighbors in H. We use $N_H[v]$ as a shorthand for $N[v] \cap H$, regardless of whether $v \in H$ or not. We start from characterizing $N_H[v]$ for every vertex v: we specify some forbidden structures not allowed to appear in a normal Helly circular-arc graph, and more importantly, we show how to find a minimal forbidden induced subgraph if one of these structures exists. The fact that they are forbidden can be easily derived from the definition and Prop. 1. Due to the lack of space, their proofs, mainly on the detection of minimal forbidden induced subgraphs, are deferred to the full version.

Lemma 1. *For every vertex v, we can in $O(d(v))$ time find either a proper sub-path of H induced by $N_H[v]$, or a minimal forbidden induced subgraph.*

We designate the ordering $h_0, h_1, h_2, \cdots$ of traversing H as *clockwise*, and the other *counterclockwise*. In other words, edges $h_0 h_1$ and $h_0 h_{-1}$ are clockwise and counterclockwise, respectively, from h_0. Now let P be the path induced by $N_H[v]$. We can assign a direction to P in accordance to the direction of H, and then we have clockwise and counterclockwise ends of P. For technical reasons, we assign canonical indices to the ends of the path P as follows.

Definition 1. *For each vertex $v \in G$, we denote by* first(v) *and* last(v) *the indices of the counterclockwise and clockwise, respectively, ends of the path induced by $N_H[v]$ in H satisfying*

- $-|H| <$ first$(v) \leq 0 \leq$ last$(v) < |H|$ *if $h_0 \in N_H[v]$; or*
- $0 <$ first$(v) \leq$ last$(v) < |H|$, *otherwise.*

It is possible that $\mathrm{last}(v) = \mathrm{first}(v)$, when $|N_H[v]| = 1$. In general, $\mathrm{last}(v) - \mathrm{first}(v) = |N_H[v]| - 1$, and $v = h_i$ or $v \sim h_i$ for each i with $\mathrm{first}(v) \le i \le \mathrm{last}(v)$. The indices $\mathrm{first}(v)$ and $\mathrm{last}(v)$ can be easily retrieved from Lem. 1, with which we can check the adjacency between v and any vertex $h_i \in H$ in constant time. Now consider the neighbors of more than one vertices in H.

Lemma 2. *Given a pair of adjacent vertices u, v s.t. $N_H[u]$ and $N_H[v]$ are disjoint, then in $O(n + m)$ time we can find a minimal forbidden induced subgraph.*

Lemma 3. *Given a set U of two or three pairwise adjacent vertices such that 1) $\bigcup_{u \in U} N_H[u] = H$; and 2) for every $u \in U$, each end of $N_H[u]$ is adjacent to at least two vertices in U, then we can in $O(n + m)$ time find a minimal forbidden induced subgraph.*

Let $T := N[h_0]$ and $\overline{T} := V(G) \setminus T$. As we have alluded to earlier, we want to duplicate T and append them to different sides of $\overline{T}$. Each edge between $v \in T$ and $u \in \overline{T}$ will be carried by only one copy of T, and this is determined by its direction specified as follows. We may assume that none of the Lems. 1, 2, and 3 applies to v or/and u, as otherwise we can terminate the algorithm by returning the forbidden induced subgraph found by them. As a result, u is adjacent to either $\{h_{\mathrm{first}(v)}, \cdots, h_{-1}\}$ or $\{h_1, \cdots, h_{\mathrm{last}(v)}\}$ but not both. The edge uv is said to be clockwise from T if $u \sim h_i$ for $1 \le i \le \mathrm{last}(v)$, and counterclockwise otherwise. Let E_c(resp., E_cc) denote the set of edges clockwise (resp., counterclockwise) from T, and let T_c (resp., T_cc) denote the subsets of vertices of T that are incident to edges in E_c (resp., E_cc). Note that $\{E_\mathrm{cc}, E_\mathrm{c}\}$ partitions edges between T and $\overline{T}$, but a vertex in T might belong to both T_cc and T_c, or neither of them. We have now all the details for the definition and construction of the auxiliary graph $\mho(G)$, which can be done in linear time.

Definition 2. *The vertex set of $\mho(G)$ consists of $\overline{T} \cup L \cup R \cup \{w\}$, where L and R are distinct copies of T, i.e., for each $v \in T$, there are a vertex v^l in L and another vertex v^r in R, and w is a new vertex distinct from $V(G)$. For each edge $uv \in E(G)$, we add to the edge set of $\mho(G)$*

- *an edge uv if neither u nor v is in T;*
- *two edges $u^l v^l$ and $u^r v^r$ if both u and v are in T; or*
- *an edge uv^l or uv^r if $uv \in E_\mathrm{c}$ or $uv \in E_\mathrm{cc}$ respectively ($v \in T$ and $u \in \overline{T}$).*

Finally, we add an edge wv^l for every $v \in T_\mathrm{cc}$.

Lemma 4. *The numbers of vertices and edges of $\mho(G)$ are upper bounded by $2n$ and $2m$ respectively. Moreover, an adjacency list representation of $\mho(G)$ can be constructed in $O(n + m)$ time.*

In an interval model, each vertex v corresponds to a closed interval $I_v = [\mathrm{lp}(v), \mathrm{rp}(v)]$. Here $\mathrm{lp}(v)$ and $\mathrm{rp}(v)$ are the left and right *endpoints* of I_v respectively, and $\mathrm{lp}(v) < \mathrm{rp}(v)$. We use unit-length circles for circular-arc models, where every point has a positive value in $(0, 1]$. Each vertex v corresponds to a closed arc $A_v = [\text{ccp}(v), \text{cp}(v)]$. Here $\text{ccp}(v)$ and $\text{cp}(v)$ are counterclockwise and clockwise endpoints of A_v respectively; $0 < \text{ccp}(v), \text{cp}(v) \leq 1$ and they are assumed to be distinct. It is worth noting that possibly $\text{cp}(v) < \text{ccp}(v)$; such an arc necessarily contains the point 1.

Lemma 5. *If G is a normal Helly circular-arc graph, then $\mho(G)$ is an interval graph.*

Fig. 3. Illustration for Lem. 5

As shown in Fig. 3, it is intuitive to transform a normal Helly circular-arc model of G to an interval model of $\mho(G)$. Note that for any vertex $v \in T$, an induced (v^l, v^r)-path corresponds to a cycle whose arcs cover the entire circle. The main thrust of our algorithm will be a process that does the reversed direction, which is nevertheless far more involved.

Theorem 5. *If $\mho(G)$ is an interval graph, then we can in $O(n + m)$ time build a circular-arc model of G.*

Proof. We can in $O(n+m)$ time build an interval model $\mathcal{I}$ for $\mho(G)$. By construction, $(wh^l_{-1}h^l_0h^l_1h_2 \cdots h_{-2}h^r_{-1}h^r_0h^r_1)$ is an induced path of $\mho(G)$; without loss of generality, assume it goes "from left to right" in $\mathcal{I}$. We may assume $\text{rp}(w) = 0$ and $\max_{u \in \overline{T}} \text{rp}(u) = 1$, while no other interval in $\mathcal{I}$ has 0 or 1 as an endpoint. Let $a = \text{rp}(h^l_0)$. We use $\mathcal{I}$ to construct a set of arcs for $V(G)$ as follows. For each $u \in \overline{T}$, let $A_u := [\text{lp}(u), \text{rp}(u)]$, which is a subset of $(a, 1]$. For each $v \in T$, let

$$A_v := \begin{cases} [\text{lp}(v^r), \text{rp}(v^l)] & \text{if } v \in T_{\text{cc}}, \\ [\text{lp}(v^l), \text{rp}(v^l)] & \text{otherwise.} \end{cases}$$

It remains to verify that the arcs obtained as such represent G, i.e., a pair of vertices u, v of G is adjacent iff A_u and A_v intersect. This holds trivially when $u, v \notin T$; hence we may assume without loss of generality that $v \in T$. By construction, $a < \text{lp}(u) < \text{rp}(u) \leq 1$ for every $u \in \overline{T}$. Note that $v^l \sim w$ and $v^r \sim \overline{T}$ for every $v \in T_{\text{cc}}$, which implies that $\text{lp}(v^l) < 0$ iff $\text{lp}(v^r) < 1$ iff $v \in T_{\text{cc}}$.

Assume first that u is also in T, then $u \sim v$ in G if and only if $u^l \sim v^l$ in $\mho(G)$. They are adjacent when both $u, v \in T_{\text{cc}}$, and since $\text{lp}(v^l), \text{lp}(u^l) < 0$, both A_u and A_v contains the point 1 and thus intersect. If neither u nor v is in T_{cc}, then $\text{lp}(v^l), \text{lp}(u^l) > 0$, and $u \sim v$ if and only if $A_u = [\text{lp}(u^l), \text{rp}(u^l)]$ and $A_v = [\text{lp}(v^l), \text{rp}(v^l)]$ intersect. Otherwise, assume, without loss of generality, that $\text{lp}(v^l) < 0 < \text{lp}(u^l)$, then $u \sim v$ in G if and only if $0 < \text{lp}(u^l) < \text{rp}(v^l)$, which implies A_u and A_v intersect (as both contain $[\text{lp}(u^l), \text{rp}(v^l)]$).

Assume now that u is not in T, and then $u \sim v$ in G if and only if either $u \sim v^l$ or $u \sim v^r$ in $\mathcal{U}(G)$. In the case $u \sim v^l$, we have $\mathrm{lp}(v^l) \leq a < \mathrm{lp}(u) \leq \mathrm{rp}(v^l)$; since both A_u and A_v contain $[\mathrm{lp}(u), \mathrm{rp}(v^l)]$, which is nonempty, they intersect. In the case $u \sim v^r$, we have $\mathrm{lp}(v^r) < \mathrm{rp}(u) \leq 1$; since both A_u and A_v contain $[\mathrm{lp}(v^r), \mathrm{rp}(u)]$, they intersect. Otherwise, $u \not\sim v$ in G and $\mathrm{lp}(v^l) < \mathrm{rp}(v^l) < \mathrm{lp}(u) < \mathrm{rp}(u) < \mathrm{lp}(v^r) < \mathrm{rp}(v^r)$, then A_u and A_v are disjoint. $\square$

We are now ready to present the recognition algorithm in Fig. 4, and prove Thm. 3. Recall that Lin et al. [6] have given a linear-time algorithm for verifying whether a circular-arc model is normal and Helly.

Algorithm **nhcag**(G)
Input: a graph G.
Output: a normal Helly circular-arc model, or a forbidden induced subgraph.

1   test the chordality of G and find a hole H if not;
     if G is chordal **then** verify whether G is an interval graph or not;
2   construct the auxiliary grpah $\mathcal{U}(G)$;
3   **if** $\mathcal{U}(G)$ is not an interval graph **then**
          call Thm. 4 to find a forbidden induced subgraph;
4   **call** 5 to build a circular-arc $\mathcal{A}$ model of G;
5   verify whether $\mathcal{A}$ is normal and Helly.

Fig. 4. The recognition algorithm for normal Helly circular-arc graphs

Proof (Thm. 3). Step 1 is clear. Steps 2-4 follow from Lem. 4, Thm. 4, and Lem. 5, respectively. If model $\mathcal{A}$ built in step 4 is not normal and Helly, then we can in linear time find a set of two or three arcs whose union covers the circle. Their corresponding vertices satisfy Lem. 3, and this concludes the proof. $\square$

It is worth noting that if we are after a recognition algorithm (with positive certificate only), then we can simply return "NO" if the hypothesis of step 3 is true (justified by Lem. 5) and the algorithm is already complete.

3   Proof of Theorem 4

Recall that Thm. 4 is only called in step 3 of algorithm nhcag; the graph is then not choral and we have a hole H. In principle, we can pick any vertex as h_0. But for the convenience of presentation, we require it satisfies some additional conditions. If some vertex v is adjacent to four or more vertices in H, i.e., $\mathrm{last}(v) - \mathrm{first}(v) > 2$, then $v \notin H$. We can thus use $(h_{\mathrm{first}(v)} v h_{\mathrm{last}(v)})$ as a short cut for the sub-path induced by $N_H[v]$, thereby yielding a strictly shorter hole. This condition, that h_0 cannot be bypassed as such, is formally stated as:

Lemma 6. *We can in $O(n + m)$ time find either a minimal forbidden induced subgraph, or a hole H such that $\{h_{-1}, h_0, h_1\} \subseteq N_H[v]$ for some v if and only if $N_H[v] = \{h_{-1}, h_0, h_1\}$.*

This linear-time procedure can be called before step 2 of algorithm nhcag, and it does not impact the asymptotic time complexity of the algorithm, which remains linear. Henceforth we may assume that H satisfies the condition of Lem. 6. During the construction of $\mho(G)$, we have checked $N_H[v]$ for every vertex v, and Lem. 1 was called if it applies. Thus, for the proof of Thm. 4 in this section, we may assume that $N_H[v]$ always induces a proper sub-path of H.

Each vertex x of $\mho(G)$ different from w is uniquely defined by a vertex of G, which is denoted by $\phi(x)$. We say that x is *derived from* $\phi(x)$. For example, $\phi(v^l) = \phi(v^r) = v$ for $v \in T$. By abuse of notation, we will use the same letter for a vertex $u \in \overline{T}$ of G and the unique vertex of $\mho(G)$ derived from u, i.e., $\phi(u) = u$ for $u \in \overline{T}$; its meaning is always clear from the context. We can mark $\phi(x)$ for each vertex of $\mho(G)$ during its construction. For a set U of vertices not containing w, we define $\phi(U) := \{\phi(v) : v \in U\}$; possibly $|\phi(U)| \neq |U|$.

By construction, if a pair of vertices x and y (different from w) is adjacent in $\mho(G)$, then $\phi(x)$ and $\phi(y)$ must be adjacent in G as well. The converse is unnecessarily true, e.g., $u \not\sim v^r$ for any vertex $v \in T_c$ and edge $uv \in E_c$, and $u^l \not\sim v^r$ and $u^r \not\sim v^l$ for any pair of adjacent vertices $u, v \in T$. We say that a pair of vertices x, y of $\mho(G)$ is a *bad pair* if $\phi(x) \sim \phi(y)$ in G but $x \not\sim y$ in $\mho(G)$. By definition, w does not participate in any bad pair, and at least one vertex of a bad pair is in $L \cup R$. Note that any induced path of length d between a bad pair x, y with $x = v^l$ or v^r can be extended to a (v^l, v^r)-path with length $d + 1$.

Figure 3 shows that if G is a normal Helly circular-arc graph, then for any $v \in T$, the distance between v^l and v^r is at least 4. We now see what happens when this necessary condition is not satisfied by $\mho(G)$. By definition of $\mho(G)$, there is no edge between L and R; for any $v \in T$, there is no vertex adjacent to both v^l and v^r. In other words, for every $v \in T$, the distance between v^l and v^r is at least 3. The following observation can be derived from Lems. 1 and 2.

Lemma 7. *Given a (v^l, v^r)-path P of length 3 for some $v \in T$, we can in $O(n + m)$ time find a minimal forbidden induced subgraph of G.*

Proof. Let $P = (v^l xyv^r)$. Note that P must be a shortest (v^l, v^r)-path, and $w \notin P$. The inner vertices x and y cannot be both in $L \cup R$; without loss of generality, let $x \in \overline{T}$. Assume first that $y \in \overline{T}$ as well, i.e., $vx \in E_c$ and $vy \in E_{cc}$. By definition, $v \in T_c \cap T_{cc}$, and then v is adjacent to both h_{-1} and h_1. If follows from Lem. 6 that $N_H[v] = \{h_{-1}, h_0, h_1\}$, and then $x \sim h_1$ and $y \sim h_{-1}$. If $x \sim h_{-1}$, i.e., $\texttt{last}(x) = |H| - 1$, then we call Lem. 2 with v and x. If $\texttt{last}(x) < \texttt{first}(y)$, then we call Lem. 1 with x and y. In the remaining case, $\texttt{first}(y) \leq \texttt{last}(x) < |H| - 1$, and $(vxh_{\texttt{last}(x)} \cdots h_{-1}v)$ is a hole of G; this hole is completely adjacent to y, and thus we find a wheel.

Now assume that, without loss of generality, $y = u^r \in R$. If $\texttt{last}(v) \geq \texttt{first}(y)$, then we call Lem. 2 with v and y. Otherwise, $(vh_{\texttt{last}(v)} \cdots h_{\texttt{first}(y)}uv)$ is a hole of G; this hole is completely adjacent to x, and thus we find a wheel. □

If G is a normal Helly circular-arc graph, then in a circular-arc model of G, all arcs for T_{cc} and T_c contain $\texttt{ccp}(h_0)$ and $\texttt{cp}(h_0)$ respectively. Thus, both T_{cc} and T_c induce cliques. This observation is complemented by

Lemma 8. *Given a pair of nonadjacent vertices $u, x \in T_{cc}$ (or T_c), we can in $O(n + m)$ time find a minimal forbidden induced subgraph of G.*

Proof. By definition, we can find $uv, xy \in E_{cc}$. We have three (possibly intersecting) chordless paths $h_0 h_1 h_2$, $h_0 uv$, and $h_0 xy$. If both u and x are adjacent to h_1, then we return $(uh_{-1}xh_1u) + h_0$ as a wheel. Hence we may assume $x \not\sim h_1$.

If $u \sim h_1$, then by Lem. 6, $N_H[u] = \{h_{-1}, h_0, h_1\}$. We consider the subgraph induced by the set of distinct vertices $\{h_0, h_1, h_2, u, v, x\}$. If v is adjacent to h_0 or h_1, then we can call Lem. 3 with u, v. By assumption, h_0, h_1, and u make a triangle; x is adjacent to neither u nor h_1; and h_2 is adjacent to neither h_0 nor u. Thus, only uncertain adjacencies in this subgraph are between v, x, and h_2. The subgraph is hence isomorphic to (1) FIS-1 if there are two edges among v, x, and h_2; (2) $\overline{C_6}$ if v, x, and h_2 are pairwise adjacent; or (3) net if v, x, and h_2 are pairwise nonadjacent. In the remaining cases there is precisely one edge among v, x, and h_2. We can return a C^*, e.g., $(vxh_0uv) + h_2$ when the edge is vx.

Assume now that u, x, and h_1 are pairwise nonadjacent. We consider the subgraph induced by $\{h_0, h_1, h_2, u, v, x, y\}$, where the only uncertain relations are between v, y, and h_2. The subgraph is thus isomorphic to (1) $K_{2,3}$ if all of them are identical; or (2) twin-C_5 if two of them are identical, and adjacent to the other. If two of them are identical, and nonadjacent to the other, then the subgraph contains a C^*, e.g., $(vuh_0xv) + h_2$ when $v = y$. In the remaining cases, all of v, y, and h_2 are distinct, and then the subgraph (1) is isomorphic to long claw if they are pairwise nonadjacent; (2) contains net $\{h_1, h_2, u, v, x, y\}$ if they are pairwise adjacent; or (3) is isomorphic to FIS-2 if there are two edges among them. If there is one edge among them, then the subgraph contains a C^*, e.g., $(vuh_0xyv) + h_2$ when the edge is vy.

A symmetrical argument applies to T_c. the runtime is clearly $O(n + m)$.   □

It can be checked in linear time whether T_{cc} and T_c induce cliques. When it is not, a pair of nonadjacent vertices can be found in the same time. By Lem. 8, we may assume hereafter that T_{cc} and T_c induce cliques. Recall that $N(w) \subseteq T_{cc}$; as a result, w is simplicial and participates in no holes.

Proposition 2. *Given a (h_0^l, h_0^r)-path nonadjacent to h_i for some $1 < i < |H| - 1$, we can in $O(n + m)$ time find a minimal forbidden induced subgraph.*

We are now ready to prove Thm. 4, which is separated into three statements, the first of which considers the case when $\mho(G)$ is not chordal.

Lemma 9. *Given a hole C of $\mho(G)$, we can in $O(n + m)$ time find a minimal forbidden induced subgraph of G.*

Proof. Let us first take care of some trivial cases. If C is contained in L or R or $\overline{T}$, then by construction, $\phi(C)$ is a hole of G. This hole is either nonadjacent or completely adjacent to h_0 in G, whereupon we can return $\phi(C) + h_0$ as a C^* or wheel respectively. Since L and R are nonadjacent, it must be one of the cases above if C is disjoint from $\overline{T}$. Henceforth we may assume that C intersects $\overline{T}$ and, without loss of generality, L; it might intersect R as well, but this fact is irrelevant in the following argument. Then we can find an edge x_1x_2 of C such that $x_1 \in L$ and $x_2 \in \overline{T}$, i.e., $x_1x_2 \in E_c$.

Let $a := \mathtt{last}(\phi(x_1))$. Assume first that $x_2 = h_a$; then we must have $a > 1$. Let x_3 and x_4 be the next two vertices of C. Note that $x_3 \notin L$, i.e., $x_3 \nsim h_0^l$; otherwise $x_1 \sim x_3$, which is impossible. If $x_3 \sim h_{a-2}$ (or h_{a-2}^l when $a = 3$), then $\phi(\{x_1, x_2, x_3\}) \cup \{h_{a-2}\}$ induces a hole of G, and we can return it and h_{a-1} as a wheel. Note that $x_4 \nsim h_a$ as they are non-consecutive vertices of the hole C. We now argue that $\mathtt{last}(\phi(x_4)) < a$. Suppose for contradiction, $\mathtt{first}(\phi(x_4)) > a$. We can extend the (x_3, x_1)-path P in C that avoids x_2 to a (h_0^l, h_0^r)-path avoiding the neighborhood of h_a, which allows us to call Prop. 2. We can call Lem. 2 with x_3 and x_4 if $\mathtt{first}(\phi(x_3)) = a$. In the remaining case, $\mathtt{first}(\phi(x_3)) = a - 1$. Let x be the first vertex in P that is adjacent to h_{a-2} (or h_{a-2}^l if $a \le 3$); its existence is clear as x_1 satisfies this condition. Then $\phi(\{x_3, \ldots, x, h_{a-2}, x_1, x_2\})$ induces a hole of G, and we can return it and h_{a-1} as a wheel.

Assume now that h_a is not in C. Denote by P the (x_2, x_1)-path obtained from C by deleting the edge x_1x_2. Let x be the first neighbor of h_{a+1} in P, and let y be either the first neighbor of h_{a-1} in the (x, x_1)-path or the other neighbor of x_1 in C. It is easy to verify that $\phi(\{x_1, \cdots, x, \cdots y, x_2\})$ induces a hole of G, which is completely adjacent to h_a, i.e., we have a wheel.   □

In the rest $\mho(G)$ will be chordal, and thus we have a chordal non-interval subgraph F of $\mho(G)$. This subgraph is isomorphic to some graph in Fig. 1, on which we use the following notation. It is immediate from Fig. 1 that each of them contains precisely three simplicial vertices (squared vertices), which are called *terminals*, and others (round vertices) are *non-terminal vertices*. In a long claw or †, for each $i = 1, 2, 3$, terminal t_i has a unique neighbor, denoted by u_i.

Proposition 3. *Given a subgraph F of $\mho(G)$ in Fig. 1, we can in $O(n + m)$ time find either all bad pairs in F or a forbidden induced subgraph of G.*

Lemma 10. *Given a subgraph F of $\mho(G)$ in Fig. 1 that does not contain w, we can in $O(n + m)$ time find a minimal forbidden induced subgraph of G.*

Proof. We first call Prop. 3 to find all bad pairs in F. If F has no bad pair, then we return the subgraph of G induced by $\phi(F)$, which is isomorphic to F. Let x, y be a bad pair with the minimum distance in F; we may assume that it is 3 or 4, as otherwise we can call Lem. 7. Noting that the distance between a pair of non-terminal vertices is at most 2, we may assume that without loss of generality, x is a terminal of F. We break the argument based on the type of F.

Long claw. We may assume that $x = t_1$ and $y \in \{u_2, t_2\}$; other situations are symmetrical. Let P be the unique (x, y)-path in F. If $\phi(t_3)$ is nonadjacent to $\phi(P)$, then we return $\phi(P) + \phi(t_3)$ as a C^*; we are thus focused on the adjacency between $\phi(t_3)$ and $\phi(P)$. If $y = t_2$, then by the selection of x, y (they have the minimum distance among all bad pairs), $\phi(t_3)$ can be only adjacent to $\phi(t_1)$ and/or $\phi(t_2)$. We return either $\phi(F)$ as an FIS-2 , or $\phi(\{t_1, t_2, t_3, u_1, u_2, u_3\})$ as a net. In the remaining cases, $y = u_2$, and $\phi(t_3)$ can only be adjacent to $\phi(u_1)$, $\phi(u_2)$, and/or $\phi(t_1)$. We point out that possibly $\phi(t_2) = \phi(t_1)$, which is irrelevant as $\phi(t_2)$ will not be used below. If $\phi(t_3)$ is adjacent to both $\phi(u_1)$ and $\phi(u_2)$ in G, then we get a $K_{2,3}$. Note that this is the only case when $\phi(t_1) = \phi(t_3)$. If $\phi(t_3)$ is adjacent to both $\phi(t_1)$ and $\phi(u_2)$ in G, then we get an FIS-1. If $\phi(t_3)$ is adjacent to only $\phi(u_2)$ or only $\phi(t_1)$ in G, then we get a domino or twin-C_5, respectively. The situation that $\phi(t_3)$ is adjacent to $\phi(u_1)$ but not $\phi(u_2)$ is similar as above.

†. Consider first that $x = t_1$ and $y = t_3$, and let $P = (t_1 u_1 u_3 t_3)$. If $\phi(t_2)$ is nonadjacent to the hole induced by $\phi(P)$, then we return $\phi(P)$ and $\phi(t_2)$ as a C^*. If $\phi(t_2)$ is adjacent to $\phi(t_3)$ or $\phi(u_1)$, then we get a domino. If $\phi(t_2)$ is adjacent to $\phi(t_1)$, then we get a twin-C_5. If $\phi(t_2)$ is adjacent to $\phi(t_1)$ and precisely one of $\{\phi(t_3), \phi(u_1)\}$, then we get an FIS-1. If $\phi(t_2)$ is adjacent to both $\phi(t_3)$ and $\phi(u_1)$, then we get a $K_{2,3}$; here the adjacency between $\phi(t_2)$ and $\phi(t_1)$ is immaterial. A symmetric argument applies when $\{t_2, t_3\}$ is a bad pair. In the remaining case, neither $\phi(t_1)$ nor $\phi(t_2)$ is adjacent to $\phi(t_3)$. Therefore, a bad pair must be in the path $F - N[t_3]$, which is nonadjacent to $\phi(t_3)$, then we get a C^*.

The whipping top and ‡ are straightforward and omitted. □

Lemma 11. *Given a subgraph F of $\mathfrak{U}(G)$ in Fig. 1 that contains w, we can in $O(n + m)$ time find a minimal forbidden induced subgraph of G.*

‖ Note that $0 \le \mathtt{last}(\phi(x_1)), \mathtt{last}(\phi(x_2)) \le 1$.
1 **if** $\mathtt{last}(\phi(x_1)) = 1$ and $y_1 \sim h_2$ **then**
    call Lem. 2 with $(y_1 \phi(x_1) h_1 h_2 y_1)$ and $\{\phi(x_2), y_2\}$;
  if $\mathtt{last}(\phi(x_1)) = 0$ and $y_1 \sim h_1$ **then**
    call Lem. 2 with $(y_1 \phi(x_1) h_0 h_1 y_1)$ and $\{\phi(x_2), y_2\}$;
  if $y_2 \sim h_{\mathtt{last}(\phi(x_2))+1}$ **then** symmetric as above;
2 **if** $\mathtt{last}(\phi(x_1)) = \mathtt{last}(\phi(x_2))$ **then**
    return $\{y_1, \phi(x_1), y_2, \phi(x_2), h_{\mathtt{last}(\phi(x_2))}, h_{\mathtt{last}(\phi(x_2))+1}\}$ as a †;
  ‖ assume from now that $\mathtt{last}(\phi(x_1)) = 1$ and $\mathtt{last}(\phi(x_2)) = 0$.
3 **if** $\phi(x_2) \sim h_2$ **then return** $(\phi(x_2) h_0 h_1 h_2 \phi(x_2)) + y_1$ as a C^*;
4 **if** $y_2 \not\sim h_{-1}$ **then return** $\{y_1, h_{-1}, \phi(x_1), y_2, \phi(x_2), h_0, h_1\}$ as a ‡;
  if $y_2 \sim h_{-1}$ **then return** $\{y_1, h_{-1}, y_2, \phi(x_2), h_0, h_1\}$ as a †.

Fig. 5. Procedure for Lem. 11

Proof. Since w is simplicial, it has at most 2 neighbors in F. If w has a unique neighbor in F, then we can use a similar argument as Lem. 10. Now let x_1, x_2 be the two neighbors of w in F. If there exists some vertex $u \in \overline{T}$ adjacent to both $\phi(x_1)$ and $\phi(x_2)$ in G, which can be found in linear time, then we can use it replace w. Hence we assume there exists no such vertex. By assumption, we can find two distinct vertices $y_1, y_2 \in \overline{T}$ such that $\phi(x_1)y_1, \phi(x_2)y_2 \in E_{cc}$; note that $\phi(x_1) \not\sim y_2$ and $\phi(x_2) \not\sim y_1$ in G. As a result, y_1 and y_2 are nonadjacent; otherwise, $\{y_1, y_2\}$ and the counterparts of $\{x_1, x_2\}$ in R induce a hole of $\mathfrak{U}(G)$, which is impossible. We then apply the procedure described in Fig. 5.

We now verify the correctness of the procedure. Since each step—either directly or by calling a previously verified lemma—returns a minimal forbidden induced subgraph of G, all conditions of previous steps are assumed to not hold in a later step. By Lem. 6, $\texttt{last}(\phi(x_1))$ and $\texttt{last}(\phi(x_2))$ are either 0 or 1. Step 1 considers the case where $y_1 \sim h_{\texttt{last}(\phi(x_1))+1}$. By Lem. 3, $y_1 \not\sim h_{\texttt{last}(\phi(x_1))}$. Thus, $(y_1\phi(x_1)h_1h_2y_1)$ or $(y_1\phi(x_1)h_0h_1y_1)$ is a hole of G, depending on $\texttt{last}(\phi(x_1))$ is 0 or 1. In the case $(y_1\phi(x_1)h_1h_2y_1)$, only $\phi(x_1)$ and h_1 can be adjacent to $\phi(x_2)$; they are nonadjacent to y_2. Likewise, in the case $(y_1\phi(x_1)h_0h_1y_1)$, vertices $\phi(x_1)$ and h_0 are adjacent to $\phi(x_2)$ but not y_2, while h_1 can be adjacent to only one of $\phi(x_2)$ and y_2. Thus, we can call Lem. 2. A symmetric argument applies when $y_2 \sim h_{\texttt{last}(\phi(x_2))+1}$. Now that the conditions of step 1 do not hold true, step 2 is clear from assumption. Henceforth we may assume without loss of generality that $\texttt{last}(\phi(x_1)) = 1$ and $\texttt{last}(\phi(x_2)) = 0$. Consequently, $\texttt{last}(y_1) = |H| - 1$ (Lem. 2). Because we assume that the condition of step 1 does not hold, $y_1 \not\sim h_2$; this justifies step 3. Step 4 is clear as y_1 is always adjacent to h_{-1}. $\qquad\square$

References

1. Cao, Y.: Linear recognition of almost (unit) interval graphs (2014) (manuscript)
2. Grippo, L., Safe, M.: On circular-arc graphs having a model with no three arcs covering the circle. arXiv:1402.2641 (2014)
3. Heggernes, P., Kratsch, D.: Linear-time certifying recognition algorithms and forbidden induced subgraphs. Nord. J. Comput. 14, 87–108 (2007)
4. Kratsch, D., McConnell, R., Mehlhorn, K., Spinrad, J.: Certifying algorithms for recognizing interval graphs and permutation graphs. SIAM J. Comput. 36, 326–353 (2006)
5. Lekkerkerker, C., Boland, J.: Representation of a finite graph by a set of intervals on the real line. Fund. Math. 51, 45–64 (1962)
6. Lin, M., Soulignac, F., Szwarcfiter, J.: Normal Helly circular-arc graphs and its subclasses. Discrete Appl. Math. 161, 1037–1059 (2013)
7. Lindzey, N., McConnell, R.M.: On finding Tucker submatrices and Lekkerkerker-Boland subgraphs. In: Brandstädt, A., Jansen, K., Reischuk, R. (eds.) WG 2013. LNCS, vol. 8165, pp. 345–357. Springer, Heidelberg (2013)
8. McConnell, R.: Linear-time recognition of circular-arc graphs. Algorithmica 37, 93–147 (2003)
9. McConnell, R., Mehlhorn, K., Näher, S., Schweitzer, P.: Certifying algorithms. Computer Science Review 5, 119–161 (2011)
10. McKee, T.: Restricted circular-arc graphs and clique cycles. Discrete Math. 263, 221–231 (2003)
11. Tarjan, R., Yannakakis, M.: Simple linear-time algorithms to test chordality of graphs, test acyclicity of hypergraphs, and selectively reduce acyclic hypergraphs. SIAM J. Comput. 13, 566–579. Addendum in the same journal 14, 254–255 (1985)

A Fixed-Parameter Approach
for Privacy-Protection with Global Recoding*

Katrin Casel

FB 4-Abteilung Informatikwissenschaften,
Universität Trier, 54286 Trier, Germany
casel@informatik.uni-trier.de

Abstract. This paper discusses a problem arising in the field of privacy-protection in statistical databases: Given a $n \times m$ $\{0,1\}$-matrix M, is there a set of mergings which transforms M into a zero matrix and only affects a bounded number of rows/columns. "Merging" here refers to combining adjacent lines with a component-wise logical AND. This kind transformation models a generalization on OLAP-cubes also called *global recoding*. Counting the number of affected lines presents a new measure of information-loss for this method. Parameterized by the number of affected lines k we introduce reduction rules and an $\mathcal{O}^*(2.618^k)$-algorithm for the new abstract combinatorial problem LMAL.

1 Introduction

With the steadily increasing amount of personal data collected for statistical research, privacy has become a matter of great federal and public interest. Statistical databases allow investigating personal records for empirical studies. This access however has to be restricted carefully to avoid disclosure of individual information. In the following, we approach this task considering parameterized complexity which was recently used for similar problems to obtain security in data-tables by entry-suppression [4,6].

In the following we discuss a method to secure access to confidential data via *OLAP-cubes*: A collection of individual records $R \subset I \times Q \times S$ characterized by x unique identifiers I, p numerical (or otherwise logically ordered) non-confidential attributes Q (*quasi-identifiers*) and numerical confidential attributes S is represented by a p-dimensional table T. Each dimension corresponds to one attribute and ranges over all of its possible values in their logical order. A cell of T with the label $(w_1, \ldots, w_p)$ contains the number of records in R with these characteristics, i.e.: $|\{r \in R : r_{x+i} = w_i \; \forall \, i = 1, \ldots, p\}|$. An example for this kind of representation is given below. Access to the confidential attribute(s) S is granted via queries on the non-confidential attributes, identifiers I are suppressed completely.

With the restriction to SUM- and COUNT- range-queries, the set of *even* range-queries (queries addressing an even number of cells) is the largest safe query-set [2]. The corresponding security-level is *l-compromise* [3], which is defined

---

* Partially funded by a PhD-scholarship of the DAAD.

J. Chen, J.E. Hopcroft, and J. Wang (Eds.): FAW 2014, LNCS 8497, pp. 25–35, 2014.
© Springer International Publishing Switzerland 2014

similarly to *k-anonymity* [10] for data-tables but does not directly suffer from the diversity-problem [9]. Unfortunately, this method is only safe for databases in which each cell of the OLAP-cube contains at least one record. To avoid this problem without major information-loss, the database can be altered by combining attribute-ranges as proposed in [1].

Table 1. Collection of records and its OLAP-cube representation

name	age	education	salary
Adam	48	Master	2500
Dave	44	College	1500
Keith	32	Bachelor	2000
Norah	34	College	1000
Tracy	48	Master	3000

$\rightarrow$

$education \backslash age$	32	34	39	44	48
College	0	1	0	1	0
Bachelor	1	0	0	0	0
Master	0	0	0	0	2

Throughout this paper, we only consider the two-dimensional version of the resulting abstract problem of transforming a Boolean matrix into a zero matrix by merging adjacent rows/columns. In this abstraction, a two-dimensional OLAP-cube is represented by a matrix $M \in \{0,1\}^{n \times m}$ where empty cells in the cube correspond to one- and non-empty cells to zero-entries. The n rows and m columns of M will always be denoted by $r_1, \ldots, r_n$ and $c_1, \ldots, c_m$, respectively. Further, since there are many statements that apply symmetrically to a row or a column, the term *line* is used to refer to both. For $M[i,j] = 1$, a one-entry in $M[i-1,j], M[i+1,j], M[i,j-1]$ or $M[i,j+1]$ is called a *neighbour*. A one-entry is called *isolated*, if it has no neighbours.

The term *merging* will be used to express the transformation of replacing two adjacent lines l_1, l_2 of M by one line l, computed by $l[i] = l_1[i] \cdot l_2[i]$. This operation, in the following also described by the term (l_1, l_2), can be seen as performing a component-wise logical AND on two adjacent lines which models combining ranges in the OLAP-cube where the resulting combined cell is empty if and only if both participating original cells are empty. This operation is commutative which allows writing the shortened term $(l_1, \ldots, l_r)$ instead of the merging-set $\{(l_1, l_2), (l_2, l_3), \ldots, (l_{r-1}, l_r)\}$.

As long as M contains at least one zero-entry, the set of all possible mergings always translates M into a zero matrix. With the original objective of minimizing information-loss, this transformation is not very reasonable. Minimizing the number of merging-operations was already discussed in [1] and [7]. A different way to measure information-loss is considering each altered original line as "lost". The idea behind this new measurement is illustrated in the example below. This objective yields the following abstract problem:

LINE-MERGING MINIMIZING AFFECTED LINES (LMAL)
Input: $M \in \{0,1\}^{n \times m}$, $k \in \mathbb{N}$.
Question: Is there a set S of operations to transform M into a zero matrix with $|\{l_i : ((l_{i-1}, l_i) \in S) \vee ((l_i, l_{i+1}) \in S)\}| \leq k$?

A solution for LMAL can also be described by the set of affected lines. A set of lines L will be called *feasible*, if $(l_{i-1} \in L) \vee (l_{i+1} \in L) \; \forall \; l_i \in L$. A feasible set of lines L is a solution for a LMAL instance (M, k), if the merging-set $S = \{(l_i, l_{i+1}): l_i, l_{i+1} \in L\}$ transforms M into a zero matrix.

Example 1. Consider the following OLAP-cube with the non-confidential attributes *age* and *education*:

education \ age	30	31	32	33	34	35	36	37	38	39
None	8	7	9	4	2	0	1	0	0	0
High-School	5	6	4	2	2	1	0	1	1	1
College	4	5	7	10	3	0	2	1	0	0
Bachelor	2	2	7	6	2	1	0	0	1	1
Master	3	3	5	4	6	0	1	0	2	1
PhD	1	2	6	8	7	2	0	1	0	0

A solution with the smallest number of mergings is $\{(r_1, r_2), (r_3, r_4), (r_5, r_6)\}$ which alters 6 original ranges. This solution creates a table with only three ranges for the attribute *age*:

education \ age	30	31	32	33	34	35	36	37	38	39
None or High-School	12	13	13	6	4	1	1	1	1	1
College or Bachelor	6	7	14	16	5	1	2	1	1	1
Master or PhD	4	5	11	12	14	2	1	1	2	1

Minimizing affected lines, an optimal solution would be merging $(c_6, \ldots, c_{10})$ which requires 4 operations but affects only 5 lines. This solution combines the dense last columns and creates the more balanced table:

education \ age	30	31	32	33	34	35–59
None	8	7	9	4	2	1
High-School	5	6	4	2	2	4
College	4	5	7	10	3	3
Bachelor	2	2	7	6	2	3
Master	3	3	5	4	6	4
PhD	1	2	6	8	7	3

The new measure for information-loss prefers neighbouring operations which seems more suitable for distributions in which empty cells tend to accumulate.

In the following we study the new problem LMAL. Section 2 discusses its complexity and rules for kernelization. Section 3 derives the parameterized algorithm with time-complexity in $\mathcal{O}^*(2.618^k)$.

2   Complexity and Reduction Rules

The previous problem-variation with the objective to minimize the number of mergings was already identified as NP-complete [7]. The new measure for information-loss does not seem to simplify this problem:

Theorem 1. *The decision-problem variation of LMAL is NP-complete.*

Proof. Membership in NP is easily seen by nondeterministically guessing k operations and checking the resulting matrix for one-entries. Reduction from vertex cover, as one of Karp's famous 21 NP-complete problems [8], proves hardness. Let (G, k) be a vertex cover instance and let $v_1, \ldots, v_n$ be the nodes and $e_1, \ldots, e_m$ the edges of G. If $k \geq n$ ((G, k) is a trivial "yes"-instance) return a zero matrix as a trivial "yes"-instance for LMAL. If $k < n$, the following construction yields a matrix $M \in \{0, 1\}^{(6n+2m) \times (3n+5m)}$ for which $(M, 2n + 2m + k)$ is a "yes"-instance for LMAL if and only if (G, k) is a "yes"-instance:

Each node v_i of G is represented by three rows $r_i, \hat{r}_i, \tilde{r}_i$. Neighbouring in this order for all nodes, these build the first $3n$ rows of M. The rows $\hat{r}_i, \tilde{r}_i$ will always be altered by a minimal solution, merging r_i corresponds to v_i being in the vertex cover. Each edge e_j is represented by three columns c_j^1, c_j, c_j^2. Neighbouring in this order for all edges, these build the first $3m$ columns of M. The columns c_j^1, c_j^2 model the connection to the incident nodes. Since just one of these nodes has to be in a cover, c_j will be forced to merge with one of its neighbours c_j^1, c_j^2 leaving just one of them to invoke row-mergings. The remaining $3n + 2m$ rows will be denoted by $h_1^r, \ldots, h_{3n+2m}^r$ and used to force the columns c_j to be altered in every minimal optimal solution. The remaining $3n + 2m$ columns $h_1^c, \ldots, h_{3n+2m}^c$ similarly trigger the choice of the rows $\hat{r}_i, \tilde{r}_i$.

Each edge $e_j = (v_i, v_l)$ (undirected but incident nodes considered given in an arbitrary, fixed order) induces a one-entry in the corresponding column c_j^1 in row r_i representing v_i, and another in column c_j^2 and row r_l representing v_l. These entries will be called edge-induced. To force the merging of c_j with one of its neighbours, this column has one-entries in all of the $3n + 2m$ additional rows. If c_j is not merged, all of these entries have to be eliminated by row-merging which requires $3n + 2m > 2n + 2m + k$ operations, hence exceeding the optimal solution. The rows $\hat{r}_i$ have one-entries in h_{2j-1}^c, $\tilde{r}_i$ in h_{2j}^c for $i = 1, \ldots, n$, $j = 1, \ldots, \lceil 3n/2 \rceil + m$. A solution that does not affect all of the rows $\hat{r}_i$ and $\tilde{r}_i$ has to merge all of the additional columns instead which again exceeds $2n + 2m + k$.

A vertex cover C for G can be translated to $L = \{\hat{r}_1, \ldots, \hat{r}_n\} \cup \{\tilde{r}_1, \ldots, \tilde{r}_n\} \cup \{c_1, \ldots, c_m\} \cup \{r_i : v_i \in C\} \cup \{c_j^1 : (e_j = (v_{i_1}, v_{i_2})) \wedge (v_{i_1} \in C)\} \cup \{c_j^2 : (e_j = (v_{i_1}, v_{i_2})) \wedge (v_{i_1} \notin C)\}$. A feasible LMAL solution L for M on the other hand can be altered to a solution of the same size that deletes at least one of the edge-induced one-entries for every edge by row-merging implying that the set $\{v_i : r_i \in L\}$ is a vertex cover for C (observe that additionally merging c_j^1 with (c_j, c_j^2) only deletes a single one-entry and can hence be replaced by covering it by the corresponding row which, by construction, always has the merging-partner $(\hat{r}, \tilde{r})$). Since covering all the additional lines of M induces a cost of $2n + 2m$, a solution of size at least $2n + 2m + k$ contains at most k of the rows r_i.   □

In the following we use the terms of parameterized complexity presented in [5]. A decision-problem can be considered *parameterized*, if its instances can be described as elements of $\Sigma^* \times \mathbb{N}$. A parameterized problem P is called *fixed parameter tractable*, if it can be solved in time $f(k)p(n)$ for every instance (I, k) of P, with $n = size(I)$ where p is an arbitrary polynomial and f an arbitrary function. An equivalent way to describe fixed parameter tractability is reduction to a *problem-kernel*, a procedure that reduces an instance (I, k) in time $p(size(I))$ to an equivalent instance (I', k') with $k', size(I') \leq f(k)$. LMAL can be interpreted as parameterized by the number of affected lines k assuming that the information-loss should be reasonably bounded.

A simple false assumption is the idea of reducing the given instance by deleting all lines without one-entries; even for two neighbouring "empty" lines, this reduction may alter the solution-size. Consider, for example, deleting the second row and column from the following matrix:

$$\begin{pmatrix} 1 & 0 & 0 & 0 \\ 0 & 0 & 0 & 0 \\ 0 & 0 & 0 & 0 \\ 0 & 0 & 0 & 1 \end{pmatrix} \rightarrow \begin{pmatrix} 1 & 0 & 0 \\ 0 & 0 & 0 \\ 0 & 0 & 1 \end{pmatrix}$$

The original matrix requires four lines to eliminate all one-entries, the reduced version can be solved with three. Counting affected lines, merging-operations with a line of zeros are not independent. Considering three neighbouring lines of zeros, the one in the middle however is never included in a minimal solution, which yields:

Reduction-Rule 1. Let (M, k) be a parameterized LMAL instance and r_{i-1}, r_i, r_{i+1} three neighbouring rows of zeros in M. Create the reduced matrix M' from M by deleting the row r_i. (M, k) is a "yes"-instance, if and only if (M', k) is.

Proof. For any set of affected lines L for a LMAL solution for M, consider the set $L' = \{r_j : r_j \in L, j < i\} \cup \{r_{j-1} : r_j \in L, j > i\}$. If L' contains the rows r_{i-1} or r_i without merging partners, delete them from L'. This creates a feasible set for M', since r_i is the only line possibly omitted from L. The operations among L' perform the same transformation on M' as L for M except for possible mergings with r_i. Since merging (r_{i-1}, r_i) or (r_i, r_{i+1}) in M does not delete any entries, the operations among L' build a solution for M' with $|L'| \leq |L|$.

Similarly, the set $L' = \{r_j : r_j \in L, j < i\} \cup \{r_{j+1} : r_j \in L, j > i\}$ (omitting r_{i-1} or r_{i+1} if they have no neighbours in L') translates affected lines of a solution for M' into affected lines of a solution for M, not increasing the size. $\qquad\square$

Rows that contain more than k entries have to be affected by the solution, since deleting its entries would otherwise affect more than k columns. This observation is similar to *Buss'* rule for vertex cover; "deleting" these rows to reduce a given instance however is not as simple as deleting nodes of large degree from a graph since the feasibility-condition for the solution requires at least one neighbouring row. The following reduction-rule introduces one possible way to reduce the number of entries by marking a reduced row with an entry in an additional column. Exchanging "row" for "column" gives the equivalent rule for columns with more than k entries, as the whole argumentation is obviously symmetrical.

Reduction-Rule 2. Let r_i be a row in M with more than k entries in a LMAL instance (M, k). Construct the reduced matrix M' from M by:

1. Delete all entries $M[i, j] = 1$ with $M[i - 1, j] = M[i + 1, j] = 0$.
2. Add two columns of zeros and the i-th unit-vector to the right border of M $(M \rightarrow [M|0|0|e_i])$.

With these transformations, (M, k) is a "yes"-instance, if and only if (M', k) is.

Proof. Let S be a solution for M that affects at most k lines. Since r_i has more than k entries, S has to at least merge either (r_{i-1}, r_i) or (r_i, r_{i+1}). Each of these operations delete the new one-entry in the last column of M'. With all other one-entries copied from M, S is a solution for M'.

A solution S for M' that merges the new columns, on the other hand, can be altered to a solution S' that affects the same number of lines and merges either (r_{i-1}, r_i) or (r_i, r_{i+1}) instead (observe that they both delete the only one-entry possibly deleted by column-merging the last column of M'). Each of these row-operations also deletes all of the entries of M which were omitted in the second step of creating M' which implies that S' is a solution for M.   □

These rules however can only yield a quadratic kernel, since even if no line contains more than k entries, the matrix-size can still be non-linear in k. Consider, for example, for a fixed $k = 4h \in \mathbb{N}$ a matrix $A \in \{0, 1\}^{(k-2) \times 6kh}$ with $A[4i+1, 3ki+3j-2]=A[4i + 2, 3k(i+h)+3j-2]=1 \ \forall \ i=0, \ldots, h-1 \ j=1, \ldots, k$.

For $k = 4$ (zero-entries denoted by ·): $\begin{pmatrix} 1 \cdot \cdot 1 \cdot \cdot 1 \cdot \cdot 1 \cdot \cdot \cdot \cdot \cdot \cdot \cdot \cdot \cdot \cdot \cdot \cdot \cdot \\ \cdot \cdot \cdot \cdot \cdot \cdot \cdot \cdot \cdot \cdot \cdot \cdot 1 \cdot \cdot 1 \cdot \cdot 1 \cdot \cdot 1 \cdot \cdot \end{pmatrix}$

Consider $M := \begin{pmatrix} A & \mathcal{O} \\ \mathcal{O} & A^T \end{pmatrix}$ with appropriately sized zero matrices $\mathcal{O}$. An optimal solution merges $(r_{4i+1}, r_{4i+2}), (c_{6kh+4i+1}, c_{6kh+4i+2}) \ \forall \ i = 0, \ldots, h - 1$, which alters k lines. No line contains more than k entries and M does not contain three neighbouring empty lines which means that none of the above reductions is applicable. With $n = m = \frac{3}{2}k^2 + k - 2$ the matrix-size is quadratic in k.

3   Parameterized Algorithm

Parameterized by the number of merged lines, branching on a single one-entry by choosing either its row or its column already provides an approach in $\mathcal{O}^*(2^k)$. Branching on both possibilities to pick a partner-line yields feasibility and preserves this running-time. Branches that consider adding one of the non-existing lines $r_0, r_{n+1}, c_0, c_{m+1}$ are to be omitted throughout this section.

Theorem 2. *Parameterized LMAL restricted to instances without neighbouring one-entries can be solved in $\mathcal{O}^*(2^k)$.*

Proof. Starting with $L = \emptyset$, compute the set of affected lines by:

$\text{while}((|L| < k) \wedge (\exists\, M[i,j] = 1 \colon (r_i \notin L) \wedge (c_j \notin L)))$ branch into:

1. if$((r_{i+1} \in L) \vee (r_{i-1} \in L))\ L = L \cup r_i$
   else branch:
   (a) $L = L \cup \{r_{i-1}, r_i\}$
   (b) $L = L \cup \{r_i, r_{i+1}\}$
2. if$((c_{j+1} \in L) \vee (c_{j-1} \in L))\ L = L \cup c_j$
   else branch:
   (a) $L = L \cup \{c_{j-1}, c_j\}$
   (b) $L = L \cup \{c_j, c_{j+1}\}$

These choices of lines arise from the obvious conditions for any feasible solution:

1. For each one-entry, either its row or column has to be included in the solution.
2. Lines in L always have a merging-partner ($r_i \in L \Rightarrow (r_{i+1} \in L) \vee (r_{i-1} \in L)$).
Branching into all sets of lines that cover a one-entry satisfying these conditions
yields all possibilities to compute a solution for M.

Since an isolated one-entry is eliminated by any adjacent operation, a feasible
set of lines covering all one-entries transforms M into a zero matrix. With this
property, any set that satisfies the conditions above gives a solution for a matrix without neighbouring one-entries. All branching-options produce a recursion
solved by $T(k) = 2^k$. $\qquad\square$

The construction above benefits from the fact that the selected lines always
merge to a vector of zeros. Unfortunately, neighbouring one-entries create ex-
ceptions which do not allow for this argumentation. The following example illustrates the problems created by general instances:

Example 2. Consider the following matrix (zero-entries omitted):

$$
\begin{array}{c}
\\
r_1 \\
r_p \\
r_{i-2} \\
\\
r_i \\
r_{i+2} \\
r_q \\
r_n
\end{array}
\begin{pmatrix}
& c_1 & \cdots & c_s & \cdots & c_j & & \cdots & c_l & & \cdots & c_t & \cdots & c_m \\
& & & & & & & & 1 & & & & & \\
& & & & & 1 & & & & & & & & \\
& & & 1 & & 1 & 1 & & & & & & & \\
& & & & & 1 & 1 & & & & & 1 & & \\
& & & & & & & & 1 & 1 & & & & 1 \\
& 1 & & & & & & & 1 & 1 & & & & \\
& & & & & 1 & & & & & & & & \\
& & & & & & & & 1 & & & & &
\end{pmatrix}
$$

With $p, s > 2$, $|p - i|, |q - i| > 3$ and $|n - q|, |m - t|, |t - l|, |l - j|, |j - s| > 2$, *the*
optimal solution is the set $\{r_{i-2}, r_{i-1}, r_i, r_{i+1}, r_{i+2}, c_j, c_{j+1}, c_l, c_{l+1}\}$. Consider
running the algorithm from the proof to theorem 2 for $k=9$. Choosing the one-
entry $M[i-2, s]$ in the first step, the only branch that arrives at the correct "yes"-
response is the one that adds the set $\{r_{i-2}, r_{i-1}\}$ (all other branches contain lines
that are not included in the optimal solution).

Branching on the one-entries $M[i+2,1]$, $M[p,j]$ and $M[1,l]$ in the following three steps always leaves just one possible successful branch as well and yields the partial solution $\{r_{i-2}, r_{i-1}, r_{i+1}, r_{i+2}, c_j, c_{j+1}, c_l, c_{l+1}\}$. This set of lines contains at least the row or the column of each one-entry and the algorithm terminates. For the remaining one-entries, both row and column are already contained in the partial solution but only the choice of the row r_i gives the optimal solution. Adding a row of zeros between r_i and r_{i+1} and deleting the columns $c_{j+2}, \ldots, c_{l-2}$ creates a matrix in which the algorithm arrives at a similar partial solution and the only optimal solution is created by adding c_{j+2}.

Larger matrices built from these structures can create instances with an arbitrarily difficult optimal set to complete an unfortunate partial solution. Efficiently computing an optimal additional set turns out to be a non-trivial task. For a parameterized approach for general instances we introduce a different branching-rule and further use the following sub-problem:

ROW-MERGING MINIMIZING AFFECTED LINES (RMAL)
Input: $M \in \{0,1\}^{n \times m}$, $k \in \mathbb{N}$.
Question: Is there a set of row-mergings that transforms M into a zero matrix and affects at most k rows?

Theorem 3. *RMAL can be solved in linear time.*

Proof. The following construction yields a solution represented by the set of affected lines:

1. Start with $L = \{r_i : \exists\, 1 \le p \le m : M[i,p] = 1\}$.
2. Collect the rows with one-entries that are not eliminated by the operations $S := \{(r_i, r_{i+1}) : r_i, r_{i+1} \in L\}$ in a new set R. Rows $r_i \in R$ are characterized by:

$$(r_i \in L) \wedge (r_{i+1} \notin L) \wedge (\exists p : \forall \min\{j : r_t \in L \vee j \le t \le i\} \le h \le i\ M[h,p] = 1).$$

These rows require another merging partner outside the current set L.
3. While $R \ne \emptyset$, set $i = \min\{1 \le j \le n : r_j \in R\}$, $R = R \setminus r_i$ and:
   (a) If $i = n$, add r_{n-1} to L.
   (b) If $i < n$, add r_{i+1} to L and delete other rows possibly covered by r_{i+1}: If $r_{i+2} \in L$ and $r_h \in R$ with $h = \max\{1 \le j \le n : \forall\ i+2 \le t \le j\ r_t \in L\}$, set $R = R \setminus r_h$.

Correctness of the (possibly not feasible) starting-set follows from the simple fact that every non-empty row has to be altered. Each row collected in R is either non-empty and without merging-partner in L ($\min\{j : r_t \in L \vee j \le t \le i\} = i$) or the row of largest index in a group that does not merge to an empty row. Transforming L into a feasible solution requires at least one row outside L as merging-partner for each row in R. Since the rows outside L are empty, any partner outside L suffices to delete all one-entries and produces an additional cost of one. With this observation, each set that contains L and at least one partner outside L for each row in R yields a feasible set that deletes all one-entries in M. Step 3 creates a set with this property.

Minimality of the constructed solution is seen as follows: Each row of zeros picked for the solution can be used to cover at most two rows in R. Always fixing the row of lowest index exploits the fact that choosing its upper neighbour can not cover another row in R. $\qquad\square$

The following final algorithm first computes a partial solution which contains the row of each not-deleted one-entry. This property allows a more efficient way of expanding it to a solution: Branching for one-entries not deleted by the partial solution, the options for row-merging can be postponed. Using the *easy* sub-problem RMAL as a final polynomial-time step, the "book-keeping" introduced in [7] allows pre-counting the cost of the postponed row-mergings to improve this algorithm.

Theorem 4. *Parameterized LMAL can be solved in* $\mathcal{O}^*(2.618^k)$.

Proof. The following three steps compute a solution L (set of affected lines):

Step 1: Starting with $L = \emptyset$, compute a feasible partial solution:

while$(\exists M[t,j] = 1\colon (r_t \notin L) \wedge (c_j \notin L) \wedge (|L| < k))$

Select a row $r_i \notin L$ with $i = argmax\{C(i,j)\colon 1 \leq i \leq n, r_i \notin L\}$ where $C(i,j)$ counts the neighbouring one-entries around c_j:
$$C(i,j) = \max\{u\colon M[i,h] = 1 \vee j \leq h \leq u\} - \min\{o\colon M[i,h] = 1 \vee o \leq h \leq j\} + 1$$
Let $c_o, \ldots, c_u$ be the $C(i,j)$ columns that contain the one-entries in r_i around $M[i,j]$. Branch into:

1. if$((r_{i+1} \in L) \vee (r_{i-1} \in L))$ $L = L \cup r_i$
   else, branch:
   (a) $L = L \cup \{r_{i-1}, r_i\}$
   (b) $L = L \cup \{r_i, r_{i+1}\}$

2. if$((c_{u+1} \in L) \vee (c_{o-1} \in L))$ $L = L \cup \{c_o, \ldots, c_u\}$
   else, branch:
   (a) $L = L \cup \{c_{o-1}, \ldots, c_u\}$
   (b) $L = L \cup \{c_o, \ldots, c_{u+1}\}$

If the corresponding operations already transform M into a zero matrix, L is a feasible solution. Else expand the partial solution with steps 2 and 3.

Step 2: Choose additional columns for L and save row-merging for step 3:

With $B_l(j) := \min\{s\colon c_s, \ldots, c_j \in L\}$, $B_r(j) := \max\{t\colon c_j, \ldots, c_t \in L\}$ for $c_j \in L$ and $B_l(j) := B_r(j) := j$ for $c_j \notin L$, the set I of one-entries which are not deleted by the operations corresponding to L can be characterized by:

$I = \{(o, u, j) \in \{1, \ldots, n\} \times \{1, \ldots, n\} \times \{1, \ldots, m\}\colon r_o, \ldots, r_u \in L, \ r_{o-1}, r_{u+1} \notin L$
and $M[h, s] = 1 \vee o \leq h \leq u, \ B_l(j) \leq s \leq B_r(j) = j\}$

Branch into either row- or column-merging for these one-entries and save row-merging for a final polynomial-step by collecting rows in a set R (initially $R = \emptyset$):

while$((I \neq \emptyset) \wedge (|L| + \frac{1}{2}|R| < k))$

Choose the column c_j with $j = \min\{t: \exists o, u: (o, u, t) \in I\}$ and branch into:
1. Column-merging:
    if$(c_j \in L)$
        if$(j < n)$ $L = L \cup c_{j+1}$
        if$(1 < B_l(j) < j = n)$ $L = L \cup c_{B_l(j)-1}$
    if$(c_j \notin L)$
        if$(c_{j-1}, c_{j+1} \notin L)$
            if$(j < n)$ $L = L \cup \{c_j, c_{j+1}\}$
            if$(1 < j = n)$ $L = L \cup \{c_{j-1}, c_j\}$

2. Row-merging: $R = R \cup \{(o, u): (o, u, j) \in I\}$
Recalculate I for the new set L omitting one-entries (o, u, j) with $(o, u) \in R$.

Step 3: Compute additional rows with RMAL: Choose merging-partner for the set $\{r_u: \exists o, j: (o, u, j) \in R\}$ with step 3 of the procedure used for theorem 3.

Since the branching-step considers all minimal, feasible possibilities to delete the one-entries $M[i, o], \ldots, M[i, u]$, any solution for M contains at least one of the sets constructed by the first step. All branches of the first step give a recursion with a running-time in $\mathcal{O}^*(2^k)$. The particular choice of the branching-row $i = argmax\{C(i, j): 1 \leq i \leq n, r_i \notin L\}$ in step 1 produces a partial solution S that contains the row of every one-entry that is not eliminated: Suppose a one-entry $M[t, j]$ is not deleted by S with $r_t \notin S$. Since S contains at least row or column for each one-entry, c_j is contained in S. Let r_i be the branching-row that added c_j to S among, w.l.o.g, $c_{o-1}, \ldots, c_u$. Since S applies the operation $(c_{o-1}, \ldots, c_u)$ and does not delete the one-entry $M[t, j]$, the row r_t contains the one-entries $M[t, o-1] = \ldots = M[t, u] = 1$ which would give $C(t, j) \geq u - o + 2 = C(i, j) + 1$, a contradiction to the choice of i.

One-entries not deleted by S are relics of neighbouring one-entries in rows that are contained in L which justifies the characterization of I used in step 2. This property further implies that rows outside L are transformed into rows of zeros by the column-mergings in L. This property is the crucial difference to the partial solution from the procedure for theorem 2. Since any additional row that merges with $r_o, \ldots, r_u$ deletes all remaining one-entries in these rows omitting one-entries (o, u, j) with (o, u) already chosen to be merged by the final polynomial step is justified and pre-counting the cost for each row in R is valid. Since each row added in the final step can cover at least two groups of rows, step 3 induces a cost of at least $\frac{1}{2}|R|$. Since the second branch increases $|R|$ by at least one, this book-keeping yields the recursion $T(k) \leq T(k - 1/2) + T(k - 1)$ for each branching-step which gives the stated running-time.

Always choosing the one-entry of lowest column-index allows to reduce the options in the branch that treats this one-entry by column-operations to merging it with the *right* neighbouring column (unless c_j is the last column of M). Like in the procedure for RMAL, choosing the left neighbour does not delete other one-entries and can be replaced by choosing the right neighbour instead. With the properties of L from step 1, the set I always contains *all* one-entries that are not deleted by adding at least one row for each row in R to the current partial solution. This implies that the columns c_h with $h < j$ and especially the possible left merging-partner are already empty. The cases in branch 1 consider all possibilities to add the right neighbour preserving feasibility.   □

4   Conclusion

This paper studied an abstract problem from the field of privacy-protection in statistical databases with global recoding. We studied the new problem LMAL considering parameterized complexity and presented reduction rules and a branching-algorithm based on the *easy* sub-problem RMAL.

Considering the motivating background, a generalization for matrices of higher dimension would be very interesting. While the ideas for solving the easy sub-problem remain applicable, a translation for the general parameterized algorithm seems to be more difficult. The model introduced here so far only suffices *1-compromise*. Another interesting generalization would be a generalization for *k-compromise* for larger values of k.

References

1. Branković, L., Fernau, H.: Approximability of a $\{0, 1\}$-matrix problem. In: Proceedings of the 16th Australasian Workshop on Combinatorial Algorithms, pp. 39–45 (2005)
2. Branković, L., Horak, P., Miller, M.: A Combinatorial Problem in Database Security. Discrete Applied Mathematics 91(1-3), 119–126 (1999)
3. Branković, L., Giggins, H.: Security of Statistical Databases. In: Petkovic, M., Jonker, W. (eds.) Security, Privacy and Trust in Modern Data Management, pp. 167–182. Springer (2007)
4. Bredereck, R., Nichterlein, A., Niedermeier, R.: Pattern-Guided k-Anonymity. In: Fellows, M., Tan, X., Zhu, B. (eds.) FAW-AAIM 2013. LNCS, vol. 7924, pp. 350–361. Springer, Heidelberg (2013)
5. Downey, R., Fellows, M.: Parameterized Complexity. Springer (1999)
6. Evans, P., Wareham, T., Chaytor, R.: Fixed-parameter tractability of anonymizing data by suppressing entries. Journal of Combinatorial Optimization 18(4), 362–375 (2009)
7. Fernau, H.: Complexity of a $\{0, 1\}$-matrix problem. The Australasian Journal of Combinatorics 29, 273–300 (2004)
8. Karp, R.: Reducibility among Combinatorial Problems. Complexity of Computer Computations, 85–103 (1972)
9. Machanavajjhala, A., Kifer, D., Gehrke, J., Venkitasubramaniam, M.:l-Diversity: Privacy Beyond k-Anonymity. ACM Transactions on Knowledge Discovery from Data 1(1) (2007)
10. Samarati, P.: Protecting Respondents' Identities in Microdata Release. IEEE Transactions on Knowledge and Data Engineering 13(6), 1010–1027 (2001)

A Circular Matrix-Merging Algorithm with Application in VMAT Radiation Therapy[*]

Danny Z. Chen[1], David Craft[2], and Lin Yang[1,**]

[1] Department of Computer Science and Engineering,
University of Notre Dame, Notre Dame, IN 46556, USA
{dchen,lyang5}@nd.edu
[2] Department of Radiation Oncology,
Massachusetts General Hospital and Harvard Medical School,
Boston, MA 02114, USA
dcraft@partners.org

Abstract. In this paper, we study an optimization problem, called the circular matrix-merging (CMM) problem: Given a cyclic sequence of n non-negative matrices that are associated with some locations on a circle and an operation that merges two consecutive matrices into one matrix at the expense of some merging error, merge the n matrices into the minimum number of matrices under the constraint of a threshold of the total merging error. This problem arises in Volumetric Intensity-Modulated Arc Therapy (VMAT) for radiation cancer treatment, where the circular structure represents a delivery arc path of $360°$ around the patient and the matrices represent the radiation fluence maps at selected delivery angles on that path. By computing the minimum k matrices, the CMM algorithm produces a most efficient delivery plan that meets clinical requirements. Based on dynamic programming and computing a set of k-link shortest paths, we present a polynomial time algorithm for the CMM problem, improving the quality and efficiency of VMAT delivery.

1 Introduction

In this paper, we study an optimization problem, called the circular matrix-merging (CMM) problem: Given a cyclic sequence of n non-negative matrices $M_0, M_1, \ldots, M_{n-1}$ of size $h \times w$ each that are associated with some locations on a circle and an operation that merges two consecutive matrices into a new $h \times w$ matrix at the expense of some merging error (as defined by the error function in Section 2.3), merge the n matrices into the minimum number, k, of matrices while keeping the sum of merging errors under an input error tolerance Er.

The CMM problem arises in the planning optimization of Volumetric Intensity-Modulated Arc Therapy (VMAT). VMAT is a rotational radiation cancer treatment modality that has the potential to precisely and efficiently deliver radiation doses to target tumors. A common hardware structure for VMAT is

---

[*] This research was supported in part by NSF under Grant CCF-1217906.
[**] Corresponding author.

J. Chen, J.E. Hopcroft, and J. Wang (Eds.): FAW 2014, LNCS 8497, pp. 36–47, 2014.

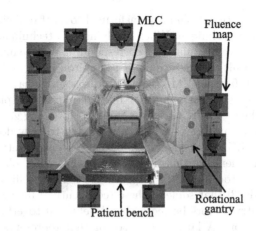

Fig. 1. A common hardware structure for VMAT (adapted from [13])

shown in Figure 1. A linear accelerator (LINAC) equipped with a multileaf colli-
mator (MLC) is mounted on a rotatable gantry. The LINAC generates radiation
beams and the MLC forms deliverable apertures to modulate the radiation beam
shapes. The rotatable gantry allows to deliver desired radiation fluence distributions along an arc path of 360° around the patient. The radiation fluence
distribution at each delivery angle is called a *fluence map* and is a non-negative
matrix. Comparing to the prevailing fixed-angle Intensity-Modulated Radiation
Therapy (fixed-angle IMRT), which uses a similar hardware structure but delivers radiation only at a set of fixed gantry angles, VMAT has more freedom in
choosing beam angles, gantry speed, dose rate, etc. Such freedom enables VMAT
to deliver more efficiently desired doses to target areas with comparable or even
better delivery quality than fix-angle IMRT [7,11,12]. In the CMM modeling of
a VMAT optimization problem, n represents the number of initial delivery angles, the matrices $M_0, M_1, \ldots, M_{n-1}$ represent the fluence map at each delivery
angle, and the circle formed by these matrices represents a 360° delivery arc.

However, such freedom of VMAT also poses new optimization problems in
VMAT planning optimization [3]. There are two main approaches for computing VMAT treatment plans [8]. The first approach, called direct aperture optimization, computes a deliverable aperture directly for each delivery angle while
taking MLC movement restrictions into account [5,6]. The second approach is a
two-stage method: The first stage divides the delivery arc into many small arc
segments and solves the fluence-map optimization (FMO) problem on each such
arc segment; the second stage uses an arc-sequencing algorithm (e.g., [2,4,14])
to compute deliverable apertures for the corresponding optimized fluence maps.

Recently, a new two-stage method, VMERGE, which offers considerable improvements on other known VMAT approaches, was proposed by Craft *et al.*
[3]. In its first stage, it divides the 360° delivery arc into 180 arc segments of
2° each, which we call *gantry sectors*. Then a multi-criteria optimization is performed to compute the optimal fluence map for each segment. In the second

stage, the desired treatment plan is transformed to a set of deliverable apertures using sliding-window technique. When applying this technique, all MLC leaves moves back and forth to modulate the corresponding fluence map for each gantry sector (for more details of this technique, see Spirou *et al.* and Svensson *et al.*'s work [9,10]). This 180-beam VMAT plan produces an "ideal" dose distribution. But, to deliver this treatment plan, each fluence map in this plan requires a full leaf sweep when the gantry rotates over the corresponding gantry sector. Thus, a large number of fluence maps require the gantry to slow down frequently in order to give the MLC leaves enough time to sweep each gantry sector. As a result, the delivery efficiency of this "ideal" 180-beam VMAT plan is unacceptable (its delivery time is too long). To improve the delivery efficiency, it was proposed in VMERGE [3] that some of these (consecutive) fluence maps and their corresponding gantry sectors be merged and delivered together at the expense of some delivery errors. A large gantry sector that represents multiple merged consecutive gantry sectors is called an *arc slice*. This VMAT paradigm [3] enables the user to explore the trade-off between the target coverage and healthy tissue sparing and to choose an appropriate compromise between delivery time and plan quality. In the examples studied, VMERGE is able to deliver a high quality plan in less than 5 minutes on average [3].

In the second stage of VMERGE, it seeks an optimal matrix merging solution that improves the delivery efficiency while keeping the delivered dose quality clinically acceptable. The smaller the number of arc slices is, the more efficient a treatment delivery becomes. However, a smaller number of arc slices also incurs a larger merging error. The CMM problem is formulated for this problem: Er represents the maximum allowed total merging error of a plan, k represents the minimum possible number of arc slices for generating a clinically acceptable plan such that the total merging error is less than Er ($1 \leq k \leq n$), and a merging solution specifies how to merging the fluence maps. VMERGE solved the CMM problem using a greedy heuristic [3]. It iteratively merges consecutive fluence maps in the cyclic sequence with the lowest merging error, and delivers the sum of these maps for each resulting arc slice [8]. Although the heuristic merging scheme of VMERGE does decrease the overall MLC leaf travel distances and increase the delivery efficiency, its greedy heuristic nature does not guarantee any optimality of the total merging error and delivery efficiency.

We develop a new approach for the CMM problem. We model this problem based on dynamic programming and computing a set of k-link shortest paths in a weighted directed acyclic graph of $O(n)$ vertices and $O(n^2)$ edges. The time complexity of our algorithm is $O(n^3 * h * w + n^3 k)$. Comparing to the greedy merging algorithm in VMERGE [3], our algorithm guarantees a globally optimal solution that allows a faster delivery time without sacrificing the plan quality.

2   Preliminaries

This section gives the CMM problem statement and discusses some key features such as the matrix merging operation and error function for merging matrices.

(a)   (b)

Fig. 2. An example of the input and output for the CMM problem: (a) An input (with $n = 8$); (b) an output (with $k = 5$)

2.1   Definition of the CMM Problem

1. INPUT: (1) A cyclic sequence of n non-negative matrices $M_0, M_1, \ldots, M_{n-1}$ of size $h \times w$ each that are associated with some locations on a circle (say, in a clockwise order around the circle). (2) The maximum total merging error tolerance Er.

2. OUTPUT: A sequence of k matrices $M_0', M_1', \ldots, M_{k-1}'$ of size $h \times w$ each which are obtained by merging the n input matrices such that k is the minimum possible number of the resulting matrices under the constraint of the total merging error $Er_{total} = \sum_{i=0}^{k-1} \varepsilon_i < Er$, where ε_i is the merging error incurred by the i-th output matrix M_i' generated by the merging process.

The error incurred by generating a matrix M (by merging multiple consecutive input matrices into M) is called the *merging error* of M. Obviously, the merging error of any input matrix M_i is 0. Figure 2 illustrates the CMM problem.

2.2   The Matrix Merging Operation

The operation of merging two consecutive matrices M_a and M_b in a cyclic sequence into a new matrix M_c replaces M_a and M_b by M_c in the resulting cyclic sequence. Such merge incurs a merging error, denoted by $\delta(M_a, M_b)$, for M_c (defined by the error function in Section 2.3). M_a and M_b can be input matrices or matrices produced by the merging process, and M_c is the sum of M_a and M_b, i.e., $M_c = M_a + M_b$. On the circle, the arc slice associated with M_c is the union of the two arc slices which are associated with M_a and M_b, respectively (e.g., see Figure 2).

2.3   Merging Errors

VMERGE uses a similarity measure for deciding which matrices to merge next in its greedy strategy [3]. We adopt this similarity measure as our error measure for merging two matrices. The merging error $\delta(M_a, M_b)$ for merging two consecutive matrices M_a and M_b is determined by the difference between M_a and M_b in the Frobenius norm. $\delta(M_a, M_b)$ is computed using Equation (1). M_a and M_b are two consecutive matrices with arc slice lengths of θ_a and θ_b respectively, where i and j are the indices of rows and columns of the matrices.

$$\delta(M_a, M_b) = (\theta_a + \theta_b) \sqrt{\sum_{i,j} (\frac{M_{ij}^a}{\theta_a} - \frac{M_{ij}^b}{\theta_b})^2} \tag{1}$$

Clearly, computing $\delta(M_a, M_b)$ takes $O(|M_a| + |M_b|) = O(h * w)$ time. Note that many other possible merging error functions can also be implemented by our CMM algorithm.

It should be pointed out that the above error function $\delta(\cdot, \cdot)$ is not associative over the matrix merging operations, i.e., the merging error of $M_a + (M_b + M_c)$ may not be the same as the merging error of $(M_a + M_b) + M_c$ (although $M_a + (M_b + M_c) = (M_a + M_b) + M_c$). This implies that for a same merging solution, the total merging errors may be different corresponding to different orders in which the matrices are merged together. Also, observe that by the definition of the CMM problem, once the minimum possible total merging error, Er_{min}, for a merging solution is smaller than Er, that solution meets the requirement of $Er_{total} < Er$. Therefore, henceforth, we always compute Er_{min} for any desired merging solution and use it to evaluate whether that solution meets the total merging error requirement. Furthermore, since the merging error incurred by generating each individual output matrix M' is merely affected by the order of merging together all the input matrices that are "spanned" by M', we only need to determine the optimal merging order for the minimum possible merging error of each candidate output matrix, judiciously select the minimum number k of output (merged) matrices, and add up their merging errors in order to compute Er_{min} for an optimal merging solution.

3   Our Algorithm

The CMM problem aims to merge the n input matrices in a cyclic sequence into the minimum possible k output matrices under the total merging error constraint of Er. Let S_{pos} denote the set of all possible candidates of output matrices. Then the CMM problem becomes a problem of selecting a subset S_{ans} of S_{pos} along a cyclic order that meets the following requirements: (1) Each input matrix is included in (or merged into) exactly one matrix in S_{ans}; (2) Er_{min} of S_{ans} is smaller than Er; (3) the total number of matrices in S_{ans}, k, is minimized.

As discussed in Section 2, we use Er_{min} to evaluate a merging solution. In order to obtain Er_{min}, we need to compute the minimum merging error of each candidate output matrix. Thus, we divide the CMM problem into two sub-problems. In the first sub-problem, the minimum merging error of each candidate matrix in S_{pos} is computed, based on a dynamic programming scheme (in Section 3.1). In the second sub-problem, a globally optimal S_{ans} is determined. Our idea is to model the second sub-problem by using a graph, and transform it to a problem of computing many k-link shortest paths in the graph (in Section 3.2).

3.1 Computing Merging Errors of All Candidate Output Matrices

The first sub-problem computes the minimum possible merging error of every candidate output matrix. Let $M(i,j)$ denote a candidate matrix in S_{pos}, which is the merged matrix of the input matrices M_i, M_{i+1}, ..., M_j along their cyclic order (say, clockwise) around the circle, and let $E(M(i,j))$ denote the minimum possible merging error of $M(i,j)$. Note that $i = j$ is possible, in which case $E(M(i,i)) = 0$ which can be computed trivially in $O(1)$ time for any i. For $i \neq j$, either $i < j$ or $i > j$. When $i < j$, $M(i,j)$ merges consecutively the input matrices M_i, M_{i+1}, ..., M_j; when $i > j$, due to the cyclic order of the input matrices, $M(i,j)$ merges consecutively the input matrices M_i, M_{i+1}, ..., M_{n-1}, M_0, M_1, ..., M_j. Since the matrix merging operation merges only two consecutive matrices in a cyclic sequence at a time, $M(i,j)$ must be obtained by merging $M(i,z)$ and $M((z+1) \mod n, j)$, for some $z \in \{i, i+1, \ldots, j-1\}$ if $i \leq j$ or $z \in \{i, i+1, \ldots, n-1, 0, 1, \ldots, j-1\}$ if $i > j$. Hence, the minimum merging error $E(M(i,j))$ is equal to the sum of $E(M(i,z))$, $E(M((z+1) \mod n, j))$, and the merging error for merging two matrices $M(i,z)$ and $M((z+1) \mod n, j)$ for some z. That is, the minimum merging error $E(M(i,j))$ can be computed using Equation (2) below.

$$E(M(i,j)) = \min\{E(M(i,z)) + E(M((z+1) \mod n, j))$$
$$+ \delta(M(i,z), M((z+1) \mod n, j))\},$$
$$\forall z \in \{i, i+1, \ldots, j-1\} \text{ if } i \leq j$$
$$\text{or } \forall z \in \{i, i+1, \ldots, n-1, 0, 1, \ldots, j-1\} \text{ otherwise} \quad (2)$$

By Equation (2), the values of $E(M(i,j))$'s for all pairs of (i,j) with $0 \leq i, j \leq n-1$ can be computed by a common dynamic programming scheme. More specifically, the dynamic programming algorithm first computes the minimum merging error of all candidate output matrices that contain only one input matrix (all such merging errors, $E(M(i,i))$'s, are simply 0), then computes the minimum merging errors of all candidate output matrices that contain two consecutive input matrices, and so forth iteratively. But before doing so, we need to deal with the cyclic order of the input matrices. That is, we like to reduce the cyclic order of the input matrices to a non-cyclic linear order, or we "linearize" the cyclic sequence of the input matrices to a non-cyclic sequence, as follows.

Based on the cyclic sequence of the n input matrices M_0, M_1, ..., M_{n-1}, we create a non-cyclic sequence S_L of $2n-1$ matrices A_0, A_1, ..., A_{n-1}, A_n, A_{n+1}, ..., A_{2n-2}, such that for each A_i in S_L with $i = 0, 1, \ldots, 2n-2$, $A_i = M_{(i \mod n)}$ (i.e., most of the input matrices M_j, except for M_{n-1}, appear exactly twice in S_L). With a little abuse of notation, we denote S_L simply as M_0, M_1, ..., M_{n-1}, M_n, M_{n+1}, ..., M_{2n-2}, with $M_{n+i} = M_i$ for each $i = 0, 1, \ldots, n-2$.

Now observe that for each $M(i,j)$ (and $E(M(i,j))$) in S_{pos}, if $i \leq j$, then we can compute $E(M(i,j))$ from the matrices $M_i, M_{i+1}, \ldots, M_j$ in S_L; otherwise ($i > j$), we can compute $E(M(i,j))$ from the matrices $M_i, M_{i+1}, \ldots, M_{n+j}$ in S_L. Therefore, using the linear sequence S_L of matrices, we can compute the minimum merging errors $E(M(i,j))$ for all candidate matrices $M(i,j)$ in S_{pos}.

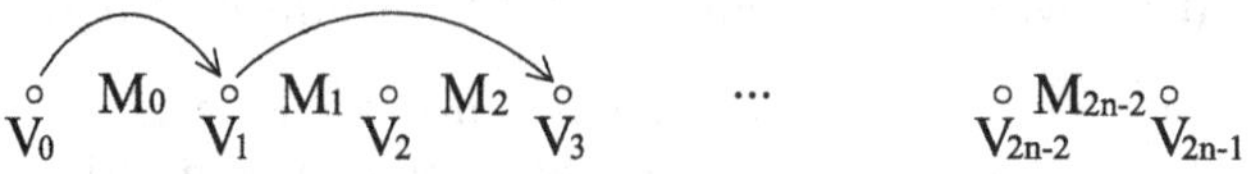

Fig. 3. Illustrating the graph $G = (V, E)$. For example, the edge (v_1, v_3) corresponds to merging the two input matrices M_1 and M_2.

Our dynamic programming algorithm, `Calc-Merge-Errors`, for computing the minimum merging errors $E(M(i, j))$ for all candidate output matrices $M(i, j)$ in S_{pos} (with $0 \leq i, j \leq n - 1$), based on the linear sequence S_L, is given below.

```
Algorithm Calc-Merge-Errors:
for i = 0 to 2n-2
  E(i,i) = 0
for number_of_matrices_merged = 2 to n
    for i = 0 to (2*n - 1 - number_of_matrices_merged)
      j = i + number_of_matrices_merged - 1
      E(i,j) = Maximum
      for k = i to (j - 1)
          if E(i,j) > E(i,k) + E(k+1,j) + Delta(M(i,k), M(k+1,j))
          then E(i,j) = E(i,k) + E(k+1,j) + Delta(M(i,k), M(k+1,j))
```

Each $\delta(M(i, k), M(k + 1, j))$ is computed in $O(h * w)$ time. Clearly, Algorithm `Calc-Merge-Errors` above computes the values of $O(n^2)$ $E(i, j)$'s for all candidate output matrices $M(i, j)$ in S_{pos}, in totally $O(n^3 * h * w)$ time. We store these $O(n^2)$ $E(i, j)$ values using $O(n^2)$ space for solving the second sub-problem.

3.2   Graph Construction

We now solve the second sub-problem: Compute an optimal subset S_{ans} of S_{pos} (i.e., the minimum number k of output matrices selected from S_{pos}) along the cyclic order such that S_{ans} meets the three requirements specified at the beginning of Section 3. We solve this sub-problem by computing a set of k-link shortest paths in a graph $G = (V, E)$. The graph G is defined based on the linear sequence S_L of matrices $M_0, M_1, \ldots, M_{2n-2}$ (as introduced in Section 3.1).

We construct $G = (V, E)$ as follows. Let $V = \{v_0, v_1, \ldots, v_{2n-1}\}$ be its vertex set. Intuitively, one may view V as a sequence of "virtual points" such that, for each $i = 1, 2, \ldots, 2n - 2$, v_i is a "virtual point" located between the two consecutive matrices M_{i-1} and M_i in S_L, and v_0 (resp., v_{2n-1}) is a "virtual point" located before (resp., after) the matrix M_0 (resp., M_{2n-2}) in S_L (see Figure 3). For each vertex v_i, there is a directed edge (v_i, v_j) for every $j = i+1, i+2, \ldots, i+n$. Let the weight $w(v_i, v_j)$ of the edge (v_i, v_j) be $E(i, j - 1)$, which is computed by Algorithm `Calc-Merge-Errors` in Section 3.1. Intuitively, a directed edge (v_i, v_j) in G corresponds to a candidate output matrix obtained by merging the matrices $M_i, M_{i+1}, \ldots, M_{j-1}$ in the linear sequence S_L (i.e., the candidate output matrix $M(i \mod n, (j - 1) \mod n)$ in S_{pos}), and its weight $w(v_i, v_j) = E(i, j - 1)$

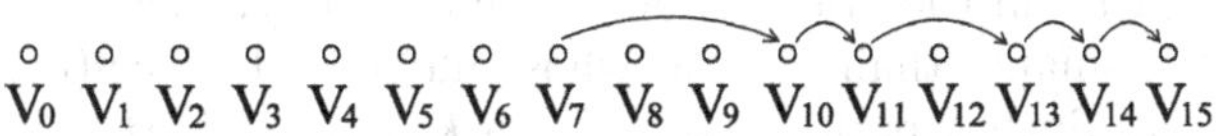

$V_0 \ V_1 \ V_2 \ V_3 \ V_4 \ V_5 \ V_6 \ V_7 \ V_8 \ V_9 \ V_{10} V_{11} V_{12} V_{13} V_{14} V_{15}$

Fig. 4. A feasible 5-link v_7-to-v_{15} path that corresponds to the merging solution in the example of Figure 2 (with $n = 8$ and $k = 5$)

represents the minimum possible merging error of $M(i \mod n, (j-1) \mod n)$ of the candidate output matrix $M(i \mod n, (j-1) \mod n)$ (see Figure 3). For example, the edge (v_0, v_1) corresponds to the candidate output matrix $M(0,0)$ (which is actually the input matrix M_0), and the edge (v_1, v_{n+1}) corresponds to the candidate output matrix $M(1,0)$ which is the merged matrix of all n input matrices $M_1, M_2, \ldots, M_0$ with the minimum merging error $E(M(1,0))$.

Clearly, the graph G thus constructed is a directed acyclic graph with non-negative edge weights, and it has $|V| = O(n)$ vertices and $|E| = O(n^2)$ edges.

Our CMM algorithm is based on some key structures of G, as stated below.

Lemma 1. *(I) For each $i = 0, 1, \ldots, n-1$, any v_i-to-v_{n+i} path $P(i, n+i)$ in the graph G corresponds to a sequence S of output matrices that satisfies requirement (1) (as specified at the beginning of Section 3); further, the number of edges in $P(i, n+i)$ corresponds to the number of matrices in S (e.g., see Figure 4).*

(II) For each $i = 0, 1, \ldots, n-1$, if a v_i-to-v_{n+i} path P in G has a total weight $W(P) < Er$, then P specifies a feasible solution for the CMM problem.

Proof. To prove (I), by the definition of the graph G, it is easy to see that such a path $P(i, n+i)$ "spans" n different input matrices. Thus, each input matrix is included by $P(i, n+i)$ exactly once (i.e., requirement (1) is satisfied). (Actually, one can also argue that each sequence of output matrices around the cyclic order satisfying requirement (1) corresponds to a v_i-to-v_{n+i} path in G.) Further, each edge of $P(i, n+i)$ corresponds to a candidate output matrix. Hence the number of edges of $P(i, n+i)$ corresponds to the number of output matrices in S.

For (II), note that each edge weight in G is equal to the merging error of a candidate output matrix. By (I), the total weight $W(P)$ of the path P is equal to the total sum of merging errors of the corresponding output matrices. Thus, if $W(P) < Er$, then P specifies a CMM solution that satisfies requirement (2) (as given at the beginning of Section 3), which is a feasible CMM solution.   ♠

Corollary 1. *For any $i \in \{0, 1, \ldots, n-1\}$, a v_i-to-v_{n+i} path P^* in the graph G whose total weight $W(P^*) < Er$ (if such a path P^* exists) and whose number of edges is as small as possible specifies a globally optimal CMM solution.*

Proof. By Lemma 1, such a path P^* satisfies all three requirements (given at the beginning of Section 3). Thus, P^* specifies a globally optimal CMM solution.   ♠

3.3   Finding an Optimal CMM Solution

Based on Lemma 1 and Corollary 1, we solve the second sub-problem optimally by finding the minimum possible number k, denotes by k_{min}, such that there is a v_i-to-v_{n+i} path P^* in the graph G, for some $i \in \{0, 1, \ldots, n-1\}$, whose total weight $W(P^*)$ is smaller than Er and whose number of edges (i.e., *links*) is no bigger than k_{min}. Such a path P^* is called a k_{min}-link shortest path in G. In general, we let $SP_k(i,j)$ denote a k-link shortest v_i-to-v_j path in G and $W_k(i,j)$ denote its total weight. It is well-known that computing a k-link shortest path in a graph $G' = (V', E')$ takes $O(k * (|V'| + |E'|))$ time. Below we show how to first find k_{min}, and then obtain an actual optimal CMM solution S^*_{ans}.

We compute k_{min} in an iterative fashion. Specifically, our algorithm computes a set of k-link shortest v_i-to-v_{n+i} paths $SP_k(i, n+i)$ in the graph G, for each $i = 0, 1, \ldots, n-1$, and for $k = 1, 2, \ldots$, until we obtain the first feasible path (i.e., its total weight is smaller than Er) among such k-link shortest paths. Once such a feasible path is found, the algorithm stops. Since our algorithm iteratively increases the value of k from $k = 1$, when it stops, the value of k must be k_{min}. The details of the algorithm are as follows.

```
Algorithm Find-Minimum-k:
for k = 1 to n
    for every i = 0, 1, ..., n - 1
        Find(SP_k(i, n + i))
        if W_k(i, n + i) < Er
            k_min = k
            S_ans = Find-Actual-Path(SP_{k_min}(i, n + i))
            Exit
```

The correctness of Algorithm `Find-Minimum-k` follows from Lemma 1 and Corollary 1. It applies the standard dynamic programming approach for computing k-link shortest paths in graphs. Note that throughout the iterative process of Algorithm `Find-Minimum-k`, we need to compute only the *total weights* of the sought k-link shortest paths, that is, their actual paths are not needed. Thus, we need not maintain the dynamic programming tables (of size $O(k * n) = O(k|V|)$ each) for the n k-link paths. Also, observe that when computing these k-link paths for every $k = 2, 3, \ldots, n$, we can utilize the results of the $(k-1)$-link shortest paths already computed at the $(k-1)$-th iteration. Hence, for each i, computing the k-link shortest v_i-to-v_{n+i} paths for all values of $k = 1, 2, \ldots, k_{min}$ takes altogether $O(k_{min} * n^2) = O(k_{min}(|V| + |E|))$ time. Since the size of the graph G is $O(n^2) = O(|V| + |E|)$, and for every $i = 0, 1, \ldots, n-1$, storing the information for computing only the *total weight* of each path $SP_k(i, n+i)$ takes only $O(n) = O(|V|)$ space, the entire Algorithm `Find-Minimum-k` uses $O(n^2)$ space. Thus, this algorithm takes totally $O(k_{min} * n^3)$ time and $O(n^2)$ space.

After k_{min} is obtained, we know exactly which value of i, say i', gives us a feasible k_{min}-link shortest path in G. Hence, we simply apply the standard k-link shortest path algorithm again to compute an actual k_{min}-link shortest $v_{i'}$-to-$v_{n+i'}$ path in G, and use this path to generate a sequence of k_{min} output matrices for an optimal CMM solution. Consequently, our entire CMM algorithm takes $O(n^3 * h * w + k_{min} * n^3)$ time and $O(n^2)$ space.

We may note that it is easy to show that in general, the concave Monge property does not hold for our graph. Thus, some common fast approaches for computing k-link shortest paths in special directed acyclic graphs (e.g., [1]) are not applicable to our problem.

Our CMM algorithm can be extended to handling other clinical considerations. These extensions will be discussed in the full version of the paper.

4   Implementation Results

We implemented the greedy merging algorithm in VMERGE [3] and our CMM algorithm using C programming language, and experimented with them using both randomly generated data and medical data obtained from the Department of Radiation Oncology, Massachusetts General Hospital. Since our CMM algorithm is optimal, the merging error generated by it is always smaller or equal to the merging error of VMERGE. Besides, for a given value k, a merging solution with smaller total merging error has a better chance to satisfy the error tolerance. Hence, we use the percentage of decrease in the total merging error, denoted by D_p, to measure the performance difference between our algorithm and VMERGE at each value k. D_p is calculated for each k using Equation (3) below, where Er_{VMERGE} and Er_{CMM} denote the errors of the two algorithms.

$$D_p = \frac{Er_{VMERGE} - Er_{CMM}}{Er_{VMERGE}}\% \qquad (3)$$

On the randomly generated data, we explore how the sizes and values of the input matrices impact the performance of these two algorithms. On the medical data, we compare the performance of our CMM algorithm with VMERGE.

4.1   On Randomly Generated Data

To explore the difference between our CMM algorithm and the merging algorithm in VMERGE, we compare their performance on randomly generated data.

On Different Matrix Sizes. Since these matrices are randomly generated, the larger their sizes are, the more stable the merging errors between such matrices become. Thus, in our first experiment, we compare the algorithm performance on matrices of different sizes. Because the values of matrix elements in the medical data are in the range of $[0, 1]$, we set the value range of elements in the randomly generated matrices as $[0, 1]$. We compare the two algorithms on matrices with sizes 1×1, 3×3, 5×5, 7×7, and 9×9. For each size, we run the algorithms on 50 different datasets, each containing 180 random matrices, and compute the average D_p. The results are given in Figure 5(a), which show that the smaller the matrix size is, the better our CMM algorithm performs.

On Different Element Values between Matrices. In the second experiment, we fix the matrix size as 9×9, and modify the range of element values of those matrices with odd indices (i.e., M_1, M_3, M_5, ...) while keeping the range of element values of the matrices with even indices unchanged (i.e., $[0, 1]$). We

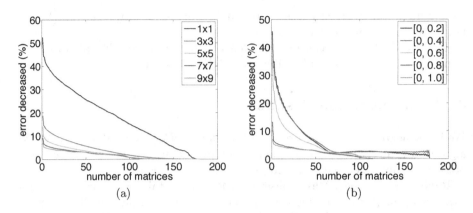

Fig. 5. Test results on random matrices

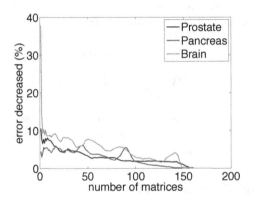

Fig. 6. Test results on medical data

set up five different cases: in the first to fifth cases, the ranges of element values of the matrices with odd indices are set, respectively, as $[0, 0.2]$, $[0, 0.4]$, $[0, 0.6]$, $[0, 0.8]$, and $[0, 1]$. In each case, we run both algorithms on 50 different datasets, each containing 180 random matrices, and compute the average D_p. The results are given in Figure 5(b), which show that the larger the value difference between consecutive matrices is, the better our CMM algorithm performs.

4.2   Medical Data

We applied the two algorithms to three medical datasets with matrix sizes 11×11, 12×10, and 16×9, respectively (for prostate, pancreas, and brain). Figure 6 shows the results. In all these cases, our algorithm runs in only a few seconds.

In [3], the final treatment plans contain about 30 output matrices. In the above cases of prostate, pancreas, and brain, our algorithm outperforms VMERGE by respectively 5.1%, 4.7%, and about 8.0%. The performance difference between our CMM algorithm and VMERGE (i.e., $Er_{VMERGE} - Er_{CMM}$) becomes larger as the number of output matrices decreases. The up and down of the curves in Figure 6 are caused by the quickly increasing values of Er_{VMERGE}.

In general, our algorithm can achieve much better results than VMERGE when the data are complicated. However, in some of the above medical datasets, the input matrices are quite similar to each other. In the future development of VMAT technology, the number of input matrices may increase and consecutive matrices may have bigger value differences for complicated cancer cases. Our CMM algorithm, which guarantees optimal solutions, may play a more effective role in such scenarios.

References

1. Aggarwal, A., Schieber, B., Tokuyama, T.: Finding a minimum weight k-link path in graphs with Monge property and applications. Discrete & Computational Geometry 12(1), 263–280 (1994)
2. Bzdusek, K., Friberger, H., Eriksson, K., Hardemark, B., Robinson, D., Kaus, M.: Development and evaluation of an efficient approach to volumetric arc therapy planning. Medical Physics 36(6), 2328–2339 (2009)
3. Craft, D., McQuaid, D., Wala, J., Chen, W., Salari, E., Bortfeld, T.: Multicriteria VMAT optimization. Medical Physics 39(2), 686–696 (2012)
4. Luan, S., Wang, C., Cao, D., Chen, D.Z., Shepard, D.M., Yu, C.X.: Leaf-sequencing for intensity-modulated arc therapy using graph algorithms. Medical Physics 35(1), 61–69 (2008)
5. Men, C., Romeijn, H.E., Jia, X., Jiang, S.B.: Ultrafast treatment plan optimization for volumetric modulated arc therapy (VMAT). Medical Physics 37(11), 5787–5791 (2010)
6. Otto, K.: Volumetric modulated arc therapy: IMRT in a single gantry arc. Medical Physics 35(1), 310–317 (2008)
7. Rao, M., Yang, W., Chen, F., Sheng, K., Ye, J., Mehta, V., Shepard, D., Cao, D.: Comparison of Elekta VMAT with helical tomotherapy and fixed field IMRT: Plan quality, delivery efficiency and accuracy. Medical Physics 37(3), 1350–1359 (2010)
8. Salari, E., Wala, J., Craft, D.: Exploring trade-offs between VMAT dose quality and delivery efficiency using a network optimization approach. Physics in Medicine and Biology 57(17), 5587–5600 (2012)
9. Spirou, S.V., Chui, C.S.: Generation of arbitrary intensity profiles by dynamic jaws or multileaf collimators. Medical Physics 21(7), 1031–1041 (1994)
10. Svensson, R., Kallman, P., Brahme, A.: An analytical solution for the dynamic control of multileaf collimators. Physics in Medicine and Biology 39(1), 37–61 (1994)
11. Teoh, M., Clark, C.H., Wood, K., Whitaker, S., Nisbet, A.: Volumetric modulated arc therapy: A review of current literature and clinical use in practice. British Journal of Radiology 84(1007), 967–996 (2011)
12. Tsai, C.L., Wu, J.K., Chao, H.L., Tsai, Y.C., Cheng, J.C.H.: Treatment and dosimetric advantages between VMAT, IMRT, and helical tomotherapy in prostate cancer. Medical Dosimetry 36(3), 264–271 (2011)
13. VMAT (Volumetric Intensity Modulated Arc Therapy), http://www.umm.uni-heidelberg.de/inst/radonk/vmat/vmat.html
14. Wang, C., Luan, S., Tang, G., Chen, D.Z., Earl, M.A., Yu, C.X.: Arc-modulated radiation therapy (AMRT): A single-arc form of intensity-modulated arc therapy. Physics in Medicine and Biology 53(22), 6291–6303 (2008)

Engineering Algorithms
for Workflow Satisfiability Problem
with User-Independent Constraints

David Cohen, Jason Crampton, Andrei Gagarin,
Gregory Gutin, and Mark Jones

Royal Holloway, University of London, UK
{D.Cohen,Jason.Crampton,Andrei.Gagarin,
G.Gutin,M.E.L.Jones}@rhul.ac.uk

Abstract. The workflow satisfiability problem (WSP) is a planning problem. Certain sub-classes of this problem have been shown to be fixed-parameter tractable. In this paper we develop an implementation of an algorithm for WSP that has been shown, in our previous paper, to be fixed-parameter for user-independent constraints. In a set of computational experiments, we compare our algorithm to an encoding of the WSP into a pseudo-Boolean SAT problem solved by the well-known solver SAT4J. Our algorithm solves all instances of WSP generated in our experiments, unlike SAT4J, and it solves many instances faster than SAT4J. For lightly constrained instances, SAT4J usually outperforms our algorithm.

1 Introduction

It is increasingly common for organizations to computerize their business and management processes. The co-ordination of the steps that comprise a computerized business process is managed by a workflow management system. Typically, the execution of these steps will be triggered by a human user, or a software agent acting under the control of a human user, and the execution of each step will be restricted to some set of authorized users. In addition, we may wish to constrain the users who execute sets of steps (even if authorized). We may, for example, require that two particular steps are executed by two different users, in order to enforce some separation-of-duty requirement.

We assume the existence of a set U of users and model a workflow as a set S of steps; a set $\mathcal{A} = \{A(u) : u \in U\}$ of authorization lists, where $A(u) \subseteq S$ denotes the set of steps for which u is authorized; and a set C of (workflow) constraints. A constraint is a pair $c = (L, \Theta)$, where $L \subseteq S$ and Θ is a set of functions from L to U: L is the *scope* of the constraint and Θ specifies those assignments of steps in L to users in U that satisfy the constraint. Given a workflow $W = (S, U, \mathcal{A}, C)$, W is said to be *satisfiable* if there exists a function (called a *plan*) $\pi : S \to U$ such that

1. for all $s \in S$, $s \in A(\pi(s))$ (each step is allocated to an authorized user);
2. for all $(L, \Theta) \in C$, $\pi|_L = \theta|_L$ for some $\theta \in \Theta$ (every constraint is satisfied).

J. Chen, J.E. Hopcroft, and J. Wang (Eds.): FAW 2014, LNCS 8497, pp. 48–59, 2014.

Evidently, it is possible to specify a workflow that is not satisfiable. Equally, an unsatisfiable workflow is of no practical use. Hence, it is important to be able to determine whether a workflow is satisfiable or not. We call this the *workflow satisfiability problem*. This problem has been studied extensively in the security community [2,5,16] and more recently as an interesting algorithmic problem [4,7].

The workflow satisfiability problem (WSP) is known to be NP-hard [16] (and is easily shown to be NP-hard even if restricted to separation-of-duty constraints). Wang and Li [16] observed that, in practice, the number k of steps is usually significantly smaller than the number n of users and, thus, suggested to parameterize WSP by k. Wang and Li [16] showed that, in general, WSP is W[1]-hard, but it is *fixed-parameter tractable (FPT)* for certain classes of constraints (i.e., it can be solved in time $O^*(f(k))$, where f is an arbitrary function of k only and O^* suppresses not only constants, but also polynomial factors; we call algorithms with such running time *fixed-parameter*). For further terminology on parameterized algorithms and complexity, see monographs [8,11,14]. Crampton *et al.* [7] extended the FPT classes of [16] and obtained significantly faster algorithms.

Cohen *et al.* recently designed a generic WSP algorithm and proved that the algorithm is fixed-parameter for WSP restricted to the class of user-independent constraints [4]. Informally, a constraint c is *user-independent* if, given a plan π that satisfies c and any permutation $\phi : U \to U$, the plan $\pi' : S \to U$, where $\pi'(s) = \phi(\pi(s))$, also satisfies c. Almost all constraints studied in [7,16] and other papers are user-independent (since the separation-of-duty constraints are user-independent, WSP restricted to the class of user-independent constraints is NP-hard).

It is well known that the gap between traditional "pen and paper" algorithmics and actually implemented and computer-tested algorithms can be very wide [3,13]. In this paper, we demonstrate that the algorithm of Cohen *et al.* is not merely of theoretical interest by developing an implementation that is able to outperform SAT4J in solving instances of WSP. In a set of computational experiments, we compare performance of our implementation with performance of the well-known pseudo-Boolean satisfiability solver SAT4J [12]. Unlike SAT4J, our implementation solves all instances of WSP generated in the experiments and usually solves the instances faster than SAT4J. However, for lightly-constrained instances, SAT4J usually outperforms our implementation.

The paper is organized as follows. In Section 2, we describe how WSP for the family of instances we are interested in can be formulated as a pseudo-Boolean SAT problem. We also describe our fixed-parameter algorithm and discuss its implementation. Section 3 describes test experiments we have conducted with synthetic data to see in which cases our implementation is more efficient and effective than SAT4J. Finally, Section 4 provides conclusions and discusses plans for future work.

2 Methods of Solving WSP

In this paper we seek to demonstrate that (i) our algorithm can be implemented in such a way that it solves certain instances of WSP in a reasonable amount of time; (ii) our implementation can have better performance than SAT4J on the same set of instances. In this section we describe concisely how WSP can be encoded as a pseudo-Boolean SAT problem, how our algorithm works, and the heuristic speed-ups that we have introduced in the implementation of our algorithm.

2.1 WSP as a Pseudo-Boolean SAT Problem

Pseudo-Boolean solvers are recognized as an efficient way to solve general constraint networks [12]. Due to the difficulty of acquiring real-world workflow instances, Wang and Li [16] used synthetic data in their experimental study. Wang and Li encoded WSP as a pseudo-Boolean SAT problem in order to use SAT4J to solve instances of WSP. Their work considered separation-of-duty constraints, henceforth called *not-equals constraints*, which may be specified as a pair (s, t) of steps; a plan π satisfies the constraint (s, t) if $\pi(s) \neq \pi(t)$. Wang and Li considered a number of other constraints in their work, which we do not use in our experimental work since they add no complexity to an instance of WSP. In the experiments of Wang and Li, SAT4J solved all generated instances, and each of them quite efficiently. We test SAT4J on a set of WSP instances of a different type than in [16]. By varying the relevant parameters, we can make our instances more difficult for SAT4J to solve, as we show in the next section.

For a step s, let $A(s) = \{u \in U : s \in A(u)\}$. For their encoding, Wang and Li introduced (0,1)-variables $x_{u,s}$ to represent those pairs (u, s) that are authorized by $\mathcal{A}$. That is, we define a variable $x_{u,s}$ if and only if $s \in A(u)$. The goal of the SAT solver is to find an assignment of values to these variables (representing a plan) where $x_{u,s} = 1$ if and only if u is assigned to step s. These variables are subject to the following constraints:

- for every step s, $\sum_{u \in A(s)} x_{u,s} = 1$ (each step is assigned to exactly one user);
- for each not-equals constraint (s, t) and user $u \in A(s) \cap A(t)$, $x_{u,s} + x_{u,t} \leqslant 1$ (no user is assigned to both s and t).

We use not-equals constraints and some relatively simple counting constraints: "at-most-3" and "at-least-3" with scopes of size 5. The former may be represented as a set $(T, \leqslant)$, where $T \subseteq S$, $|T| = 5$, and is satisfied by any plan that allocates no more than three users in total to the steps in T. The latter may be represented as $(T, \geqslant)$ and is satisfied by any plan that allocates at least three users to the steps in T. For convenience and by an abuse of notation, given a constraint c of the form $(T, \geq)$ or $(T, \leq)$, we will write $s \in c$ to denote that $s \in T$.

We encode "at-least-3" and "at-most-3" constraints as part of the pseudo-Boolean SAT instance in the following way. For each "at-least-3" constraint c

and user u, we introduce a $(0,1)$-variable $z_{u,c}$. The variables $z_{u,c}$ will be such that if $z_{u,c} = 1$ then u performs a step in c, and will be used to satisfy a lower bound on the number of users performing steps in c. For each "at-most-3" constraint c and user u, we introduce a $(0,1)$-variable $y_{u,c}$. The variables $y_{u,c}$ will be such that if u performs a step in c then $y_{u,c} = 1$, and will be used to satisfy an upper bound on the number of users performing steps in c.

The variables are subject to the following constraints:

- for each "at-least-3" constraint c and user u: $z_{u,c} \leq \sum_{s \in c \cap A(u)} x_{u,s}$;
- for each "at-least-3" constraint c: $\sum_{u \in U} z_{u,c} \geqslant 3$.
- for each "at-most-3" constraint c, $s \in c$ and $u \in A(s)$: $x_{u,s} \leq y_{u,c}$;
- for each "at-most-3" constraint c: $\sum_{u \in U} y_{u,c} \leqslant 3$.

It is possible to prove the following assertion. The proof is omitted due to the space limit.

Lemma 1. *A WSP instance has a solution if and only if the corresponding pseudo-Boolean SAT problem has a solution.*

2.2   Fixed-Parameter Algorithm

We proceed on the basis of the assumption that the number k of steps is significantly smaller than the number n of users. Any function $\pi : T \to X$ with $T \subseteq S$ and $X \subseteq U$, is called a *partial plan*. We order the set of users and incrementally construct "patterns" for partial plans that violate no constraints for the first i users. At each iteration we assign a set of steps (which may be empty) to user $i + 1$. An assignment is *valid* if it violates no constraints and extends at least one existing pattern for users $1, \ldots, i$. Under certain conditions, we dynamically change the ordering of the remaining users.

Patterns are encodings of equivalence classes of partial plans and are used to reduce the number of partial plans the algorithm needs to consider. In particular, we define an equivalence relation on the set of all possible plans. This equivalence relation is determined by the particular set of constraints under consideration [4]. In the case of user-independent constraints, two partial plans $\pi : T \to X$ and $\pi' : T' \to X'$ are equivalent, denoted by $\pi \approx \pi'$, if and only if $T = T'$ and for all $s, t \in T$, $\pi(s) = \pi(t)$ if and only if $\pi'(s) = \pi'(t)$.

A pattern is a representation of an equivalence class, and there may be many possible patterns. In the case of $\approx$, we assume an ordering $s_1, \ldots, s_k$ of the set of steps. Then the encoding of an equivalence class for $\approx$ is given by $(T, (x_1, \ldots, x_k))$, where $T \subseteq S$ and, for some π in the equivalence class, we define

$$x_i = \begin{cases} 0 & \text{if } s_i \notin T, \\ x_j & \text{if } \pi(s_i) = \pi(s_j) \text{ and } j < i, \\ \max\{x_1, \ldots, x_{i-1}\} + 1 & \text{otherwise.} \end{cases}$$

We must ensure that there exist efficient algorithms for searching and inserting elements into the set of patterns. Cohen *et al.* [4] show that such algorithms exist for user-independent constraints, essentially because the set of patterns admits a natural lexicographic order.

The overall complexity of the algorithm is determined by k, n, and the number w of equivalence classes for a pair (U_i, T), where U_i denotes the set of the first i users considered by the algorithm. Cohen *et al.* show that the algorithm has run-time $O^*(3^k w \log w)$ [4, Theorem 1]. Thus, we have an FPT algorithm when w is a function of k. For user-independent constraints, $w \leqslant B_k$, where B_k is the kth Bell number; $B_k = 2^{k \log k(1-o(1))}$ [1].

2.3   Implementing the Algorithm

In this section, we describe our algorithm in more detail, via a pseudo-code listing (Algorithm 1), focusing on the heuristic speed-ups we have introduced in our implementation. The algorithm iterates over the set of users constructing a set Π of patterns for valid partial plans (that is partial functions from S to U that violate no constraints or authorizations). For each user u, we attempt to find the set Π_u of valid plans that extend a pattern in Π and assign some non-empty subset of unassigned steps to u.

We assume a fixed ordering of the elements of S and that the users are ordered initially by the cardinality of their respective authorization lists (with ties broken according to the lexicographic ordering of the authorized steps). The ordering we impose on the set of users allows us to introduce a heuristic speed-up based on the idea of a useless user. A user v is *useless* if there exists a user u such that $A(v) \subseteq A(u)$, u has already been processed, and $\Pi_u = \emptyset$. Each iteration of the algorithm considers assigning steps to a particular user u and constructs a set Π_u of extended plans that includes this user. If, having examined all patterns in Π, we have $\Pi_u = \emptyset$, then there is no subset of steps (for which u is authorized) that can be added to Π without violating some constraint. Since all constraints are user-independent, we may now remove all useless users from the list of remaining users.

Much of the work of the algorithm is done in line 8. The time taken to check the validity of a plan can be reduced by considering the constraints in the following order: (1) not-equals constraints; (2) at-most-3 constraints; (3) at-least-3 constraints. The intuition underlying this design choice is that we should consider constraints that violate the most plans first. In line 8 we consider a plan with a prescribed pattern and test whether its extension (by the assignment of steps in T' to user u) is valid. In line 12 we have to compute the pattern for the extended plan and add it to the list Π_u of extended plans. As we noted in the previous section, results by Cohen *et al.* [4] assert that these subroutines can be computed efficiently.

Finally, we are able to propagate information about the current state of any at-most-3 constraints. Suppose ℓ steps from the 5 steps in an at-most-3 constraint $(T, \leqslant)$ have been assigned to two distinct users, where $\ell \in \{2, 3\}$. Then the remaining $5 - \ell$ steps must be assigned to a single user. Hence we may discard the pattern immediately if there is no user that is authorized for all the $5 - \ell$ remaining steps in T. Similarly, if there are any not-equals constraints defined on

Algorithm 1. Algorithm for WSP with user-independent constraints

```
 1 begin
 2 │  Initialize the set Π of patterns to the zero-pattern (∅, (0, . . . , 0))
 3 │  foreach  u ∈ U do
 4 │  │  Initialize Πᵤ = ∅;
 5 │  │  foreach  pattern p = (T, (x₁, . . . , xₖ))  in Π do
 6 │  │  │  Tᵤ ← A(u) \ T;
 7 │  │  │  foreach  ∅ ≠ T' ⊆ Tᵤ do
 8 │  │  │  │  if  plan π ∪ (T' → u) is valid (where π is a plan with pattern p)
    │  │  │  │  then
 9 │  │  │  │  │  if  T ∪ T' = S then
10 │  │  │  │  │  │  return SATISFIABLE and π ∪ (T' → u);
11 │  │  │  │  │  end
12 │  │  │  │  │  Compute the pattern for π ∪ (T' → u) and add it to Πᵤ if
    │  │  │  │  │  not present in Π;
13 │  │  │  │  end
14 │  │  │  end
15 │  │  end
16 │  │  if  Πᵤ = ∅ then
17 │  │  │  Remove useless users
18 │  │  else
19 │  │  │  Π ← Π ∪ Πᵤ
20 │  │  end
21 │  │  Re-order the list of remaining users according to propagation of
    │  │  at-most-3 constraints;
22 │  end
23 │  return UNSATISFIABLE;
24 end
```

any pair of the $5 - \ell$ remaining steps in T, then we know the pattern (encoding a partial plan) cannot possibly generate a total valid plan, and we may discard it at that point. Moreover, in an effort to determine whether the constraint can be satisfied as quickly as possible, we identify users who are authorized for all the $5 - \ell$ remaining steps in T and move one such user to the beginning of the list of remaining users at the end of each iteration accordingly (line 21). This determines a dynamic ordering of users.

3   Experiments

We used C++ to write an implementation of our algorithm for WSP with user-independent constraints. We then generated a number of instances of WSP and compared the performance of our implementation with that of SAT4J when solving those instances. All our experiments used a MacBook Pro computer having a 2.6 GHz Intel Core i5 processor, 8 GB 1600 MHz DDR3 RAM[1] and running Mac OS X 10.9.2.

---

[1] Our computer is more powerful than the one used by Wang and Li [16].

3.1   Testbed

Based on what might be expected in practice, we used values of $k = 16, 20, 24$, and set $n = 10k$. The number c_1 of at-most-3 constraints plus the number c_2 of at-least-3 constraints was set equal to $20, 40, 60$, and 80 (for $k = 24$, we did not consider $c_1 + c_2 = 20$ as the corresponding instances are normally easy to solve by SAT4J). We varied the "not-equals constraint density," i.e. the number of not-equals constraints as a percentage of $\binom{k}{2}$ (the maximum possible number), by using values in the set $\{10, 20, 30\}$. We also assumed that every user was authorized for at least one step but no more than $k/2$ steps; that is, $1 \leqslant |A(u)| \leqslant k/2$. While not-equals constraints and authorizations were generated for each instance separately, the at-most-3 constraints and at-least-3 constraints were kept the same for the corresponding three different values of densities of not-equal constraints and the same value of k.

We adopt the following convention to label our test instances in Tables 1 and 3: $c_1.c_2.d$ denotes an instance with c_1 at-most-3 constraints, c_2 at-least-3 constraints and constraint density d. Informally, we would expect the difficulty of solving instances $c_1.c_2.d$ for fixed k, c_1 and c_2 would increase as d increases (as the problem becomes "more constrained"). Similarly, at-most-3 constraints are more difficult to satisfy than at-least-3 constraints, so, for fixed k and d, instances $c_1.c_2.d$ would become harder to solve as c_1 increases. We would also expect that the time taken to solve an instance would depend on whether the instance is satisfiable or not, with unsatisfiable instances requiring all possible plans to be examined.

Assuming the number c_2 of at-least-3 constraints does not influence satisfiability too much ($c_2 \leq 80$), we varied c_1 trying to generate WSP borderline instances, i.e. instances which have a good chance of being both satisfiable and unsatisfiable. Some instances we generated are lightly constrained, i.e. have a relatively large number of valid plans, while others are highly constrained, i.e. unsatisfiable or have a relatively small number of valid plans. The minimum and maximum values of c_1 used in the experiments to generate Tables 1–4 correspond to instances which we view as mainly borderline. In other words, we started with instances that are experimentally not clearly lightly constrained and stopped at instances which are likely to be highly constrained as the corresponding three instances for the same values of k, c_1, c_2 are unsatisfiable.

Counting constraints were generated by first enumerating all 5-element subsets of S using an algorithm from Reingold et $al.$ [15]. We then used Durstenfeld's version of the Fisher-Yates random shuffle algorithm [9,10] to select independently at random c_1 constraint sets for at-most-3 constraints and c_2 constraint sets for at-least-3 constraints, respectively. The random shuffle algorithm was also used to select steps for which each user is authorized (the list of authorization sets was generated randomly subject to the cardinality constraints above). Finally, the random shuffle was used to select steps for each not-equals constraint.

Table 1. Experimental test results for $k = 20$

Instance	SAT4J		Algorithm 1			
	Output	CPU Time (s)	Output	CPU Time (s)	Users	Patterns
15.5.10	Y	0.974	Y	43.853	6	2,204,316
15.5.20	Y	3.888	Y	26.983	16	208,816
15.5.30	N	1,624.726	N	106.959	200	151,646
20.0.10	Y	1.022	Y	23.777	12	244,494
20.0.20	Y	354.240	Y	25.333	64	34,326
20.0.30	N	987.137	N	15.332	200	12,911
15.25.10	Y	1.092	Y	26.167	8	870,082
15.25.20	Y	12.008	Y	68.890	35	257,808
15.25.30	N	1,886.133	N	75.804	200	112,561
20.20.10	Y	3.997	Y	63.146	33	232,886
20.20.20	N	3,014.967	N	38.623	200	34,310
20.20.30	N	178.208	N	12.091	200	6,093
25.15.10	Y	13.425	Y	79.267	34	153,833
25.15.20	N	1,611.904	N	27.582	200	19,332
25.15.30	N	3,157.095	N	16.070	200	8,116
30.10.10	Y	1.244	Y	20.399	21	63,540
30.10.20	N	2,002.985	N	19.225	200	11,279
30.10.30	N	2,413.049	N	8.393	200	3,416
35.5.10	Y	11.363	Y	23.035	32	20,381
35.5.20	?	2,406.241	N	11.771	200	6,363
35.5.30	N	2,061.416	N	3.409	200	2,107
40.0.10	N	2,734.843	N	28.124	200	13,582
40.0.20	?	3,720.915	N	6.543	200	3,570
40.0.30	N	844.309	N	3.712	200	1,483
15.45.10	Y	1.039	Y	41.948	10	1,201,221
15.45.20	Y	19.302	Y	109.774	31	512,416
15.45.30	Y	97.056	Y	2.123	10	25,433
20.40.10	Y	218.154	Y	58.158	26	209,141
20.40.20	?	3,290.306	N	46.538	200	47,382
20.40.30	N	777.610	N	12.490	200	10,165
25.35.10	Y	5.075	Y	37.573	15	119,829
25.35.20	N	1,984.297	N	30.570	200	26,063
25.35.30	N	3,710.115	N	19.508	200	10,332
30.30.10	Y	317.208	Y	52.735	50	57,123
30.30.20	?	3,514.502	N	14.979	200	8,148
30.30.30	N	489.492	N	7.217	200	5,088
35.25.10	?	7,099.028	N	24.928	200	13,493
35.25.20	N	2,293.669	N	6.365	200	4,072
35.25.30	N	783.961	N	4.229	200	1,577
15.65.10	Y	2.296	Y	35.941	6	2,305,213
15.65.20	Y	13.504	Y	40.761	12	770,002
15.65.30	N	2,351.671	N	156.593	200	274,532
20.60.10	Y	68.804	Y	51.024	27	209,564
20.60.20	?	3,380.584	N	82.120	200	94,197
20.60.30	N	1,438.178	N	13.085	200	10,736
25.55.10	?	3,197.404	N	183.198	200	124,643
25.55.20	N	630.904	N	17.206	200	11,102
25.55.30	N	600.167	N	7.010	200	4,896

3.2 Results

In our experiments we compare the run-times and performance of SAT4J and our algorithm. For the number k of steps equal to 20 and 24, we provide Tables 1 and 3, respectively, which give detailed results of our experiments. We record whether an instance was solved, and the response if it was solved. Thus, we report 'Y', 'N', or '?', indicating, respectively, a satisfiable instance, an unsatisfiable instance, or an instance for which the algorithm terminated without reaching a decision. Note that our algorithm reached a conclusive decision ('Y' or 'N') in all cases, whereas SAT4J failed to reach such a decision for some instances, typically because the machine ran out of memory. For Algorithm 1, we record the number of patterns generated before a valid plan was obtained or the instance was recognized as unsatisfiable, as well as the number of users considered before the algorithm terminated. We also record the time taken for the algorithms to run on each instance.

For lightly constrained instances, SAT4J performs better than our algorithm. This is to be expected, because many of the (large number of) potential patterns are valid. Thus, the number of possible patterns explored by our iterative algorithm is rather large, even when the number of users required to construct a valid plan is relatively small. In contrast, SAT4J simply has to find a satisfying assignment for (all) the variables. The tables also exhibit the expected correlation between the running time of our algorithm and two numbers: the number of patterns generated by the algorithm and the number of users considered, which, in turn, is related to the number of constraints and constraint density.

However, the situation is rather different for highly constrained instances, whether they are satisfiable or not. For such instances, SAT4J will have to consider very many possible valuations for the variables and the running times increase dramatically as a consequence. In contrast, our algorithm has to consider far fewer patterns and this more than offsets the fact that we may have to consider every user (for those cases that are unsatisfiable). Table 2 shows the summary statistics for the running times in Table 1 (to two decimal places). The statistics are based on running times for instances where both algorithms were able to return conclusive decisions. Note that the average time taken by our algorithm for satisfiable and unsatisfiable instances is of a similar order of magnitude; the same cannot be said for SAT4J. Note also the variances of the running times for the two algorithms, indicating that the running time of SAT4J varies quite significantly between instances, unlike our algorithm.

Table 2. Summary statistics for $k = 20$

Output	SAT4J		Algorithm 1	
	Mean	Variance	Mean	Variance
Satisfiable	60.30	11570.72	43.73	585.10
Unsatisfiable	1708.04	896901.01	28.62	1360.75

Table 3. Experimental test results for $k = 24$

	SAT4J		Algorithm 1			
Instance	Output	CPU Time (s)	Output	CPU Time (s)	Users	Patterns
30.10.10	Y	15.88	Y	1,596.07	35	1,351,463
30.10.20	?	2,044.95	N	295.33	240	83,153
30.10.30	N	1,406.89	N	122.50	240	23,201
35.5.10	?	2,867.81	N	2,651.70	240	682,217
35.5.20	?	2,416.26	N	326.75	240	66,962
35.5.30	N	133.53	N	74.61	240	12,296
40.0.10	Y	42.57	Y	639.62	33	472,122
40.0.20	?	2,172.18	N	233.77	240	40,080
40.0.30	N	989.68	N	44.69	240	7,646
30.30.10	Y	4.11	Y	2,666.69	41	2,505,089
30.30.20	?	2,487.96	N	380.47	240	96,073
30.30.30	?	2,506.33	N	119.31	240	22,237
35.25.10	Y	277.69	Y	842.69	43	493,585
35.25.20	?	3,307.86	N	369.22	240	78,041
35.25.30	?	2,782.98	N	85.79	240	14,700
40.20.10	?	2,610.55	N	1,878.97	240	394,284
40.20.20	?	2,596.38	N	283.73	240	47,707
40.20.30	N	3,269.74	N	65.44	240	10,521
30.50.10	?	4,298.72	Y	7,151.83	99	2,582,895
30.50.20	N	197.10	N	604.01	240	175,348
30.50.30	N	1,004.96	N	194.36	240	36,095
35.45.10	?	3,911.16	N	1,106.80	240	243,817
35.45.20	?	3,141.84	N	251.76	240	46,863
35.45.30	?	2,642.05	N	74.00	240	13,959

In Table 4, we summarize the results of our algorithm and SAT4J for all three values of k. As k increases, SAT4J fails more frequently, and was unable to reach a conclusive decision for over half the instances when $k = 24$. This is unsurprising, given that the number of variables will grow quadratically as k and n (which equals $10k$) increase. In Table 4 we report the average run-times for those instances in which both algorithms were able to reach a conclusive decision. We also report (in brackets) the average run-time of our algorithm over all instances. For $k = 16$, the average run-time of our algorithm was two orders of magnitude better than that of SAT4J. As for the other values of k, our algorithm was much faster than SAT4J for unsatisfiable instances. However, for satisfiable instances, the picture was often more favourable towards SAT4J. Overall, for larger values of k, the average run-time advantage of our algorithm over SAT4J decreases, but the relative number of instances solved by SAT4J decreases as well.

It is interesting to note the way in which the mean running time $\hat{t}$ varies with the number of steps. In particular, $\hat{t}$ for our algorithm grows exponentially with k (with a strong correlation between k and $\log \hat{t}$), which is consistent with

Table 4. Test results for $k \in \{16, 20, 24\}$

				SAT4J		Algorithm 1	
Steps k	Min c_1	Sums $c_1 + c_2$	Output	Number	Mean Time	Number	Mean Time
16	5, 10	20, 40, 60, 80	Sat	38	3.98	38	1.27
			Unsat	28	408.20	28	0.77
20	15	20, 40, 60, 80	Sat	19	60.30	19	43.73
			Unsat	22	1,708.04	29	28.62 (34.47)
			Unknown	7		0	
24	30	40, 60, 80	Sat	4	85.06	5	1,436.27 (2,579.38)
			Unsat	6	1,166.98	19	184.27 (482.27)
			Unknown	14		0	

the theoretical running time of our algorithm ($O^*(2^{k \log k})$). The running time of SAT4J is also dependent on k, with a strong correlation between k and $\log \hat{t}$, which is consistent with the fact that there are $O(n^k)$ possible plans to consider. However, it is clear that the running time of SAT4J is much more dependent on the number of variables (determined by the number of users, authorizations, and constraints), than it is on k, unlike the running time of our algorithm.

4 Concluding Remarks

In this paper, we describe the implementation of a fixed-parameter algorithm designed to solve a specific hard problem known as the workflow satisfiability problem (WSP) for user-independent constraints. In theory, there exists an algorithm that can solve WSP for user-independent constraints in time $O^*(2^{k \log k})$ in the worst case. However, WSP is a practical problem with applications in the design of workflows and the design of access control mechanisms for workflow systems [6]. Thus, it is essential to demonstrate that theoretical advantages can be transformed into practical computation advantages by concrete implementations.

Accordingly, we have developed an implementation using our algorithm as a starting point. In developing the implementation, it became apparent that several application-specific heuristic improvements could be made. In particular, we developed specific types of propagation and pruning techniques for counting constraints.

We compared the performance of our algorithm with that of SAT4J—an "off-the-shelf" SAT solver. In order to perform this comparison, we extended Wang and Li's encoding of WSP as a pseudo-Boolean satisfiability problem. The results of our experiments suggest that our algorithm does, indeed, have an advantage over SAT4J when solving WSP, although this advantage does not extend to lightly constrained instances of the problem. The results also suggest that those advantages could be attributable to the structure of our algorithm, with its focus on the small parameter (in this case the number of workflow steps).

We plan to continue working on algorithm engineering for WSP. In particular, we plan to continue developing ideas presented in this paper and in [4] to develop an efficient implementation of a modified version of our algorithm. We hope to obtain a more efficient implementation than the one presented in this paper. We also plan to try different experimental setups. For example, in this paper, we have used a uniform random distribution of authorizations to users with an upper bound at 50% of the number of steps for which any one user can be authorized. In some practical situations, a few users are authorized for many more steps than others. We have only considered counting constraints, rather than a range of user-independent constraints. In some ways, imposing these constraints enables us to make meaningful comparisons between the two different algorithms, but we would still like to undertake more extensive testing to confirm the initial results that we have obtained for this particular family of instances of WSP.

Acknowledgment. This research was supported by an EPSRC grant EP/K005162/1.

References

1. Berend, D., Tassa, T.: Improved bounds on Bell numbers and on moments of sums of random variables. Probability and Mathematical Statistics 30(2), 185–205 (2010)
2. Bertino, E., Ferrari, E., Atluri, V.: The specification and enforcement of authorization constraints in workflow management systems. ACM Trans. Inf. Syst. Secur. 2(1), 65–104 (1999)
3. Chimani, M., Klein, K.: Algorithm engineering: Concepts and practice. In: Bartz-Beielstein, T., Chiarandini, M., Paquete, L., Preuss, M. (eds.) Experimental Methods for the Analysis of Optimization Algorithms, pp. 131–158 (2010)
4. Cohen, D., Crampton, J., Gagarin, A., Gutin, G., Jones, M.: Iterative plan construction for the workflow satisfiability problem. CoRR abs/1306.3649 (2013)
5. Crampton, J.: A reference monitor for workflow systems with constrained task execution. In: Ferrari, E., Ahn, G.J. (eds.) SACMAT, pp. 38–47. ACM (2005)
6. Crampton, J., Gutin, G.: Constraint expressions and workflow satisfiability. In: Conti, M., Vaidya, J., Schaad, A. (eds.) SACMAT, pp. 73–84. ACM (2013)
7. Crampton, J., Gutin, G., Yeo, A.: On the parameterized complexity and kernelization of the workflow satisfiability problem. ACM Trans. Inf. Syst. Secur. 16(1), 4 (2013)
8. Downey, R.G., Fellows, M.R.: Parameterized Complexity. Springer (1999)
9. Durstenfeld, R.: Algorithm 235: Random permutation. Communications of the ACM 7(7), 420 (1964)
10. Fisher, R.A., Yates, F.: Statistical tables for biological, agricultural and medical research, 3rd edn. Oliver and Boyd (1948)
11. Flum, J., Grohe, M.: Parameterized Complexity Theory. Springer (2006)
12. Le Berre, D., Parrain, A.: The SAT4J library, release 2.2. J. Satisf. Bool. Model. Comput. 7, 59–64 (2010)
13. Myrvold, W., Kocay, W.: Errors in graph embedding algorithms. J. Comput. Syst. Sci. 77(2), 430–438 (2011)
14. Niedermeier, R.: Invitation to Fixed-Parameter Algorithms. Oxford U. Press (2006)
15. Reingold, E.M., Nievergelt, J., Deo, N.: Combinatorial algorithms: Theory and practice. Prentice Hall (1977)
16. Wang, Q., Li, N.: Satisfiability and resiliency in workflow authorization systems. ACM Trans. Inf. Syst. Secur. 13(4), 40 (2010)

The Complexity of Zero-Visibility Cops and Robber

Dariusz Dereniowski[1,*], Danny Dyer[2],
Ryan M. Tifenbach[2,**], and Boting Yang[3,***]

[1] Dept. of Algorithms and System Modeling,
Gdańsk University of Technology, Poland
[2] Dept. of Mathematics and Statistics,
Memorial University of Newfoundland, Canada
[3] Dept. of Computer Science, University of Regina, Canada
boting.yang@uregina.ca

Abstract. In this work we deal with the computational complexity aspects of the zero-visibility Cops and Robber game. We provide an algorithm that computes the zero-visibility copnumber of a tree in linear time and show that the corresponding decision problem is NP-complete even for the class of starlike graphs.

1 Introduction

There are many pursuit-evasion models in graph theory where a collection of agents, known as cops, move through a graph attempting to capture an evader, known as the robber. In the classic cops and robber game of Nowakowski/Winkler/Quillot [13,15], the cops and robber have full information about each other's location and the structure of the graph, then alternate turns moving from a vertex to an adjacent vertex. In this game, the cops win if a strategy exists whereby they can guarantee they occupy that same vertex as the the robber in a finite number of moves. Otherwise, the robber wins. For a thorough survey of this game, see [2].

The *zero-visibility cops and robber game* is a variant of this game in which the robber is invisible; that is, the cops have no information at any time about the location of the robber. The game for the cops becomes to guarantee that after some finite time they must have occupied the same position as the robber. Moreover, the goal becomes to do so with the fewest cops possible. This is the *zero-visibility copnumber* of G. This model was introduced by Tošić [17], who

---

[*] D. Dereniowski has been partially supported by Narodowe Centrum Nauki under contract DEC-2011/02/A/ST6/00201, and by a scholarship for outstanding young researchers founded by the Polish Ministry of Science and Higher Education.

[**] R.M. Tifenbach has been supported by a postdoctoral fellowship from the Atlantic Association for Research in the Mathematical Sciences.

[***] B. Yang has been partially supported by an NSERC Discovery Research Grant.

characterized those graphs with copnumber one, and computed the copnumber of paths, cycles, complete graphs and complete bipartite graphs.

Due to the invisibility of the robber, this pursuit-evasion game bears similarities to the edge-searching model of Parsons [14], which employs an "arbitrarily fast" invisible robber. Here, the minimum number of cops needed to capture a robber is essentially the pathwidth of the graph [6,10], and allowing recontamination (a robber being allowed to return to an edge that has previously been cleared) does not reduce the number of cops needed [1]. The authors have similarly bounded the zero-visibility case involving pathwidth, but have shown the surprising result that recontamination does help; that is, zero-visibility cops and robber is not monotonic [5,4].

Determining if the number of cops needed to catch the robber in the edge-searching model is less than k is NP-complete for arbitrary graphs and linear for trees [12]. In the classic cops and robber model, since one cop is sufficient to catch a robber on any tree, the decision problem is trivial. However, it is a much more recent result that the corresponding decision problem for arbitrary graphs is actually EXPTIME-complete [11]. For zero-visibility cops and robber, a quadratic time algorithm for trees is known [16]. We improve on this result by finding a linear time algorithm.

In Section 2 we introduce the notation used in this work. Our algorithmic result is divided into two parts. The first part, given in Section 3, is a constructive characterisation of trees having a given copnumber. This characterisation is then directly used in Section 4 to obtain our algorithm for computing the zero-visibility copnumber of a tree. In Section 5 we analyze a computationally hard case by proving that computing a zero-visibility copnumber of a starlike graph is NP-hard.

2   Preliminaries

We consider only *simple* graphs which contain no loops or multiple edges. For graph theoretic terminology, we follow [18].

We consider the *zero-visibility cops and robber* game previously examined in [5,4]. The game is played by two players, the *cop* and the *robber,* on a graph G; we refer to both players and their respective pieces as cops and robbers. The game begins with the cop player placing one or more cop pieces on vertices of G followed by the robber placing a single robber piece on a vertex of G unknown to the cop player. Beginning with the cop, the players then alternate turns; on a player's turn, he may move each of his pieces along an edge to a vertex adjacent to its current position, or leave any pieces where they are. The game ends with victory for the cop player if the robber piece and a cop piece ever simultaneously occupy the same vertex; the robber's goal is to avoid this situation indefinitely. The position of the robber piece is kept secret from the cop player until the game ends.

By a *copwalk* we mean a walk ℓ that describes a movement of a cop piece, i.e., $\ell(s)$ is the position of the piece after its controller has taken s turns. Let G be a connected graph. A *strategy* on G of length T and order k is a collection $\mathcal{L} = \{\ell_i\}_{i=1}^{k}$ of k copwalks of length T in G, with ℓ_i the copwalk of cop i. We also denote $\mathcal{L}_s = \{\ell_1(s), \ldots, \ell_k(s)\}$ for $s \geq 0$. A strategy of order k and length T gives us a sequence of prescribed moves for a cop player utilising k cops; we might imagine that the cop player following such a strategy forfeits if he has not won after T turns. We refer to a strategy as *successful* if it guarantees victory for the cop player.

Evidently, a strategy $\mathcal{L}$ is successful if and and only if for every walk α of length $T - 1$ in G, there is $\ell_i \in \mathcal{L}$ and $s \leq T - 1$ such that $\alpha(s) \in \{\ell_i(s), \ell_i(s + 1)\}$ (the robber that follows α may be caught by either moving onto a cop, which occurs when $\alpha(s) = \ell_i(s)$, or by having a cop move onto it, which occurs when $\alpha(s) = \ell_i(s + 1)$).

Rather than tracking individual moves by a robber piece, we will view this game as an exercise in graph cleaning, played by a single player. At the beginning of the game, a number of cop pieces are placed on vertices in the graph. Every occupied vertex is marked as "clean" and every unoccupied vertex is marked as "dirty". The cop player has a series of turns, where the cops can move as above. Every time a dirty vertex is occupied by a cop it becomes clean. In between each of the cop player's turns, every clean vertex that is unoccupied and is adjacent to a dirty vertex becomes dirty.

At each point during the game, the dirty vertices are those vertices that could contain the robber, given that he has not yet been caught. We refer to the point in between the cop's turns where clean vertices may become dirty as *recontamination*. We note that for any graph G, a strategy $\mathcal{L}$ on G of length T is successful if and only if following $\mathcal{L}$ results in every vertex being clean after T turns.

The *zero-visibility copnumber* of a graph G, denoted by $c_0(G)$, is the minimum number of cops required to guarantee capture of the robber in a finite number of turns. Thus, $c_0(G) = k$ if there is a successful strategy of order k on G and there is no successful strategy of order $k - 1$ on G.

3   A Characterisation of the Zero-Visibility Copnumber of a Tree

We start by recalling the following result:

Lemma 1 ([16]). *Let G be a graph and let H be an isometric subgraph of G; then $c_0(H) \leq c_0(G)$.*

We now define a tree construction that plays a crucial role in our arguments. Let $k \geq 1$ and let T_1, T_2, T_3 be trees such that $c_0(T_i) = k$ for each $i \in \{1, 2, 3\}$. A tree T is obtained in the manner shown in Figure 1, where the edge that joins each T_i to the remainder of T can be incident to any vertex in T_i. The vertex x will be called the *central vertex* of T and y_i is the *connecting vertex* of T_i, $i \in \{1, 2, 3\}$. We then say that T is *derived* from T_1, T_2 and T_3.

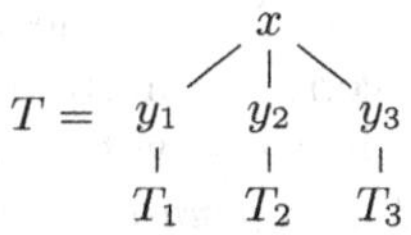

$$T = \begin{array}{ccc} & x & \\ \diagup & | & \diagdown \\ y_1 & y_2 & y_3 \\ | & | & | \\ T_1 & T_2 & T_3 \end{array}$$

Fig. 1. The tree T derived from T_1, T_2, T_3

We now describe a cleaning procedure, given in Algorithm 1, that we often use. The input to the procedure is a tree T, a vertex x of T with degree l adjacent to the vertices $y_1, y_2, \ldots, y_l$, and the number of cops k. We refer to the procedure as a *standard cleaning* of T from x. Note that the procedure does not prescribe

---

Algorithm 1. Standard cleaning of T from x with k cops.

Initially place all k cops on x.
for $i := 1, \ldots, l$ **do**
  Let the k-th cop start vibrating on xy_i.
  for each tree H in $T - N_T[x]$ having a vertex adjacent to y_i **do**
    Clean H using the first $k - 1$ cops.
  end for
  Let the first $k - 1$ cops return to x.
  Wait (at most one turn) so that all cops are on x.
end for

---

how the subtrees H are cleaned. We will refer to this procedure usually to obtain some upper bounds on the number of cops some trees require.

We have the following observation.

Observation 1. *Let T be a tree, let $v \in V_T$ and let $k \geq 1$. If $c_0(H) \leq k - 1$ for each connected component H of $T - N_T[v]$, then a standard cleaning of T from v with k cops produces a successful strategy on T.*   □

This method of cleaning leads to the following theorem, paralleling the development of edge-searching [14].

Theorem 2. *Let T be a tree and let $k \geq 1$. Then, $c_0(T) \geq k + 1$ if and only if there is $x \in V_T$ such that at least three neighbours of x are adjacent to a connected component H of $T - N_T[x]$ with $c_0(H) \geq k$.*

Recall that a *minor* of a graph G is a graph H that can be obtained from G via some sequence of edge contractions, edge deletions and/or vertex deletions. When G is a tree, any connected minor of G is, itself, a tree and can be obtained by a sequence of edge contractions.

Theorem 3. *Let T be a tree and let H be a connected minor of T. Then, $c_0(H) \leq c_0(T)$.*

This follows by induction on $n = |V_T|$. Using Theorem 2, it can be shown that if H is formed by at most one edge contraction from T, then $\mathsf{c}_0(H) \leq \mathsf{c}_0(T)$; this, together with Lemma 1, is sufficient to show the claim.

Let T be a tree. We say that T is *copwin-critical* if the only connected minor of T that has zero-visibility copnumber equal $\mathsf{c}_0(T)$ is T itself.

We define a sequence of families of trees that will characterise copwin-critical trees. The family $\mathcal{T}_1$ consists of the tree on one vertex. For $k \geq 1$, the family $\mathcal{T}_{k+1}$ consists of all trees T derived from $T_1, T_2, T_3 \in \mathcal{T}_k$ (not necessarily distinct).

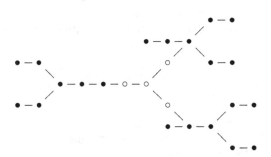

Fig. 2. One of the 10 trees contained in $\mathcal{T}_3$. The 3 copies of the tree in $\mathcal{T}_2$ are shown with dots; the root and its children are shown with circles.

The family $\mathcal{T}_2$ consists of a single tree, the star $K_{1,3}$ with each edge subdivided. There are three distinct (nonisomorphic) ways in which the tree in $\mathcal{T}_2$ could be connected to another tree by a single edge – this edge could be joined to one of the 3 leaves, to one of the 3 vertices adjacent to the leaves, or to the central vertex. Thus, there are 10 distinct trees contained in $\mathcal{T}_3$. We show one of the members of $\mathcal{T}_3$ in Figure 2; in this tree, we see each of the 3 possible ways of joining the tree in $\mathcal{T}_2$. The number of trees in $\mathcal{T}_k$ grows very rapidly with k; an involved counting argument shows that $\mathcal{T}_4$ consists of 204156 distinct trees.

Theorem 4 says that the zero-visibility copnumber of a tree T is the largest k such that $\mathcal{T}_k$ contains a minor of T.

Theorem 4. *Let T be a tree; then, $\mathsf{c}_0(T) \geq k$ if and only if there is $S \in \mathcal{T}_k$ such that S is a minor of T.* $\qquad\Box$

4   An Algorithm for Finding the Zero-Visibility Copnumber of a Tree

We give a modified version of the algorithm first presented in [6]. We will utilise, extensively, the concept of a rooted tree in this section. A *rooted tree* is a tree T with some vertex $\mathbf{r}(T) \in V_T$ as root. For each $v \in V_T$, the *descendent subtree* $T[v]$ is the rooted subtree of T with root v and whose vertices are the descendents of v along with v itself.

We denote $\overline{T} = T - r(T)$ for a rooted tree T. Thus, each tree in $\overline{T}$ is rooted at the vertex adjacent to $r(T)$ in T.

A vertex v in a rooted tree T is a *k-pre-branching* if $c_0(T[v]) = k$ and v has a child u with $c_0(T[u]) = k$; *weakly k-branching* if $c_0(T[v]) = k$ and v has at least two children that are k-pre-branchings; and *k-branching* if $c_0(T[v]) = k$ and v has a weakly k-branching child.

Let T' be a graph whose each connected component is a rooted tree. If T' has a k-branching, then $b^k(T')$ equals any of its k-branchings; otherwise we write $b^k(T') = \perp$ for brevity. We define the following counters for k-branchings and k-pre-branchings of T':

$$\#_b^k(T') = \left|\{T \mid T \text{ is a connected component of } T' \text{ and } b^k(T) \neq \perp\}\right|;$$

$$\#_{pb}^k(T') = \left|\{T \mid T \text{ is a connected component of } T' \right.$$
$$\left. \text{ and } T \text{ contains a } k\text{-pre-branching}\}\right|;$$

$$\mathbb{1}_{wb}^k(T') = \begin{cases} 1, & \text{if } T' \text{ has a weakly } k\text{-branching}, \\ 0, & \text{otherwise}; \end{cases}$$

$$\#_{c_0}^k(T') = \left|\{T \mid T \text{ is a connected component of } T' \text{ and } c_0(T) = k\}\right|.$$

Note that all four above functions give 0 or 1, i.e., they act as 'indicator functions', if T' is a tree. For an integer $p \geq 0$, we define $\mathbb{1}(p) = 0$ if $p = 0$ and $\mathbb{1}(p) = 1$ if $p > 0$.

With these definitions, Lemma 2 follows from Theorems 2 and 3, and provides a constructive method for the computation of the label of the root of a rooted tree, which will give the desired zero-visibility copnumber.

Lemma 2. *Let T be a rooted tree on two or more vertices. Denote $k = \max\{c_0(T') \mid T' \text{ is a connected component of } \overline{T}\}$ and $v = b^k(\overline{T})$. Then:*

(i) *If $\#_b^k(\overline{T}) > 1$, then $c_0(T) = k+1$ and $\#_b^{k+1}(T) = \mathbb{1}_{wb}^{k+1}(T) = \#_{pb}^{k+1}(T) = 0$.*

(ii) *If $\#_b^k(\overline{T}) = 1$ and $c_0(T - V_{T[v]}) \geq k$, then $c_0(T) = k+1$ and $\#_{pb}^{k+1}(T) = \mathbb{1}_{wb}^{k+1}(T) = \#_b^{k+1}(T) = 0$.*

(iii) *If $\#_b^k(\overline{T}) = 1$ and $c_0(T - V_{T[v]}) < k$, then $c_0(T) = k$, $b^k(T) = v$ and $\#_b^k(T) = \mathbb{1}_{wb}^k(T) = \#_{pb}^k(T) = 1$.*

(iv) *If $\#_b^k(\overline{T}) = 0$, then:*

    (1) *If $\mathbb{1}_{wb}^k(\overline{T}) = 1$ and $\#_{c_0}^k(\overline{T}) > 1$, then $c_0(T) = k+1$ and $\#_b^{k+1}(T) = \mathbb{1}_{wb}^{k+1}(T) = \#_{pb}^{k+1}(T) = 0$.*

    (2) *If $\mathbb{1}_{wb}^k(\overline{T}) = 1$ and $\#_{c_0}^k(\overline{T}) \leq 1$, then $c_0(T) = k$, $b^k(T) = r(T)$ and $\#_b^k(T) = \mathbb{1}_{wb}^k(T) = \#_{pb}^k(T) = 1$.*

    (3) *If $\mathbb{1}_{wb}^k(\overline{T}) = 0$, then:*

        (3a) *If $\#_{pb}^k(\overline{T}) > 2$, then $c_0(T) = k+1$ and $\#_b^{k+1}(T) = \mathbb{1}_{wb}^{k+1}(T) = \#_{pb}^{k+1}(T) = 0$.*

(3b) If $\#^k_{\mathrm{pb}}(\overline{T}) = 2$, then $\mathsf{c}_0(T) = k$, $\#^k_{\mathrm{b}}(T) = 0$ and $\mathbb{1}^k_{\mathrm{wb}}(T) = \#^k_{\mathrm{pb}}(T) = 1$. ($\mathbf{r}(T)$ is a weakly k-branching.)

(3c) If $\#^k_{\mathrm{pb}}(\overline{T}) < 2$, then $\mathsf{c}_0(T) = k$, $\#^k_{\mathrm{b}}(T) = \mathbb{1}^k_{\mathrm{wb}}(T) = 0$ and $\#^k_{\mathrm{pb}}(T) = 1$ ($\mathbf{r}(T)$ is a k-pre-branching).

We will use Lemma 2 to directly obtain a dynamic programming algorithm that computes the zero-visibility copnumber of a given tree T. Roughly speaking, the algorithm roots T at any vertex (we refer by T to the rooted tree in the following), and then T is 'processed' in a bottom-up fashion. In particular, for each vertex v, a label of $T[v]$ (and possibly a label of some subtree of $T[v]$) is computed. (See below for the definition of a label.) The label of $T[v]$ gives the zero-visibility copnumber of $T[v]$, and it can be computed based on the labels of the children of v. For this reason the label contains also some other entries besides $\mathsf{c}_0(T[v])$. Once the labels of all vertices are computed, the label of $\mathbf{r}(T)$ gives us the desired $\mathsf{c}_0(T)$.

Formally, we define a *label* of any rooted tree T, as $L(T) = \big(k, \#^k_{\mathrm{b}}(T),$ $\mathbb{1}^k_{\mathrm{wb}}(T), \#^k_{\mathrm{pb}}(T), v\big)$ where $k = \mathsf{c}_0(T)$, and v is a k-branching if T has one, or v is undefined otherwise. We use for brevity the symbol $\perp$ in the last entry when v is undefined. Then, for any $k \geq \mathsf{c}_0(T)$, a *k-label* of T is $L_k(T) = L(T)$ if $k = \mathsf{c}_0(T)$, and $L_k(T) = (k, 0, 0, 0, \perp)$ otherwise.

We start with a claim that allows us to determine a k-label of a tree T in case when $\#^k_{\mathrm{b}}(\overline{T}) \neq 1$.

Lemma 3. *Let T be a rooted tree. Denote the trees in $\overline{T}$ by $T_1, \ldots, T_l$. Let $k = \max\{\mathsf{c}_0(T_j) \mid j \in \{1, \ldots, l\}\}$. If the k-labels of $T_1, \ldots, T_l$ are given and $\#^k_{\mathrm{b}}(\overline{T}) \neq 1$, then the k'-label of T can be computed in time $O(l)$, where $k' \geq \mathsf{c}_0(T)$.*

Proof. We use Lemma 2 to state the formulas for the particular entries of the label of T, $L(T) = \big(m, \#^m_{\mathrm{b}}(T), \mathbb{1}^m_{\mathrm{wb}}(T), \#^m_{\mathrm{pb}}(T), v\big)$. Since $\#^k_{\mathrm{b}}(\overline{T}) \neq 1$, we have:

$$
m = \mathsf{c}_0(T) = \begin{cases} k+1, & \text{if } \#^k_{\mathrm{b}}(\overline{T}) > 1 \vee \big(\mathbb{1}^k_{\mathrm{wb}}(\overline{T}) = 1 \wedge \#^k_{\mathrm{c}_0}(\overline{T}) > 1\big) \\ & \vee \big(\mathbb{1}^k_{\mathrm{wb}}(\overline{T}) = 0 \wedge \#^k_{\mathrm{pb}}(\overline{T}) > 2\big) \\ k, & \text{otherwise} \end{cases}.
$$

Note that the following formulas allow us to find $\#^k_{\mathrm{b}}(\overline{T})$, $\#^k_{\mathrm{pb}}(\overline{T})$, $\mathbb{1}^k_{\mathrm{wb}}(\overline{T})$ and $\#^k_{\mathrm{c}_0}(\overline{T})$:

$$
\#^k_{\mathrm{b}}(\overline{T}) = \sum_{j=1}^{l} \#^k_{\mathrm{b}}(T_j), \quad \#^k_{\mathrm{pb}}(\overline{T}) = \sum_{j=1}^{l} \#^k_{\mathrm{pb}}(T_j),
$$

$$
\#^k_{\mathrm{c}_0}(\overline{T}) = \big|\{j \mid \mathsf{c}_0(T_j) = k\}\big|, \quad \mathbb{1}^k_{\mathrm{wb}}(\overline{T}) = \mathbb{1}\big(\sum_{j=1}^{l} \mathbb{1}^k_{\mathrm{wb}}(T_j)\big).
$$

The values of $\#^k_{\mathrm{b}}(T_j)$, $\#^k_{\mathrm{pb}}(T_j)$, $\mathsf{c}_0(T_j)$ and $\mathbb{1}^k_{\mathrm{wb}}(T_j)$ are taken directly from the k-label of T_j for each $j \in \{1, \ldots, l\}$.

If $m = k + 1$, then the label of T is $(m, 0, 0, 0, \perp)$. Thus, the k'-label of T is $L_{k'}(T) = (k', 0, 0, 0, \perp)$ and the proof is completed. Hence, let $m = k$ in the following.

We have $\#_b^k(T) = \mathbb{1}(q + \#_b^k(\overline{T}))$, where $q = 1$ if $\mathbf{r}(T)$ is a k-branching and $q = 0$ otherwise. Lemma 2, $m = k$ and $\#_b^k(\overline{T}) \neq 1$ imply that $q = 1$ if and only if: $\#_b^k(\overline{T}) = 0$, $\mathbb{1}_{wb}^k(\overline{T}) = 1$ and $\#_{co}^k(\overline{T}) \leq 1$.

Then, $\mathbb{1}_{wb}^k(T) = \mathbb{1}(q' + \mathbb{1}_{wb}^k(\overline{T}))$, where $q' = 1$ if $\mathbf{r}(T)$ is a weakly k-branching and $q = 0$ otherwise. Lemma 2, $m = k$ and $\#_b^k(\overline{T}) \neq 1$ imply that $q' = 1$ if and only if: $\#_b^k(\overline{T}) = 0$, $\mathbb{1}_{wb}^k(\overline{T}) = 0$ and $\#_{pb}^k(\overline{T}) = 2$.

Similarly, $\#_{pb}^m(T) = \mathbb{1}(q'' + \#_{pb}^k(\overline{T}))$, where $q'' = 1$ if $\mathbf{r}(T)$ is a k-pre-branching and $q = 0$ otherwise. This follows from the fact that if the root of T is a k-branching or a weakly k-branching, then one of the subtrees T_j's contains a k-pre-branching. Again, Lemma 2, $m = k$ and $\#_b^k(\overline{T}) \neq 1$ imply that $q'' = 1$ if and only if: $\#_b^k(\overline{T}) = 0$, $\mathbb{1}_{wb}^k(\overline{T}) = 0$ and $\#_{pb}^k(\overline{T}) < 2$.

It remains to determine the vertex v in $L(T)$ in case when $m = k$. Thus, $\#_b^k(\overline{T}) = 0$ and hence $v = \mathbf{r}(T)$ if $q = 1$ and $v = \perp$ otherwise. Finally, we obtain that $L_{k'}(T) = L(T)$ when $k' = k$ and $L_{k'}(T) = (k', 0, 0, 0, \perp)$ otherwise.

Theorem 5. *There exists a linear time algorithm that computes the zero-visibility copnumber of any tree.*

Our proof is constructive, i.e., we describe an algorithm computing the zero-visibility copnumber of an input tree T. We start with its informal description that points out the main ideas. First, T is rooted at any vertex. The vertices of T are ordered as $v_1, \ldots, v_n$ so that each vertex appears in it prior to its parent. The vertices are then processed according to this order. The processing of a vertex u leads to finding the label of the subtree $T[v]$ and, if this subtree has a $c_0(T[u])$-branching v, then the $c_0(T)$-label of the subtree $T[u] - V_{T[v]}$. Denote by k the maximum zero-visibility copnumber among the subtrees descendent from the children of u, and denote by $v_1, \ldots, v_l$ the children of u. Note that the number of k-branchings in the subtrees $T[v_1], \ldots, T[v_l]$, i.e., $\#_b^k(\overline{T[u]})$, can be computed on the basis of the labels of those subtrees. Then, two cases are considered. In the first case $\#_b^k(\overline{T[u]}) = 1$ where in order to compute the label of $T[u]$ we need to know the zero-visibility copnumber of the subtree $T[u] - V_{T[v]}$, where $v = \mathbf{b}^k(T[u])$. (See Lemma 2(ii) and 2(iii).) In order to compute the desired $c_0(T[u] - V_{T[v]})$, we compute the k-label of this subtree. Note that, for the computation of the latter label, we do not use 2(ii) and 2(iii) since, by assumption, there is no k-branching in $T[u] - V_{T[v]}$. In the second case $\#_b^k(\overline{T[u]}) \neq 1$ where we use Lemma 2(i) and 2(iv) for the label computation.

We now give a sketch of the algorithm in the form of a pseudo-code in Algorithm 2. This sketch gives the order of computations of labels of selected subtrees. We omit the details, but it follows by induction and applications of Lemma 2 and Lemma 3 that lines 6, 9, 10 and 12 can be implemented in linear time.

Algorithm 2. Computing the zero-visibility copnumber of an input tree T.

1: Root T at any vertex.
2: Take a permutation $u_1, \ldots, u_n$ of vertices of T such that $u_i \prec u_j$ implies $i < j$ for
   all $i \neq j$.
3: **for** $i := 1$ to n **do**
4:     Let $v_1, \ldots, v_l$ be the children of u_i in T.
5:     $k := \max\{c_0(T[v_j]) \mid j \in \{1, \ldots, l\}\}$
6:     Compute $\#_b^k(T[u_i])$.
7:     **if** $\#_b^k(T[u_i]) = 1$ **then**
8:         Find the index $s \in \{1, \ldots, l\}$ such that $\#_b^k(T[v_s]) = 1$.
9:         Compute the k-label of $T[u_i] - V_{T[x_i]}$, where $x_i = b^k(T[v_s])$.
10:        Compute the label of $T[u_i]$.
11:    **else**
12:        Compute the label of $T[u_i]$.
13:    **end if**
14: **end for**
15: **return** $c_0(T)$ that is stored in its label.

---

5   NP-Hardness of Zero-Visibility Cops and Robber

Let G be an n-vertex graph with vertex set $V_G = \{v_1, \ldots, v_n\}$. Define $\xi(G)$ to be
the graph obtained as follows. The vertex set of $\xi(G)$ is $V_{\xi(G)} = C \cup \bigcup_{v_i v_j \in E_G} V_{i,j}$
where $C = \{c_1, \ldots, c_n\}$, and $V_{i,j}$ is of size n for each i, j such that $v_i v_j \in E_G$.
Moreover, C and all $V_1, \ldots, V_m$ are pairwise disjoint. Then, the edge set of $\xi(G)$
is defined so that C induces a clique and $V_{i,j} \cup \{c_i, c_j\}$ induces a clique of order
$n + 2$ for each edge $v_i v_j$ of G. Denote $\mathcal{V}_G = \{V_{i,j} \mid v_i v_j \in E_G\}$. For any $n > 0$
define $\mathcal{G}_n = \{\xi(G) \mid |V_G| = n\}$. These graphs $\mathcal{G}_n$ are starlike graphs.

Theorem 6 ([8]). *Given $G \in \mathcal{G}_n$, where n is even, and an even integer k^*,
$2 \leq k^* \leq n - 4$, the problem of deciding whether $pw(G) \leq n + k^* - 1$ is NP-
complete.*

In the remaining part of this section we assume that the integers n and k^* are
even and $k^* \leq n - 4$.

Now we recall the node search problem that we will use in our reduction.
Given any graph G that contains a fugitive, a *node k-search* for G is a sequence
of moves such that each move is composed of the following two actions: place
a number of searchers on the vertices of G, which results in at most k vertices
being occupied by the searchers; or remove from G a subset of searchers that
are present on the vertices of G. The fugitive is invisible and fast, i.e., it can
traverse at any moment along an arbitrary path that is free of searchers. We
define *contaminated, successful,* and *node search number* of G (denoted $ns(G)$)
analagously to the cops and robber model.

It is well known that for any graph G, $pw(G) = ns(G) - 1$, and that a successful
node search strategy may be derived for any path decomposition. Then we know

that given $G \in \mathcal{G}_n$ and k^*, the problem of deciding whether $\mathbf{ns}(G) \leq n + k^*$ is NP-complete.

Note that the node search strategy derived from a path decomposition is monotonic. Then using the methods of [5], we obtain the following lemma.

Lemma 4. *Let $G \in \mathcal{G}_n$ and let k^* be given. If $\mathbf{ns}(G) \leq n + k^*$, then $\mathbf{c}_0(G) \leq n/2 + k^*/2$.*

The reverse implication is also true. That is, for a graph and given k^*, $G \in \mathcal{G}_n$ $\mathbf{c}_0(G) \leq n/2 + k^*/2$ implies $\mathbf{ns}(G) \leq n + k^*$. The key idea that we use is an observation that, due to the small diameter and the clique structure of the graphs in $\mathcal{G}_n$, the recontamination spreads, informally speaking, 'as quickly' in the cops and robber game as in the node search. This discussion, and Lemma 4 give the following.

Theorem 7. *Given $G \in \mathcal{G}_n$ and an integer $k > 0$, the problem of deciding whether $\mathbf{c}_0(G) \leq k$ is NP-complete.* □

6   Conclusions and Future Directions

Having shown that the zero-visibility cops and robber game has a linear time algorithm for trees (and is NP-complete for starlike graphs), it is natural to consider other special classes of graphs for which similar complexity results are known for edge and node searching. Particularly, interval graphs (which can be essentially searched from left to right) and cycle-disjoint graphs (which are 'tree-like') are good candidates [7,19].

Outside of complexity, it also remains open to characterize those graphs with $\mathbf{c}_0(G) \leq 2$ [17,9]. This is a natural question, as characterizations exist for graphs whose edge search number is at most 3 [12]. As well, graphs with copnumber one have long been characterized [13], and there is a recent characterization for larger copnumbers [3].

References

1. Bienstock, D., Seymour, P.: Monotonicity in graph searching. Journal of Algorithms 12, 239–245 (1991)
2. Bonato, A., Nowakowski, R.: The Game of Cops and Robbers on Graphs. Student Mathematical Library, vol. 61. American Mathematical Society (2011)
3. Clarke, N.E., MacGillivray, G.: Characterizations of k-copwin graphs. Discrete Mathematics 312, 1421–1425 (2012)
4. Dereniowski, D., Dyer, D., Tifenbach, R.M., Yang, B.: Zero-visibility cops and robber game on a graph. In: Fellows, M., Tan, X., Zhu, B. (eds.) FAW-AAIM 2013. LNCS, vol. 7924, pp. 175–186. Springer, Heidelberg (2013)
5. Dereniowski, D., Dyer, D., Tifenbach, R.M., Yang, B.: Zero-visibility cops & robber and the pathwidth of a graph. Journal of Combinatorial Optimization (2014)
6. Ellis, J.A., Sudborough, I.H., Turner, J.S.: The vertex separation and search number of a graph. Information and Computing 113, 50–79 (1994)

7. Fomin, F.V., Heggernes, P., Mihai, R.: Mixed search number and linear-width of interval and split graphs. In: Brandstädt, A., Kratsch, D., Müller, H. (eds.) WG 2007. LNCS, vol. 4769, pp. 304–315. Springer, Heidelberg (2007)
8. Gustedt, J.: On the pathwidth of chordal graphs. Discrete Applied Mathematics 45, 233–248 (1993)
9. Jeliazkova, D.: Aspects of the cops and robber game played with incomplete information. Master's thesis, Acadia University (2006)
10. Kinnersley, N.G.: The vertex separation number of a graph equals its path-width. Information Processing Letters 42, 345–350 (1992)
11. Kinnersley, W.B.: Cops and robbers is EXPTIME -complete. arXiv.1309.5405 [math.CO] (2013)
12. Megiddo, N., Hakimi, S.L., Garey, M., Johnson, D., Papadimitriou, C.H.: The complexity of searching a graph. Journal of the ACM 35, 18–44 (1988)
13. Nowakowski, R., Winkler, P.: Vertex-to-vertex pursuit in a graph. Discrete Mathematics 43, 235–239 (1983)
14. Parsons, T.: Pursuit-evasion in a graph. In: Proceedings of the International Conference on the Theory and Applications of Graphs. Lecture Notes in Mathematics, vol. 642, pp. 426–441. Springer (1978)
15. Quilliot, A.: Problèmes de jeux, de point fixe, de connectivité et de représentation sur des graphes, des ensembles ordonnés et des hypergraphes. PhD thesis, Université de Paris VI (1983)
16. Tang, A.: Cops and robber with bounded visibility. Master's thesis, Dalhousie University (2004)
17. Tošić, R.: Inductive classes of graphs. In: Proceedings of the Sixth Yugoslav Seminar on Graph Theory, pp. 233–237. University of Novi Sad (1985)
18. West, D.B.: Introduction to Graph Theory, 2nd edn. Pearson (2000)
19. Yang, B., Zhang, R., Cao, Y.: Searching cycle-disjoint graphs. In: Dress, A.W.M., Xu, Y., Zhu, B. (eds.) COCOA. LNCS, vol. 4616, pp. 32–43. Springer, Heidelberg (2007)

Combining Edge Weight and Vertex Weight for Minimum Vertex Cover Problem

Zhiwen Fang[1], Yang Chu[1], Kan Qiao[2], Xu Feng[1], and Ke Xu[1]

[1] State Key Lab. of Software Development Environment,
Beihang University, Beijing 100191, China
{zhiwenf,mottotime,isaiah.feng}@gmail.com,
kexu@nlsde.buaa.edu.cn
[2] Department of Computer Science,
Illinois Institute of Technology, Chicago 60616, USA
kqiao@iit.edu

Abstract. The Minimum Vertex Cover (MVC) problem is an important NP-hard combinatorial optimization problem. Constraint weighting is an effective technique in stochastic local search algorithms for the MVC problem. The edge weight and the vertex weight have been used separately by different algorithms. We present a new local search algorithm, namely VEWLS, which integrates the edge weighting scheme with the vertex weighting scheme. To the best of our knowledge, it is the first time to combine two weighting schemes for the MVC problem. Experiments over both the DIMACS benchmark and the BHOSLIB benchmark show that VEWLS outperforms NuMVC, the state-of-the-art local search algorithm for MVC, on 73% and 68% of the instances, respectively.

1 Introduction

Given an undirected graph $G = (V, E)$, a vertex cover of G is a subset $V' \subseteq V$ such that every edge in G has at least one endpoint in V'. The Minimum Vertex Cover (MVC) problem consists of finding a vertex cover of smallest possible size. MVC is an important NP-hard combinatorial optimization problem and its decision version is one of Karp's 21 NP-complete problems. It appears in a variety of real-world applications, such as the network security, VLSI designs and bioinformatics. MVC is equivalent to two other well-known NP-hard combinatorial optimization problems: the Maximum Clique (MC) problem and the Maximum Independent Set (MIS) problem, which have a mass of applications in the areas of information retrieval, social networks and coding theory. Due to the equivalence, algorithms for one of these problems can be applied to the others in practice.

A huge amount of effort has been devoted to solve the MVC (MC, MIS) problem over the last decades in the AI community. Generally, algorithms to solve the MVC (MC, MIS) problem fall into two types: exact algorithms mainly including the branch-and-bound search [13, 20, 19, 10, 11] and heuristic algorithms including the stochastic local search [6, 17, 14, 16, 2–4]. Exact approaches guarantee the optimality, but cost exponential time in the worst case and thus fail

J. Chen, J.E. Hopcroft, and J. Wang (Eds.): FAW 2014, LNCS 8497, pp. 71–81, 2014.
© Springer International Publishing Switzerland 2014

to give solutions for large instances within reasonable time. Although heuristic approaches cannot guarantee the optimality, they can provide solutions which are optimal or near-optimal. Therefore, stochastic local search algorithms are frequently used to solve large and hard MVC (MC, MIS) instances in practice. Typically, stochastic local search algorithms solve the MVC problem by searching a vertex cover of size k iteratively. To solve the k-vertex cover problem, i.e., finding a vertex cover of size k, one starts from a subset of vertices with size k and then swaps two vertices until a vertex cover is obtained, where *swap* means to remove a vertex from the subset and add a vertex outside into it without changing its size.

How to escape from the local optimal solution is a challenging problem in local search. Constraint weighting, which is first used in the satisfiability (SAT) problem and the constraint satisfaction problem (CSP) [12, 8, 18] and then introduced to the MVC problem, is a powerful technique to refrain from being stuck at a local optimal point. The edge weighting scheme and the vertex weighting scheme have been applied separately by different MVC solvers. COVER [17] introduces the edge weight to MVC for the first time. It is an iterative best improvement algorithm and updates the edge weight at each step. EWLS [2] also uses the edge weight. However, it is iterated local search and updates the edge weight only in the case of being stuck at local optimal points. EWCC [3] improves EWLS by combining the edge weight and configuration checking. NuMVC [4] introduces the forgetting technique, which allows one to decrease the weight of edges when necessary. Besides, it uses a more effective two-stage exchange strategy to achieve state-of-the-art performance. To the best of our knowledge, NuMVC is the dominating local search algorithm for MVC on the BHOSLIB benchmark. Dynamic Local Search (DLS) [14] is designed to solve the MC problem first. DLS alternates between phases of the iterative improvement and the plateau search. During the iterative improvement phase, suitable vertices are added to the current clique. While during the plateau search phase, vertices in the current clique are swapped with vertices outside. Which vertex to add or remove is guided by vertex penalties, which can be regarded as a vertex weighting scheme. Its improved version named the Phased Local Search (PLS) algorithm [15] has no instance-dependent parameters and performs well for the MC problem over a large range of DIMACS instances. Its author extends PLS algorithm to MIS and MVC problems [16]. To our best knowledge, PLS achieves the best performance on most instances from the DIMACS benchmark.

As mentioned above, none of algorithms dominates all others over different kinds of benchmarks. For example, PLS achieves a better performance on Brock* instances than NuMVC and is competitive with NuMVC on most instances from the DIMACS benchmark, while NuMVC is the dominating algorithm on the BHOSLIB benchmark. The different weighting schemes are the main reason for their different performance. Taking advantages of the techniques used in other algorithms allows one to achieve a better performance. In this paper, we combine the edge weight with the vertex weight and propose a new local search algorithm named VEWLS for MVC. To our best knowledge, it is the first time that both the edge weight and the vertex weight are used together in one solver. Experimental results over widely used DIMACS and BHOSLIB benchmarks show that VEWLS is competitive with NuMVC and outperforms it on most instances from both benchmarks.

The remainder of the paper is organized as follows. In the next section, we introduce some necessary background knowledge. Then we describe the VEWLS algorithm in Section 3. Experimental results are presented in Section 4. Finally, we conclude our main contributions and propose some directions for future work.

2 Preliminaries

An undirected graph $G = (V, E)$ comprises a set $V = \{v_1, v_2, ..., v_n\}$ of n vertices together with a set E of m edges, where each edge $e = \{u, v\}$ has two endpoints, namely u and v. The set of neighbor vertices of a vertex v is denoted by $N(v) = \{u | \{u, v\} \in E\}$ and $|N(v)|$ is the degree of v. A vertex cover of G is a subset $V' \subseteq V$ such that every edge $\{u, v\}$ has at least one endpoint (u or v or both) in V' and the vertex cover with k vertices is a k-vertex cover. The Minimum Vertex Cover (MVC) problem calls for finding a vertex cover of the smallest possible size. Note that more than one minimum vertex covers may exist in a graph. Local search algorithms for MVC always maintain a partial vertex cover, where a partial vertex cover is defined as a subset $P \subseteq V$, which covers a subset $E' \subseteq E$ and an edge is covered by P if at least one endpoint of the edge is contained in P.

The MVC problem is equivalent to two other important optimization problems: the Maximum Independent Set (MIS) problem and the Maximum Clique (MC) problem. If V' is a vertex cover of G, then $V \setminus V'$ is an independent set of G as well as a clique in the complement graph of G. Therefore, algorithms for any of the three problems can be used to solve the others in practice. MVC is NP-hard and its associated decision problem is NP-complete. Therefore, there will be no polynomial algorithms for MVC unless $P = NP$. Moreover, Dinur and Safra prove that the MVC problem cannot be approximated with a factor of 1.3606 for any sufficiently large vertex degree unless $P = NP$ [5] and it cannot even be approximated within any constant factor better than 2 if the unique games conjecture is true [9]. Therefore, one usually wants to resort to heuristic methods to solve large and hard instances.

Constraint weighting is a powerful technique in local search algorithms, which is widely used in SAT [12, 8, 18] and then applied to MVC by state-of-the-art local search algorithms, such as Cover [13], PLS [15, 16], EWLS [2], EWCC [3] and NuMVC [4]. Generally, two types of weighting schemes have been used for MVC. One is the vertex weighting scheme, where each vertex is associated with a non-negative weight and the other is the edge weighting scheme, where every edge has a non-negative weight. As both schemes are used in this paper, we would like to give a brief introduction of the vertex weight and the edge weight before we describe our algorithm.

The state-of-the-art algorithm for MVC based on the vertex weighting scheme is PLS [15, 16], which is an improved version of DLS [14]. DLS is a dynamic local search algorithm for MC and it refers to the vertex weight as the vertex penalty. DLS alternates between the iterative improvement phase and the plateau search phase with vertex penalties which are updated dynamically during the search. The iterative improvement phase starts with a clique consisting of a randomly selected vertex and successively extends the clique by adding vertices connected to all vertices in the clique. When the clique cannot be extended, i.e., a maximal clique is found, the plateau search starts. The plateau search phase tries to find a vertex v that is adjacent to all but one of the vertices in the current clique then swaps the vertex v with the only vertex in the clique that is not connected to v. A new clique with the same size as the old one is obtained after the plateau search phase, which makes further extension of the current possible solution.

Which vertex to add or remove is an essential problem in the algorithm. In DLS, vertices are selected based on their vertex penalties. The vertex penalty works as follows: Every vertex has a vertex penalty with an initial value of zero at the beginning. Vertex penalties are updated by increasing the penalty values of all vertices in the current clique by one at the end of the plateau search phase. DLS chooses the vertex with the minimum penalty to add or remove, and choses uniformly at random if more than one such vertex exists. The selected vertex becomes unavailable for subsequent selections until penalties have been updated and perturbations have been performed. The purpose of vertex penalties is to prevent the search process from repeatedly visiting the same search space. A single control parameter named penalty delay, which depends on the input instance, is used to avoid vertex penalties getting too large. Fortunately, its improved version, PLS [14], has no instance-dependent parameters and achieves the state-of-the-art performance over a large range of instances from the DIMACS benchmark when it is extended for MVC and MIS by its author [16].

On the other hand, Cover, EWLS, EWCC and NuMVC use the edge weighting scheme. An edge weighted graph is a graph $G = (V, E)$ combined with a weight function $w : E \rightarrow R^+$ such that each edge e is associated with a positive value $w(e)$ as its weight. Given a graph $G = (V, E)$ and a partial vertex cover P of G, let w be the edge weight function for G, the cost of P is defined as follows.

Definition 1

$$cost(G, P) = \sum_{e \in E \text{ and } e \text{ not covered by } P} w(e).$$

$cost(G, P)$ indicates the total weight of edges uncovered by P and P becomes a vertex cover when its cost equals 0. EWLS, EWCC and NuMVC all prefer to maintain a partial vertex cover with lower cost. To distinguish vertices by edge weight, decreasing score($dscore$) is calculated for each vertex v as follows.

Definition 2

$$dscore(v) = \begin{cases} cost(G, P) - cost(G, P \setminus \{v\}) & \text{if } v \in P \\ cost(G, P) - cost(G, P \cup \{v\}) & \text{if } v \notin P \end{cases}$$

EWCC and NuMVC select vertices depending on their *dscore*. It is obvious that $dscore(v) \leq 0$ if v is in P and the greater $dscore(v)$ is, the less loss to remove v out of P is. In contrast, $dscore(v) \geq 0$ if v is not in P and the greater $dscore(v)$ means the more gain to add v into P.

3   VEWLS: A New Algorithm for MVC

The pseudocode of VEWLS is presented in Algorithm 1. We introduce how to use the edge weight and the vertex weight as well as how to update both weights. Then we describe how VEWLS works.

The operation *select_add_vertex* selects a vertex with the highest dscore and breaks ties with a smaller timestamp to add, where the timestamp of v is the time when it is added into the partial vertex cover most recently. To remove a vertex, *select_remove_vertex* only considers vertices with *confChange* values equaling 1 and selects a vertex with the highest dscore and breaks ties with a smaller timestamp. *confChange* is defined for configuration checking, which is proposed in EWCC [3] to handle the cycle problem in local search. As shown in line 18, the value of *confChange*[u] is set to 1 when the state of its neighbours changes. Please note that the vertex to add always has a positive *dscore* while the vertex to remove has a negative *dscore*. To reconstruct a vertex cover, VEWLS prefers to select a vertex with higher *dscore* and breaks ties in favor of a smaller penalty.

The weight of each uncovered edge increases by 1 after each *swap*. To avoid the edge weight getting too large, we make use of the forgetting technique in [4]. The edge weight should be decreased by a coefficient ρ when the average of edge weights exceeds a threshold γ. In VEWLS, we set $\gamma = |V|/2$ and $\rho = 0.3$, where $|V|$ is the number of vertices. Let P be the current partial vertex cover and *best* be the best solution found so far. The penalty of each vertex in P should be increased by 1 when a better solution is found or a solution as good as *best* is found. We use at most one edge being uncovered by P as the condition for increasing vertex penalties, because P has one vertex less than *best*, P will become a better solution if all edges are covered and when only one edge is uncovered by P, $P \cup \{v\}$, where v covers the only uncovered edge, will become a vertex cover with the same size as *best*.

VEWLS works as follows: It starts with an initial vertex cover which is constructed by selecting a vertex covering more edges greedily. In VEWLS, the initial a vertex cover is constructed by adding the vertex to cover as many edges as possible each time iteratively. When a vertex cover of size k is found, it tries to find a solution with $k-1$ vertices by removing a vertex from the current solution and swapping a pair of vertices until the partial vertex cover becomes a vertex cover. The edge weights will increase after each *swap* so that VEWLS restarts when the average value of weights exceeds the threshold. To restart, VEWLS makes use of both the edge weight and the vertex penalty to reconstruct a new vertex cover.

Algorithm 1. VEWLS(G, *cutoff*), a new local search algorithm for MVC

Input: An undirected graph $G=(V,E)$ and cutoff time *cutoff*
Output: Vertex cover with the smallest size found so far

1 **begin**
2   construct a vertex cover P greedily
3      $best \leftarrow P$
4   **foreach** $e \in E$ **do**
5          $weight[e] \leftarrow 1$
6   **foreach** $v \in V$ **do**
7          $penalty[v] \leftarrow 0$
8   compute $dscore[v]$ by Definition 2
9          $confChange[v] \leftarrow 1$
10          $timestamp[v] \leftarrow 0$
11   **while** *not reach cutoff time* **do**
12   **while** P *is a vertex cover* **do**
13              $best \leftarrow P$
14              $v \leftarrow$ select_remove_vertex($dscore$, $timestamp$)
15   remove v from P
16          $v_1 \leftarrow$ select_remove_vertex($dscore$, $timestamp$)
17   **foreach** u *in* $N(v_1)$ **do**
18              $confChange[v] \leftarrow 1$
19          $v_2 \leftarrow$ select_add_vertex($dscore$, $timestamp$, $confChange$)
20   swap v_1 and v_2
21   update $timestamp[v_2]$
22   **foreach** $e = \{v, u\}$ *is uncovered* **do**
23              $weight[e] \leftarrow weight[e] + 1$
24   update $dscore[v]$ and $dscore[u]$ by Definition 2
25   **if** *number of uncovered edge is less than 1* **then**
26   **foreach** *vertex* v *in* P **do**
27                  $penalty[v] \leftarrow penalty[v]+1$
28   **if** *average edge weight* $\geq$ *threshold* **then**
29   **foreach** *edge* $e = \{u, v\}$ **do**
30                  $weight[e] \leftarrow \varphi \times weight[e]$
31   update $dscore[u]$ and $dscore[v]$
32   reconstruct P greedily by $dscore$ and break tie in favor of smaller penalty
33   **return** $best$

The difference between VEWLS and NuMVC is that VEWLS adds the restart phase by making use of the vertex penalty. In fact, restart is an effective technique, which has been widely used in local search algorithms for SAT and CSP [7, 1]. Experiments, presented in next section, show that it allows one to improve the performance of NuMVC on a huge range of instances.

4 Empirical Evaluation

We evaluate the performance of VEWLS on standard benchmarks for the MVC (MIS, MC) problem, namely the DIMACS[1] benchmark and the BHOSLIB (Benchmarks with Hidden Optimum Solutions)[2] benchmark. We compare VEWLS with NuMVC, which dominates other algorithms, such as Cover, EWLS and EWCC [4]. In addition, NuMVC is the best local algorithm for MVC on the BHOSLIB benchmark.

Table 1. Comparing VEWLS with NuMVC over the DIMACS benchmark

Instance		NuMVC			VEWLS		
Graph	MVC	$Suc(\%)$	$Time$(s)	$Step$	$Suc(\%)$	$Time$(s)	$Step$
brocks400_2	371	38	329.26	159323577	**44**	188.88	93389005
brocks400_4	367	100	**15.65**	7653271	100	15.67	7706357
DSJC1000.5.mis	985	100	1.10	142883	100	**1.04**	135109
keller6	3302	100	7.49	399533	100	**7.30**	390686
p_hat1500-1	1488	100	9.26	449552	100	**8.02**	387999
MANN_a45	690	**96**	180.88	78768223	92	163.86	70856196
MANN_a81	2221	4	325.64	49751930	4	**310.73**	46898081
C1000.9	932	100	**3.73**	1194201	100	3.74	1238411
C2000.5	1984	100	6.02	271050	100	**5.65**	243584
C2000.9	1921	14	228.06	37797183	20	**204.19**	33686264
C4000.5	3982	76	267.27	4471005	76	**261.86**	4168885

The DIMACS benchmark is taken from the Second DIMACS Implementation Challenge for the Maximum Clique problem (1992-1993). The 80 DIMACS instances are generated from real-world problems in coding theory, fault diagnosis problems, Kellers conjecture and the Steiner triple problem, in addition to randomly generated graphs and graphs where the maximum clique has been hidden by incorporating low-degree vertices. These instances range in size from less than 50 vertices and 1000 edges to greater than 3300 vertices and 5,000,000 edges. The BHOSLIB benchmark is a suit of hard random instances based on the CSP model RB [21, 22]. BHOSLIB instances are much harder than most DIMACS instances as well as an optimum solution can be hidden in the instance without reducing the difficulty to solve it. The BHOSLIB benchmark has been widely used to evaluate the performance algorithms for SAT, CSP, MC and MVC. We use 40 instances with the size from 450 vertices to 1534 vertices.

---

[1] ftp://dimacs.rutgers.edu/pub/challenges

[2] http://www.nlsde.buaa.edu.cn/~kexu/benchmarks/graph-benchmarks.htm

Table 2. Comparing VEWLS with NuMVC over the BHOSLIB benchmark

Instance		NuMVC			VEWLS		
Graph	MVC	Suc(%)	Time(s)	Step	Suc(%)	Time(s)	Step
frb30-15-1	420	100	0.08	35996	100	**0.07**	33943
frb30-15-2	420	100	0.10	49871	100	**0.09**	45558
frb30-15-3	420	100	0.33	167344	100	**0.30**	151613
frb30-15-4	420	100	0.09	45425	100	0.09	**42820**
frb30-15-5	420	100	**0.18**	89159	100	0.20	98466
frb35-17-1	560	100	**1.04**	447412	100	1.05	450567
frb35-17-2	560	100	0.90	384553	100	**0.87**	372007
frb35-17-3	560	100	**0.31**	129988	100	0.32	134120
frb35-17-4	560	100	0.94	402533	100	**0.89**	378461
frb35-17-5	560	100	0.46	**194745**	100	0.46	195662
frb40-19-1	720	100	**0.54**	195147	100	0.62	218655
frb40-19-2	720	100	**8.57**	3138719	100	10.04	3685558
frb40-19-3	720	100	2.39	875947	100	**2.30**	842605
frb40-19-4	720	100	8.93	3279988	100	**8.42**	3071682
frb40-19-5	720	100	24.19	8901977	100	**22.72**	8293541
frb45-21-1	900	100	8.91	2738077	100	**7.96**	2425589
frb45-21-2	900	100	13.14	4072965	100	**12.91**	3981183
frb45-21-3	900	100	39.08	12186665	100	**33.00**	10251540
frb45-21-4	900	100	**7.90**	2462682	100	9.71	2981122
frb45-21-5	900	100	20.95	6498414	100	**20.87**	6446523
frb50-23-1	1100	90	199.19	17243802	**96**	204.68	17710399
frb50-23-2	1100	**34**	273.38	23286406	28	291.15	24955502
frb50-23-3	1100	18	261.92	22275787	**20**	375.30	32788511
frb50-23-4	1100	100	66.91	5661385	100	**64.43**	5489741
frb50-23-5	1100	100	139.89	12055015	100	**136.15**	11845047
frb53-24-1	1219	**16**	383.00	93946856	10	166.47	41231273
frb53-24-2	1219	**76**	199.59	48531856	62	221.86	53881695
frb53-24-3	1219	100	123.72	30025024	100	**111.85**	27162736
frb53-24-4	1219	44	216.97	52832600	**64**	257.66	62427079
frb53-24-5	1219	100	117.16	28768079	100	**97.66**	23565609
frb56-25-1	1344	40	226.26	52569948	**54**	298.45	68745658
frb56-25-2	1344	28	331.97	76976027	**30**	254.53	58653134
frb56-25-3	1344	**90**	217.31	49723429	80	195.66	45068725
frb56-25-4	1344	100	97.29	22183589	100	**86.90**	19958391
frb56-25-5	1344	100	71.11	16264565	100	**59.89**	13642697
frb59-26-1	1475	**8**	345.38	22837670	4	183.20	12145443
frb59-26-2	1475	2	538.61	35519067	**6**	574.01	38090645
frb59-26-3	1475	**16**	338.09	22679122	10	231.79	15455807
frb59-26-4	1475	4	334.04	22197811	**8**	224.94	14965750
frb59-26-5	1475	36	166.06	11012989	**74**	210.58	13957374

VEWLS is implemented in C++. We obtain the source code of NuMVC from one of its authors. Both solvers are compiled by g++ with "-O2" option. All experiments are running on the machine with Intel Xeon CPU E5405 @ 2.00GHz and 8GB RAM under CestOS 5.7. The cutoff time for each instance is 600s (10$mins$) and we set $\gamma = |V|/2$ and $\rho = 0.3$ for all experiments. We run each instance for 50 times independently. We consider the best solution(s) found during 50 runnings as the minimum vertex cover for the instance and the size of the best solution is denoted by MVC in Table 1 and Table 2. A run is considered successful when it gives the same size as the best solution. The Success Rate (Suc for short) is computed as $T/50 * 100\%$, where T is the number of successful runs. Time (in seconds) and Step is the average time and the average steps of all successful runs, respectively. To compare the performance of solvers, we compare Success Rate first, then compare Time if Success Rates are the same and compare Step at last.

Table 1 shows the results on the DIMACS benchmark. All 80 instances are used in our experiments and we do not display the instances solved by both solvers within 1s or instances solved by neither of solvers within the cutoff time. VEWLS outperforms NuMVC on 8 instances out of 11. VEWLS improves the success rate with less time consumption on brocks400_2 and C2000.9 and it cuts down the time and steps on other 6 instances. VEWLS performs extremely close to NuMVC on brocks400_4 and C1000.9. On the other hand, NuMVC performs much better than VEWLS does only on MANN_a45.

Table 2 presents the performance of VEWLS and NuMVC on the BHOSLIB benchmark. It shows that VEWLS outperforms NuMVC on 27 instances out of 40. For the 20 hard instances with size larger than 50*23, VEWLS improves the success rates for 8 instances of them and reduces time consumptions for all instances on which VEWLS shares the same success rate with NuMVC.

Based on the empirical evaluation, we can conclude that integrating the vertex penalty into NuMVC can improve its performance on both DIMACS and BHOSLIB benchmarks. Combining the vertex weight and the edge weight takes advantages of both weighting schemes, which allows one to obtain a more effective local search algorithm for MVC.

5   Conclusion and Future Work

We present a new local search algorithm for the MVC problem, namely VEWLS, which combines the vertex weight and the edge weight. We evaluate the performance of VEWLS on both DIMACS and BHOSLIB benchmarks and compare it with NuMVC, which is the state-of-the-art local search algorithm for MVC and the dominating solver on the BHOSLIB benchmark. Experimental results show that VEWLS outperforms NuMVC on most instances from both benchmarks. To our best knowledge, it is the first time to combine the vertex weighting scheme and the edge weighting scheme for the MVC problem and it allows one to improve the performance of the best algorithm. However, the method to integrate two weights used in our paper is quite simple and we believe that the combination can be much more powerful. How to combine them more tightly and take more advantages of both weighting schemes will be part of our future work.

Acknowledgment. We would like to thank Jichang Zhao, Ge Zheng and anonymous reviewers for their helpful comments and suggestions. This research was partly supported by the fund of SKLSDE (Grant No. SKLSDE-2013ZX-06), and Research Fund for the Doctoral Program of Higher Education of China (Grant No. 20111102110019).

References

1. Biere, A.: Adaptive restart strategies for conflict driven SAT solvers. In: Kleine Büning, H., Zhao, X. (eds.) SAT 2008. LNCS, vol. 4996, pp. 28–33. Springer, Heidelberg (2008)
2. Cai, S., Su, K., Chen, Q.: EWLS: A New Local Search for Minimum Vertex Cover. Proc. of AAAI-2010, pp. 45–50 (2010)
3. Cai, S., Su, K., Sattar, A.: Local search with edge weighting and configuration checking heuristics for minimum vertex cover. Artif. Intell. 175(9-10), 1672–1696 (2011)
4. Cai, S., Su, K., Luo, C., Sattar, A.: NuMVC: An Efficient Local Search Algorithm for Minimum Vertex Cover. J. Artif. Intell. Res., 687–716 (2013)
5. Dinur, I., Safra, S.: On the hardness of approximating minimum vertex cover. Ann. of Math. 162(1) (2005)
6. Evans, I.: An evolutionary heuristic for the minimum vertex cover problem. In: Proc. of EP 1998, pp. 377–386 (1998)
7. Gomes, C., Selman, B., Kautz, H.: Boosting Combinatorial Search Through Randomization. In: Proc. of AAAI 1998, pp. 431–438 (1998)
8. Hutter, F., Tompkins, D.A.D., Hoos, H.H.: Scaling and probabilistic smoothing: Efficient dynamic local search for SAT. In: Van Hentenryck, P. (ed.) CP 2002. LNCS, vol. 2470, pp. 233–248. Springer, Heidelberg (2002)
9. Khot, S.: On the power of unique 2-Prover 1-Round games. In: Proc. 34th ACM Symp. on Theory of Computing, STOC, pp. 767–775 (May 2002)
10. Li, C., Quan, Z.: An efficient branch-and-bound algorithm based on maxsat for the maximum clique problem. In: Proc. of AAAI 2010, pp. 128–133 (2010)
11. Li, C., Fang, Z., Xu, K.: Combining maxsat reasoning and incremental upper bound for the maximum clique problem. In: Proc. of ICTAI 2013, pp. 939–946 (2013)
12. Morris, P.: The breakout method for escaping from local minima. In: Proc. of AAAI 1993, pp. 40–45 (1993)
13. Ostergard, P.R.J.: A fast algorithm for the maximum clique problem. Discrete Applied Mathematics 120, 197–207 (2002)
14. Pullan, W., Hoos, H.H.: Dynamic local search for the maximum clique problem. J. Artif. Intell. Res (JAIR), 159–185 (2006)
15. Pullan, W.: Phased local search for the maximum clique problem. J. Comb. Optim 12(3), 303–323 (2006)
16. Pullan, W.: Optimisation of unweighted/weighted maximum independent sets and minimum vertex covers. Discrete Optimization 6, 214–219 (2009)

17. Richter, S., Helmert, M., Gretton, C.: A stochastic local search approach to vertex cover. In: Hertzberg, J., Beetz, M., Englert, R. (eds.) KI 2007. LNCS (LNAI), vol. 4667, pp. 412–426. Springer, Heidelberg (2007)

18. Thornton, J.: Clause weighting local search for sat. J. Autom.Reasoning 35(1-3), 97–142 (2005)

19. Tomita, E., Kameda, T.: An Efficient Branch-and-bound Algorithm for Finding a Maximum Clique with Computational Experiments. Journal of Global Optimization 37, 95–111 (2007)

20. Tomita, E., Seki, T.: An Efficient Branch-and-Bound Algorithm for Finding a Maximum Clique. In: Calude, C.S., Dinneen, M.J., Vajnovszki, V. (eds.) DMTCS 2003. LNCS, vol. 2731, pp. 278–289. Springer, Heidelberg (2003)

21. Xu, K., Li, W.: Exact phase transitions in random constraint satisfaction problems. J. Artif. Intell. Res (JAIR) 12, 93–103 (2000)

22. Xu, K., Boussemart, F., Hemery, F., Lecoutre, C.: Random constraint satisfaction: Easy generation of hard (satisfiable) instances. Artif. Intell. 171(8-9), 514–534 (2007)

Randomized Parameterized Algorithms for Co-path Set Problem[*]

Qilong Feng[1], Qian Zhou[1], and Shaohua Li[1,2]

[1] School of Information Science and Engineering,
Central South University,
Changsha 410083, P.R. China
[2] Information School,
Guangdong University of Commercial Studies,
Guangzhou 510320, P.R. China

Abstract. Co-path Set problem is of important applications in mapping unique DNA sequences onto chromosomes and whole genomes. Given a graph G, the parameterized version of Co-path Set problem is to find a subset F of edges with size at most k such that each connected component in $G[E \backslash F]$ is a path. In this paper, we give a kernel of size $9k$ for the problem, and a randomized algorithm of running time $O^*(2.29^k)$ is presented for the Parameterized Co-path Set problem.

1 Introduction

The Minimum Co-path Set problem is to find a minimum set of edges in a given graph G such that after deleting those edges, each component in the remaining graph is a path, which has lots of applications in radiation hybrid mapping. In studying the relationship between DNA sequences and chromosomes, genomes, the radiation hybrid mapping technique can be used to map unique DNA sequences onto chromosomes and whole genomes [2],[8],[3], [9], where using gamma radiation, small DNA fragments can be obtained by breaking chromosomes.

A genetic marker is a gene or DNA sequence with a known location on a chromosome. Assume that $M = \{1, 2, \cdots, n\}$ is a set of markers, and $S_i \subseteq M$ ($1 \leq i \leq h$) be a subset of markers being present together from DNA fragments. In order to get a radiation hybrid mapping, for each $S_i \subseteq M$ ($1 \leq i \leq h$), a linear ordering of the markers in S_i should be given, i.e., the markers in S_i are consecutive. However, there might exist errors for the markers in S_i, which will affect the finding of the linear ordering in S_i. Under this case, the problem is that how to remove the minimum number of subsets from $\{S_1, S_2, \cdots, S_h\}$ such that a linear ordering can be found in the remaining subsets, which can be transformed to Minimum Co-path Set problem when $|S_i| = 2$, $1 \leq i \leq h$, that is, for a simple undirected graph $G = (V, E)$, where each marker in M is seen as a vertex in V, and for a subset $S_i = \{u, w\}$ ($1 \leq i \leq h$), an edge (u, w) is in E,

---

[*] This work is supported by the National Natural Science Foundation of China under Grant (61232001, 61103033, 61173051).

J. Chen, J.E. Hopcroft, and J. Wang (Eds.): FAW 2014, LNCS 8497, pp. 82–93, 2014.

and the objective is to find a minimum subset $F \subseteq E$, such that after deleting the edges in F, each component in the remaining graph is a path.

Based on the relationship between the number of edges deleted and the degrees of vertices, an approximation algorithm of ratio 2 was given in [2]. Recently, Chen et al. [1] gave an approximation algorithm with ratio $10/7$ using local search and dynamic programming. In this paper, we are mainly focused on the Parameterized Co-path Set problem, which is defined as follows:

Parameterized Co-path Set problem: Given a graph $G = (V, E)$ and an integer k, find a subset $F \subseteq E$ of size k such that after deleting the edges in F, each connected component in the remaining graph is a path, or report that no such subset exists.

Garey and Johnson [4] proved that Parameterized Co-path Set problem is NP-complete. Our objective is to design fixed parameter tractable algorithm for Parameterized Co-path Set problem, i.e., an algorithm with running time $O(f(k)n^c)$, where f is a computable function only related to k, n is the size of the input, and c is a constant. In the literature, Jiang et al. [6] presented a "weak kernel" (size of the potential search space) of size $5k$ for the Parameterized Co-path Set problem, and gave a parameterized algorithm of running time $O(2^{3.61k}(n + k))$ based on the weak kernel. Recently, Zhang et al. [10] gave a kernel of size $22k$ and a parameterized algorithm of running time $O^*(2.45^k)$ for the Parameterized Co-path Set problem, which are the current best results for the problem.

In this paper, we give a kernel of size $9k$ and randomized algorithm of running time $O^*(2.29^k)$ for the Parameterized Co-path Set problem. The randomized methods solving the Parameterized Co-path Set problem give a new sight to some branching based algorithms, which are of certain promising to be used to solve other problems.

2 Kernelization for Parameterized Co-path Set Problem

In this section, we study the kernelization algorithm for Parameterized Co-path Set problem. For an instance of the Parameterized Co-path Set problem (G, k), assume that there exists a subset $F \subseteq E$, $|F| \leq k$ such that after deleting the edges in F, each component in graph $G[E \backslash F]$ (the subgraph induced by the edges in $E \backslash F$) is a path, and F is called a *co-path set* of graph G. We assume that the co-path set F is a minimal co-path set, i.e., the removal of any edge in F makes F not a co-path set.

In graph G, a vertex with degree larger than or equal to three is called a degree$_{\geq 3}$ vertex, and the vertices with degree one, degree two are called degree$_1$ vertices and degree$_2$ vertices respectively. Let $G[X], X \subseteq V$, be a connected component consisting of only degree$_1$ and degree$_2$ vertices. Obviously $G[X]$ is a path or a cycle. If $G[X]$ is a path, we can just remove it from G. If $G[X]$ is a cycle, we can arbitrarily put an edge of $G[X]$ into F to make $G[X]$ a path. We can get the following reduction rule.

Rule 1. For a set $X \subseteq V$ such that $G[X]$ is a connected component consisting of only degree$_1$ and degree$_2$ vertices. If $G[X]$ is a path, remove X from V. If $G[X]$ is a cycle, arbitrarily put an edge in $G[X]$ into F, remove X from V, and decrease k by one.

In the following, we will bound the number of degree$_{\geq 3}$, degree$_2$, and degree$_1$ vertices respectively to get a kernel for the Parameterized Co-path Set problem. For a vertex v, let $H(v)$ denote the set of edges incident to v.

We first discuss the number of degree$_{\geq 3}$ vertices in G.

Lemma 1. *The number of degree$_{\geq 3}$ vertices in G is bounded by $2k$.*

Proof. Assume that v is a degree$_{\geq 3}$ vertex in G. Since the vertices in $G[E\backslash F]$ have degree either one or two, at least one edge in $H(v)$ must be deleted. Since the deletion of one edge can reduce the degrees of two vertices and at most k edges can be deleted, the number of degree$_{\geq 3}$ is at most $2k$, otherwise, (G, k) is not a yes-instance of the Parameterized Co-path Set problem.   □

We now study the number of degree$_2$ vertices in G. For a path $P = (u, w_1, w_2, \cdots, w_i, v)$, including the case that $v = u$, if all vertices in $\{w_1, w_2, \cdots, w_i\}$ have degree two and u, v do not have degree two, P is called a *degree$_2$ path*. We first study the properties of degree$_2$ path.

Lemma 2. *For a degree$_2$ path P, if an edge e in P is contained in a co-path set of graph G, then there is a co-path set F containing e such that e is only from $\{(u, w_1), (w_i, v)\}$, i.e., all edges in $\{(w_j, w_{j+1})|1 \leq j \leq i - 1\}$ are not contained in F.*

Proof. Assume that edge $e = (w_j, w_{j+1}), 1 \leq j \leq i-1$, is contained in a minimal co-path set F'. We prove the lemma by the following two cases.

Case (1). Vertices w_j, w_{j+1} are contained in different paths in $G[E\backslash F']$, denoted by P_1, P_2 respectively. The removal of edge e from F' makes the paths P_1, P_2 into one path, contradicting the fact that F' is a minimal co-path set.

Case (2). Vertices w_j, w_{j+1} are contained in one path, denoted by P_3. When no vertex of $\{u, v\}$ is a degree$_1$ vertex, then let $F = (F' \cup \{(u, w_1)\}) \backslash \{(w_j, w_{j+1})\}$. It is easy to see that F is a co-path set of the Parameterized Co-path Set problem. On the other hand, if one of $\{u, v\}$ is a degree$_1$ vertex (assume that v is a degree$_1$ vertex), then $F = (F' \cup \{(u, w_1)\}) \backslash \{(w_j, w_{j+1})\}$ is still a co-path set of the Parameterized Co-path Set problem.   □

Rule 2. For a degree$_2$ path $P = (u, w_1, w_2, \cdots, w_i, v)$, contract all the edges in $\{(w_j, w_{j+1})|1 \leq j \leq i - 1\}$, and denote the contracted path by $P' = (u, w_j, v)$. If there exist multiple edges after contraction, replace the multiple edges with a simple edge and decrease k by one. If one of $\{u, v\}$ is a degree$_1$ vertex (assume that v is a degree$_1$ vertex, and the contracted path is (u, w', v)), then contract the edge (u, w').

After using reduction Rules 1,2, denote the graph by $G' = (V', E')$. We now bound the number of degree$_2$ vertices in G'.

Lemma 3. *The number of degree$_2$ vertices in graph G' is bounded by $5k$.*

Proof. For a degree$_{\geq 3}$ vertex v in G', let $P(v)$ denote the set of degree$_2$ paths with end vertex v. If $|P(v)| \geq 3$, at least one of degree$_2$ paths (denoted as P) in $P(v)$ should be destroyed, i.e., at least one edge from P is contained in co-path set F.

From above discussion, we can see that for each degree$_{\geq 3}$ vertex v, there are at most two degree$_2$ paths in $P(v)$ without being destroyed. Since the number of degree$_{\geq 3}$ vertices is bounded by $2k$, the number of degree$_2$ paths undestroyed is bounded by $4k$. For each destroyed degree$_2$ path P', at least one edge of P' should be contained in co-path set F. Therefore, in graph G', the number of destroyed degree$_2$ paths is bounded by k. Since each degree$_2$ path has one degree$_2$ vertex after applying reduction Rule 2, the total number of degree$_2$ vertices in G' is bounded by $5k$. $\qquad\qquad\square$

We now study the number of degree$_1$ vertices in graph G'.

Lemma 4. *In graph G', if a degree$_{\geq 3}$ vertex v is adjacent to at least three degree$_1$ vertices (denoted by $\{v_1, \cdots, v_h\}$, $h \geq 3$), then there exists a co-path set F' of the Parameterized Co-path Set problem such that $h - 2$ edges from $\{(v, v_i)|1 \leq i \leq h\}$ are contained in F', and any $h-2$ edges of $\{(v, v_i)|1 \leq i \leq h\}$ can be chosen to put into F'.*

Proof. Since in $G'[E'\backslash F']$, each vertex has degree at most two, at least $h - 2$ edges from $H(v)$ should be contained in F'. Let $(v, v') \in \{(v, v_i)|1 \leq i \leq h\}$ be an edge contained in F', and let $(v, v'') \in \{(v, v_i)|1 \leq i \leq h\}$ be an edge in $G'[E'\backslash F']$. Let $F'' = (F' - \{(v, v')\}) \cup \{(v, v'')\}$. Then, F'' is still a co-path set of the Parameterized Co-path Set problem. Therefore, any $h - 2$ edges of $\{(v, v_i)|1 \leq i \leq h\}$ can be chosen to put into F'. $\qquad\square$

> **Rule 3.** In graph G', if a degree$_{\geq 3}$ vertex v is adjacent to at least three degree$_1$ (denoted by $\{v_1, \cdots, v_h\}$, $h \geq 3$), choose any $h - 2$ edges from $\{(v, v_i)|1 \leq i \leq h\}$ to delete, $k = k - (h - 2)$.

Lemma 5. *In graph G', if a degree$_{\geq 3}$ vertex v is adjacent to exactly two degree$_1$ vertices u, w, then there exists a co-path set F' of the Parameterized Co-path Set problem such that path (u, v, w) is a path in $G'[E'\backslash F']$.*

Proof. Assume that edge set F'' ($|F''| \leq k$) is a minimal co-path set of the Parameterized Co-path Set problem, i.e., in graph $G'[E'\backslash F'']$, each connected component is a path. We prove the lemma by the following two cases.

(1) One edge of $\{(u, v), (v, w)\}$ is contained in F''.

Without loss of generality, assume that edge (u, v) is contained in F''. Then, there must exist a path P' in $G'[E'\backslash F'']$ such that w is one endpoint of P' and P' passes through v. Let z be the other neighbor of v in the path P' such that (z, v) is in P'. Let $F' = (F''\backslash\{(u, v)\}) \cup \{(z, v)\}$. Then, each connected component in $G'[E'\backslash F']$ is a path and path (u, v, w) is in $G'[E'\backslash F']$.

(2) Edges (u, v) and (v, w) are contained in F''.

In this case, we first prove that in graph $G'[E'\backslash F'']$, vertex v must not be an endpoint of a path or be an isolated vertex, otherwise, $F''\backslash\{(u,v)\}$ is still a co-path set of the Parameterized Co-path Set problem, contradicting the fact that F'' is a minimal co-path set. Thus, v is an internal vertex of a path in $G'[E'\backslash F'']$. Assume that v is contained in path P', and edge $(v,x),(v,y)$ are in P'. Let $F' = (F''\backslash\{(u,v),(v,w)\}) \cup \{(v,x),(v,y)\}$. Then, each connected component in $G'[E'\backslash F']$ is a path and path (u,v,w) is in $G'[E'\backslash F']$.   □

Based on the Lemma 5, we can get the following reduction rule.

Rule 4. In graph G', if a degree$_{\geq 3}$ vertex v is adjacent to exactly two degree$_1$ vertices u, w (assume that the number of edges with endpoint v is x), then delete all the edges with endpoint v except edges $(u,v),(v,w)$, $k = k - x + 2$, and remove path (u,v,w) from G'.

After applying reduction Rules 3,4, denote the new reduced graph by G''. We can get the following lemma.

Lemma 6. *In graph G'', the number of degree$_1$ vertices is bounded by $2k$.*

Proof. By applying reduction Rules 1,2, each degree$_1$ vertex is adjacent to one degree$_{\geq 3}$ vertex. After applying reduction Rules 3,4, it is easy to see that each degree$_{\geq 3}$ vertex is adjacent to at most one degree$_1$ vertex. Since the number of degree$_{\geq 3}$ vertices is bounded by $2k$, the number of degree$_1$ vertices is bounded by $2k$.   □

Lemma 7. *The Parameterized Co-path Set problem admits a kernel of size $9k$.*

Proof. By applying reduction Rules 1-4 exhaustedly, the number of degree$_1$ and degree$_2$ vertices can be bounded. By Lemma 1, Lemma 3, Lemma 6, the number of degree$_{\geq 3}$, degree$_2$, and degree$_1$ vertices is bounded by $2k$, $5k$, $2k$ respectively, which means that the total number of vertices in reduced graph is bounded by $2k + 5k + 2k = 9k$.   □

3   The Randomized Algorithm for Parameterized Co-path Set Problem

In this section, we give a randomized parameterized algorithm with running time $O^*(2.29^k)$ for the Parameterized Co-path Set problem. The running time of our randomized algorithm dependents mainly on the time dealing with vertices with degree bounded by three, and the NP-completeness of Parameterized Co-path Set in cubic graph can be proven by getting a reduction from the Hamiltonian path problem.

Given an instance (G,k) of the Parameterized Co-path Set problem, assume that F is a co-path set of the instance, i.e., after deleting the edges in F, each component in the remaining graph is a path. For a vertex v of degree d $(d \geq 3)$, at least $d - 2$ edges need to be deleted to make v have degree at most two in $G[E\backslash F]$, which is the main idea to solve the Parameterized Co-path Set problem.

The major part of the process to find set F is to reduce the degrees of all degree$_{\geq 3}$ vertices. A vertex u with degree at least four is called a degree$_{\geq 4}$ vertex. For a degree$_{\geq 3}$ vertex v in G, let $H(v)$ denote the set of edges incident to v, and let $H'(v)$ be the set of edges in $H(v)$ that are contained in co-path set. For a vertex v, let $N(v)$ denote the set of neighbors of v. We will study the randomized algorithm for Parameterized Co-path Set problem by dealing with the following two different kinds of vertices.

(1). Degree$_{\geq 4}$ vertices.

For a degree$_{\geq 4}$ vertex v, let d_v be the degree of v. Randomly choose an edge e from $H(v)$ to put into F. The probability that e is from $H'(v)$ is at least: $(d_v - 2)/d_v \geq 1/2$.

(2). Degree$_3$ vertices.

We apply different random methods to analyze the degree$_3$ vertices.

For a degree$_3$ vertex v, if at least two edges of $H(v)$ are contained in F, then by randomly choosing an edge e from $H(v)$ to put into F, the probability that e is from $H'(v)$ is at least $2/3$. Now we are mainly focused on the degree$_3$ vertex with only one edge from $H(v)$ contained in F. Assume that the three neighbors of v are x, y, z. We discuss the probability of choosing edges from $H(v)$ based on the degrees of x, y, z, as follows.

Case 1. One vertex in $\{x, y, z\}$ has degree one.

Without loss of generality, assume that vertex x has degree one.

Lemma 8. *For the three edges $\{e_1, e_2, e_3\}$, where $e_1 = (v, x), e_2 = (v, y), e_3 = (v, z)$, if e_1 is contained in F, then there exists an edge e' in $\{e_2, e_3\}$ such that $(F - \{e_1\}) \cup \{e'\}$ is still a co-path set of the Parameterized Co-path Set problem.*

Proof. Let $P = \{P_1, P_2, \cdots, P_h\}$ be the set of components in $G[E\backslash F]$, where each $P_i (1 \leq i \leq h)$ is a path. By our above assumption, only one edge of $H(v)$ is contained in F, thus, vertices y, z are contained in the same connected component in $G[E\backslash F]$.

In $G[E\backslash F]$, vertex x is an isolated vertex, and assume that y and z are contained in path P_i in $G[E\backslash F]$. Let u, w be the two endpoints of path P_i satisfying that the subpath from u to z passing through y. Let $F' = (F - \{e_1\}) \cup \{e_2\}$. Then, in $G[E\backslash F']$, a path P' can be constructed by the subpath from w to z, and edges $(x, v), (v, z)$. Moreover, the subpath from y to u of P_i is contained in $G[E\backslash F']$. Therefore, in $G[E\backslash F']$, each connected component is a path, i.e., F' is still a co-path set of the Parameterized Co-path Set problem.   □

Lemma 9. *For the Case 1, by randomly choosing an edge e from $\{e_2, e_3\}$, the probability that e is from $H'(v)$ is 1/2.*

Proof. By our assumption, exactly one edge of $H(v)$ is contained in F. By Lemma 8, for the three edges $\{e_1, e_2, e_3\}$, where $e_1 = (v, x), e_2 = (v, y), e_3 = (v, z)$, if e_1 is contained in F, then there exists an edge e' in $\{e_2, e_3\}$ such that $(F - \{e_1\}) \cup \{e'\}$ is still a co-path set of the Parameterized Co-path Set problem. Therefore, one edge of $\{e_2, e_3\}$ is contained in co-path set of the Parameterized Co-path Set problem. Then, by randomly choosing an edge e from $\{e_2, e_3\}$, the probability that e is from $H'(v)$ is 1/2.   □

Case 2. The vertices x, y, and z all have degree two.

Lemma 10. *For the three edges $\{e_1, e_2, e_3\}$, where $e_1 = (v, x), e_2 = (v, y), e_3 = (v, z)$, if e_i is contained in F, then there exists an edge e' in $\{e_1, e_2, e_3\}\backslash\{e_i\}$ such that $(F - \{e_i\}) \cup \{e'\}$ is still a co-path set of the Parameterized Co-path Set problem.*

Proof. Without loss of generality, assume that e_1 is contained in F. Let $P = \{P_1, P_2, \cdots, P_h\}$ be the set of components in $G[E\backslash F]$, where each $P_i(1 \leq i \leq h)$ is a path. By our above assumption, only one edge of $H(v)$ is contained in F, thus, vertices y, z are contained in the same connected component in $G[E\backslash F]$. We prove the lemma by the following two cases.

1) x and y, z are in different paths in $G[E\backslash F]$.

Assume that x is contained in path P_i, and y, z are contained in path P_j in $G[E\backslash F]$, where $i \neq j, 1 \leq i \leq h, 1 \leq j \leq h$. Then, x is one endpoint of P_i. Let t be the other endpoint of path P_i, and let u, w be the two endpoints of path P_j satisfying that the subpath from u to z passing through y. Let $F' = (F - \{e_1\}) \cup \{e_2\}$. Then, in $G[E\backslash F']$, a path P' can be constructed by the subpath from w to z, the supbath from x to t, and edges $(x, v), (v, z)$. Moreover, the subpath from y to u of P_j is contained in $G[E\backslash F']$. Therefore, in $G[E\backslash F']$, each connected component is a path, i.e., F' is still a co-path set of the Parameterized Co-path Set problem.

2) x, y, z are in the same path in $G[E\backslash F]$.

Assume that x, y, z are in path P_i. Then, x is one endpoint of P_i and assume that w is the other endpoint of P_i. Without loss of generality, assume that from vertex x, along the path P_i, y is the first vertex of $\{y, z\}$ encountered. Let $F' = (F - \{e_1\}) \cup \{e_2\}$. Then, in $G[E\backslash F']$, a path P' can be constructed by the subpath from w to z, the supbath from x to y, and edges $(x, v), (v, z)$. Therefore, in $G[E\backslash F']$, each connected component is a path, i.e., F' is still a co-path set of the Parameterized Co-path Set problem. □

Based on the Lemma 10, we can get the following lemma.

Lemma 11. *For the Case 2, by randomly choosing an edge e from $H(v)$ to put into F, the probability that e is from $H'(v)$ is 2/3.*

Proof. By our assumption, exactly one edge of $H(v)$ is contained in F. By Lemma 10, for the three edges $\{e_1, e_2, e_3\}$, where $e_1 = (v, x), e_2 = (v, y), e_3 = (v, z)$, if e_i is contained in F, then there exists an edge e' in $\{e_1, e_2, e_3\}\backslash\{e_i\}$ such that $(F - \{e_i\}) \cup \{e'\}$ is still a co-path set of the Parameterized Co-path Set problem. Then, by randomly choosing an edge e from $H(v)$ to put into F, the probability that one of $\{e_i, e'\}$ is chosen is 2/3. □

Case 3. Two vertices of $\{x, y, z\}$ are degree$_2$ vertices.

Without loss of generality, assume that vertices x, y have degree two and vertex z has degree three. We can get the following lemma.

Lemma 12. *For the three edges $\{e_1, e_2, e_3\}$, where $e_1 = (v, x), e_2 = (v, y), e_3 = (v, z)$, if an edge e_i from $\{e_1, e_2\}$ is contained in F, then $(F - \{e_i\}) \cup (\{e_1, e_2\} - \{e_i\})$ is still a co-path set of the Parameterized Co-path Set problem.*

The proof of Lemma 12 is similar to the proof of Lemma 10, which is neglected here.

Lemma 13. *For the Case 3, by randomly choosing an edge e from $\{e_2, e_3\}$ (or $\{e_1, e_3\}$), the probability that e is from $H'(v)$ is $1/2$.*

Proof. By our assumption, exactly one edge of $H(v)$ is contained in F. By Lemma 12, for the three edges $\{e_1, e_2, e_3\}$, where $e_1 = (v, x), e_2 = (v, y), e_3 = (v, z)$, if an edge e_i from $\{e_1, e_2\}$ is contained in F, then $(F - \{e_i\}) \cup (\{e_1, e_2\} - \{e_i\})$ is still a co-path set of the Parameterized Co-path Set problem. Therefore, one edge of $\{e_2, e_3\}$ is contained in a co-path set of the Parameterized Co-path Set problem. Then, by randomly choosing an edge e from $\{e_2, e_3\}$, the probability that e is from $H'(v)$ is $1/2$.   □

Case 4. Only one vertex of $\{x, y, z\}$ has degree two.

Without loss of generality, assume that vertex x has degree two and vertices y, z have degree three. We can get the following lemma.

Lemma 14. *For the three edges $\{e_1, e_2, e_3\}$, where $e_1 = (v, x), e_2 = (v, y), e_3 = (v, z)$, if edge e_1 is contained in F, then there exists an edge e' in $\{e_2, e_3\}$ such that $(F - \{e_1\}) \cup \{e'\}$ is still a co-path set of the Parameterized Co-path Set problem.*

The proof of Lemma 14 is similar to the proof of Lemma 12, which is neglected here.

Lemma 15. *For the Case 4, by randomly choosing an edge e from $\{e_2, e_3\}$, the probability that e is from $H'(v)$ is $1/2$.*

Proof. By our assumption, exactly one edge of $H(v)$ is contained in F. By Lemma 14, for the three edges $\{e_1, e_2, e_3\}$, where $e_1 = (v, x), e_2 = (v, y), e_3 = (v, z)$, if edge e_1 is contained in F, then there exists an edge e' in $\{e_2, e_3\}$ such that $(F - \{e_1\}) \cup \{e'\}$ is still a co-path set of the Parameterized Co-path Set problem. Therefore, one edge of $\{e_2, e_3\}$ is contained in a co-path set of the Parameterized Co-path Set problem. Then, by randomly choosing an edge e from $\{e_2, e_3\}$, the probability that e is from $H'(v)$ is $1/2$.   □

Case 5. The three vertices $\{x, y, z\}$ all have degree three.

In this case, just randomly choose an edge e from $H(v)$ to put into F. Then, with probability $1/3$, e is from $H'(v)$. Without loss of generality, assume that edge (v, z) is put into F. Then, after deleting the edge (v, z), the degrees of vertices x, y become two, which satisfies one of cases from Case 2 to Case 4.

Based on the above cases, the specific randomized algorithm solving the Parameterized Co-path Set problem is given in Figure 1.

Theorem 1. *The Parameterized Co-path Set problem can be solved randomly in time $O^*(2.29^k)$.*

Proof. First note that if the input instance is a no-instance, step 2 could not find a subset $F \subseteq E$ with size at most k such that in graph $G[E \backslash F]$, each component is a path, which is rightly handled by step 3.

Now suppose that a subset $F \subseteq E$ can be found in G such that each component is a path in $G[E \backslash F]$. Then, there must exist two subsets $F', F'' \subseteq F$ ($F' \cup F'' = F$) such that in $G[E \backslash F']$, each vertex has degree at most two, and after deleting the edges in F'', all the cycles in $G[E \backslash F']$ are destroyed. Thus, there must exist k_1, k_2 with $k_1 + k_2 = k$ such that $|F'| \leq k_1$, $|F''| \leq k_2$. Moreover, we assume that in F', F'_1 ($|F'_1| = k_{11}$) is the set of edges deleted to reduce the degree$_{\geq 4}$ vertices (at least one endpoint of each edge in F'_1 has degree larger than or equal to four), and F'_2 ($|F'_2| = k_{12}$) is the set of edges deleted to reduce the remaining degree$_3$ vertices (at least one endpoint of each edge in F'_2 has degree three), where $k_{11} + k_{12} = k_1$.

Step 2.2 is to reduce the degrees of degree$_{\geq 4}$ vertices. For each degree$_{\geq 4}$ vertex v, with probability $(d_v - 2)/d_v \geq 1/2$, one edge of $H'(v)$ is added in F. Since the size of F'_1 is k_{11}, the probability that all edges in F'_1 are deleted is at least $(1/2)^{k_{11}}$.

For a vertex v with degree three, let x, y, z be the three neighbors of v in G, and at least one edge of $H(v)$ must be deleted. If one vertex of $\{x, y, z\}$ has degree one, by Lemma 9, step 2.4 can be rightly done with probability at least $1/2$. If vertices x, y, z all have degree two, by Lemma 11, step 2.7 can be rightly done with probability at least $2/3$. If two vertices of $\{x, y, z\}$ have degree two, by Lemma 13, step 2.8 can be rightly done with probability at least $1/2$. If only one vertex of $\{x, y, z\}$ has degree two, by Lemma 15, step 2.9 can be rightly done with probability at least $1/2$.

If there does not exist a degree$_3$ vertex with at least one neighbor having degree less than three, then step 2.10 can be done with probability at least $1/3$. We now prove the following claim.

Claim 1. The number of executions of step 2.10 is bounded by $k/3$.

It is easy to see that for each execution of step 2.10, one edge of $H(u)$ is deleted. Let a, b, c be three neighbors of u. Then, the degrees of a, b, c are also three. By deleting one edge (denoted by (u, x)) from $H(u) = \{(u, a), (u, b), (u, c)\}$, two degree$_2$ vertices u, x can be obtained. In terms of the number of vertices adjacent to u and x, there are three different cases. Obviously, each of the vertices adjacent to u and x is a degree$_3$ vertex, otherwise x or u can be done by former steps. We now prove the claim by the following three cases:

Case (1). There are four vertices in $N(u) \cup N(v)$, which are denoted by v_1, v_2, v_3, v_4. For a vertex $v_i, 1 \leq i \leq 4$, it can be incident to at most two vertices of $\{v_1, v_2, v_3, v_4\} \backslash \{v_i\}$. Therefore, there are at least two vertices $v_i, v_j, 1 \leq i, j \leq 4$ not adjacent to each other, which means in this case at least two nonadjacent degree$_3$ vertices with neighbors having degree two can be found

---

Algorithm R-MCP(G, k)
Input: a graph G, and parameter k
Output: a subset $F \subseteq E$ of size at most k such that each component in $G[E \backslash F]$
is a path, or report no such subset exists.
1. **for** each k_1, k_2 with $k_1 + k_2 = k$ **do**
2.   **loop** $c \cdot 2.29^{k_1}$ times
2.1 $F = \emptyset$;
2.2   **for** any vertex v with degree larger than or equal to four in G **do**
          randomly choose an edge e from $H(v)$, delete it and add it to F;
2.3   **while** there exists a vertex with degree three in G **do**
2.4   **if** there exists a degree$_3$ vertex v with one neighbor having degree one
          then
            let u be the neighbor of v with degree one;
            randomly choosing an edge e from $H(v) \backslash \{(u, v)\}$ to delete, and
            $F = F \cup \{e\}$;
          else
2.5   **if** there exists a degree$_3$ vertex v with at least one neighbor having
            degree two **then**
2.6   let x, y, z be the three neighbors of v;
2.7   **if** vertices x, y, z all have degree two **then**
              randomly choosing an edge e from $H(v)$ to delete, and
              $F = F \cup \{e\}$;
2.8   **if** two vertices of $\{x, y, z\}$ have degree two **then**
              let u be any vertex from $\{x, y, z\}$ with degree two;
              randomly choosing an edge e from $H(v) \backslash \{(u, v)\}$ to delete, and
              $F = F \cup \{e\}$;
2.9   **if** only one vertex of $\{x, y, z\}$ has degree two **then**
              let u be the vertex from $\{x, y, z\}$ with degree two;
              randomly choosing an edge e from $H(v) \backslash \{(u, v)\}$ to delete, and
              $F = F \cup \{e\}$;
2.10   **else**
            find a vertex u with three neighbors having degree three;
            randomly choose an edge e from $H(u)$ to delete, and $F = F \cup \{e\}$;
2.11   **if** $|F| \le k_1$ **then**
2.12   denote the remaining graph by G';
2.13   **if** the number of cycles in G' is at most k_2 **then**
2.14   **for** each cycle C in G' **do**
2.15   delete an edge e'' from C, and add e'' to F;
2.16   return(F); break;
3.   return("no such subset exists").

Fig. 1. A randomized algorithm for the Parameterized Co-path Set problem

after the execution of step 2.10. Then, at least two more edges need to be deleted to get a co-path set.

Case (2). There are three vertices in $N(u) \cup N(v)$, that is, u, x share a common neighbor. We denote the three $degree_3$ vertices in $N(u) \cup N(v)$ by v_1, v_2, v_3 and assume that v_1 is the common neighbor of u, x. As v_1 is a $degree_3$ vertex, v_1 can be adjacent to at most one vertex of $\{v_2, v_3\}$. In this case at least two nonadjacent $degree_3$ vertices v_1 and $v_i, i \in \{2, 3\}$ with neighbors having degree two can be found. Therefore at least two more edges need to be deleted to get a co-path set.

Case (3). There are two vertices in $N(u) \cup N(v)$, i.e., u, x share two common $degree_3$ neighbors, which are denoted by v_1, v_2 respectively. If v_1 and v_2 are not adjacent, v_1, v_2 are the two nonadjacent $degree_3$ vertices with neighbors having degree two and this can be done similarly as Cases 1,2. If v_1 and v_2 are adjacent, the vertices in $\{u, x, v_1, v_2\}$ induce an isolated K_4(a complete graph with 4 vertices). We can delete edges $(v_1, v_2), (v_1, x), (x, u)$ to get a co-path set.

Combining all cases discussed above, by deleting one edge in $H(u)$, there are least two extra edges needing to be deleted to get a co-path set. If the number of executions of step 2.10 is more than $k/3$, the number of extra edges deleted to get a co-path set is at least $2k/3$, a contradiction. This completes the proof of the Claim.

Since the number of executions of step 2.10 is bounded by $k/3$, and the number of $degree_3$ vertices with neighbors having degree less than three in the whole process is at least $2k/3$, the probability that all the degrees of $degree_3$ vertices is reduced is at least $(1/3)^{k_{12}/3}(1/2)^{2k_{12}/3} = 1/2.29^{k_{12}}$. Then, the probability that all edges in F_1' and F_2' are deleted is at least $(1/2)^{k_{11}} \cdot (1/2.29)^{k_{12}} \geq (1/2.29)^{k_{11}+k_{12}} = (1/2.29)^{k_1}$. Therefore, the probability that the edges in $F_1' \cup F_2'$ are not rightly deleted is $1 - (1/2.29)^{k_1}$. Therefore, after $c \cdot 2.29^{k_1}$ operations, none of the executions of steps 2.2-2.10 can rightly handle the edges in $F_1' \cup F_2'$ is $(1 - (1/2.29)^{k_1})^{c \cdot 2.29^{k_1}} = ((1 - (1/2.29)^{k_1})^{2.29^{k_1}})^c \leq (1/e)^c$. Therefore, after $c \cdot 2.29^{k_1}$ operations, the algorithm can correctly handle the edges in $F_1' \cup F_2'$ with probability larger than $1 - (1/e)^c$.

Step 2.2 can be done in time $O(n + m)$, and step 2.3 can be done in $O(n(m + n))$, where m, n are the number of edges and vertices respectively. Therefore, algorithm R-MCP runs in time $O(2.29^{k_1}n(n + m)) = O^*(2.29^k)$. □

4   Conclusion

In this paper, kernelization and improved parameterized algorithm for the Parameterized Co-path Set problem are studied. Based on the structure of the problem, a kernel of size $9k$ is given. Finally, the random property of the problem is analyzed, and a randomized algorithm of running time $O^*(2.29^k)$ is given. Our future research is to apply the random analysis method in the paper to other parameterized problems.

References

1. Chen, Z., Lin, G., Wang, L.: An approximation algorithm for the minimum co-path set problem. Algorithmica 60, 969–986 (2011)
2. Cheng, Y., Cai, Z., Goebel, R., Lin, G., Zhu, B.: The radiation hybrid map construction problem: recognition, hardness, and approximation algorithms (2008) (unpublished manuscript)
3. Cox, D.R., Burmeister, M., Price, E.R., Kim, S., Myers, R.M.: Radiation hybrid mapping: a somatic cell genetic method for constructing high resolution maps of mammalian chromosomes. Science 250, 245–250 (1990)
4. Garey, M.R., Johnson, D.S.: Computers and Intractability: A Guide to the Theory of NP Completeness. W.H. Freeman (1979)
5. Garey, M.R., Johnson, D.S., Tarjan, R.E.: The planar hamiltonian circuit problem is NP-complete. SIAM J. Comput. 5(4), 704–714 (1976)
6. Jiang, H., Zhang, C., Zhu, B.: Weak kernels. ECCC Report, TR10-005 (2010)
7. Kanj, I., Xia, G., Zhu, B.: The Radiation Hybrid Map Construction Problem Is FPT. In: Cai, Z., Eulenstein, O., Janies, D., Schwartz, D. (eds.) ISBRA 2013. LNCS, vol. 7875, pp. 5–16. Springer, Heidelberg (2013)
8. Richard, C.W., Withers, D.A., Meeker, T.C., Maurer, S., Evans, G.A., Myers, R.M., Cox, D.R.: A radiation hybrid map of the proximal long arm of human chromosome 11, containing the multiple endocrine neoplasia type 1 (MEN-1) and bcl-1 disease loci. American J. of Human Genetics 49, 1189–1196 (1991)
9. Slonim, D., Kruglyak, L., Stein, L., Lander, E.: Building human genome maps with radiation hybrids. J. of Computational Biology 4, 487–504 (1997)
10. Zhang, C., Jiang, H., Zhu, B.: Radiation Hybrid Map Construction Problem Parameterized. In: Lin, G. (ed.) COCOA 2012. LNCS, vol. 7402, pp. 127–137. Springer, Heidelberg (2012)

Improved LP-rounding Approximations for the k-Disjoint Restricted Shortest Paths Problem*

Longkun Guo

College of Mathematics and Computer Science,
Fuzhou University, China

Abstract. Let $G = (V, E)$ be a given (directed) graph in which every edge is with a cost and a delay that are nonnegative. The k-disjoint restricted shortest path (kRSP) problem is to compute k (edge) disjoint minimum cost paths between two distinct vertices $s, t \in V$, such that the total delay of these paths are bounded by a given delay constraint $D \in \mathbb{R}_0^+$. This problem is known to be NP-hard, even when $k = 1$ [4]. Approximation algorithms with bifactor ratio $(1 + \frac{1}{r}, r(1 + \frac{2(\log r + 1)}{r})(1 + \epsilon))$ and $(1 + \frac{1}{r}, r(1 + \frac{2(\log r + 1)}{r}))$ have been developed for its special case when $k = 2$ respectively in [11] and [3]. For general k, an approximation algorithm with ratio $(1, O(\ln n))$ has been developed for a weaker version of kRSP, the k bi-constraint path problem of computing k disjoint st-paths to satisfy the given cost constraint and delay constraint simultaneously [7].

In this paper, an approximation algorithm with bifactor ratio $(2, 2)$ is first given for the kRSP problem. Then it is improved such that for any resulted solution, there exists a real number $0 \le \alpha \le 2$ that the delay and the cost of the solution is bounded, respectively, by α times and $2 - \alpha$ times of that of an optimal solution. These two algorithms are both based on rounding a basic optimal solution of a LP formula, which is a relaxation of an integral linear programming (ILP) formula for the kRSP problem. The key observation of the two ratio proofs is to show that, the fractional edges of a basic solution to the LP formula will compose a graph in which the degree of every vertex is exactly 2. To the best of our knowledge, it is the first algorithm with a single factor polylogarithmic ratio for the kRSP problem.

Keywords: LP rounding, flow theory, k-disjoint restricted shortest path problem, bifactor approximation algorithm.

1 Introductions

This paper addresses on the k restricted shortest path problem, whose definition is formally as in the following:

* This research was supported by Natural Science Foundation of China (#61300025), Doctoral Funds of Ministry of Education of China for Young Scholars (#20123514120013), Natural Science Foundation of Fujian Province (#2012J05115), and Fuzhou University Development Fund (2012-XQ-26).

J. Chen, J.E. Hopcroft, and J. Wang (Eds.): FAW 2014, LNCS 8497, pp. 94–104, 2014.

Definition 1. *(The k restricted shortest path problem, kRSP) Let $G = (V, E)$ be a (directed) graph with a pair of distinct vertices $s, t \in V$. Assume that $c : E \to \mathbb{R}_0^+$ and $d : E \to \mathbb{R}_0^+$ are a cost function and a delay function on the edges of E respectively. The k restricted shortest path problem is to compute k disjoint st-paths $P_1, \ldots, P_k$, such that $\sum\limits_{i=1,\ldots,k} c(P_i)$ is minimized while $\sum\limits_{i=1,\ldots,k} d(P_i) \leq D$ holds for a given delay bound $D \in \mathbb{R}_0^+$.*

The kRSP problem has broad applications in industry, e.g., end-to-end video transmission with delay constraints, construction of minimum cost time-sensitive survivable networks, design of minimum cost fault tolerance systems subjected to given energy consumption constraint (or other additive constraints) and etc. Before the technique paragraphs, we would like to give the statement of the bifactor approximation algorithms for the kRSP problem first: An algorithm $\mathcal{A}$ is a bifactor (α, β)-approximation for the kRSP problem iff for every instance of kRSP, $\mathcal{A}$ computes k disjoint st-paths whose delay sum and cost sum are bounded by αD and $\beta c(OPT)$ respectively, where OPT is an optimum solution to the kRSP problem and $c(OPT) = \sum_{e \in OPT} c(e)$. We shall use bifactor $(1, \beta)$-approximation and β-approximation interchangeably in the text while no confusion arises.

1.1 Related Work

The kRSP problem have been studied for some fixed positive integral k. When $k = 1$, the problem becomes the restricted shortest path problem (RSP) of finding a single shortest path that satisfies a given QoS constraint. The RSP problem is known as one of Karp's 21 NP-hard problems [4] and admits full polynomial time approximation scheme (FPTAS) [9]. As a generalization of the RSP problem, the single Multiple Constraint Path (MCP) problem of computing a path subjected to multiple given QoS constraints is still attracting interest in the research community. For MCP, the $(1 + \epsilon)$-approximation developed by Xue et al [16,10] is the best result in the current state of the art. When $k = 2$, approximation algorithms with bifactor ratio $(1 + \frac{1}{r}, r(1 + \frac{2(\log r + 1)}{r})(1 + \epsilon))$ and $(1 + \frac{1}{r}, r(1 + \frac{2(\log r + 1)}{r}))$ have been developed respectively in [11] and [3]. To the best of our knowledge, there exists no non-trivial approximation that strictly obeys the delay constraint in the literature. However, for general k, the author, together with Shen and Liao, have developed approximation algorithms with bifactor ratio $(1, O(\ln n))$ for a weaker version of the kRSP problem, namely the k bi-constraint path problem, in which the goal is to compute k disjoint paths satisfying a given cost constraint and a given delay constraint simultaneously [7].

There are also some other interesting results that addressed on other special cases of the kRSP problem. When all edges are with delay 0, this problem becomes the min-sum problem of computing k disjoint paths with the total cost minimized, which is known polynomial solvable [13,14]. The min-min and min-max problems are two problems which are close related to the min-sum problem. The former problem is to find two paths with the length of the shorter one minimized, while the latter is to make the length of the longer one minimized. Our previous work, together with Xu et al's and Bhatia et al's [6,15,2], show that the min-min problem is NP-complete and doesn't admit K-approximation for any $K \geq 1$. Moreover, the edge-disjoint min-min problem remains NP-complete and admits no polynomial time approximation scheme in planar digraphs [5]. The min-max problem is also *NP*-complete. But unlike the min-min problem, it admits a best possible approximation ratio 2 in digraphs [8], which can be achieved immediately by employing Suurballe and Tarjan's algorithm for the min-sum problem [13,14]. In addition, as a variant of the min-max problem, the length bounded disjoint path problem of computing two disjoint paths whose lengths are both bound a given constraint, is also known *NP*-complete [8].

1.2   Our Technique and Results

In this paper, we first give a LP formula for the kRSP problem. Then by rounding the value of fractional edges of a solution to the LP formula, two approximation algorithms are developed. The first algorithm uses traditional rounding method, and computes solutions with a bifactor ratio of $(2, 2)$. The second algorithm improves the first rounding approach to an approximation with a pseudo ratio of $(\alpha, 2 - \alpha)$ for $0 \leq \alpha \leq 2$. That is, for any output solution of the improved approximation algorithm, there always exists $0 \leq \alpha \leq 2$, such that the delay and cost of the solution are bounded by α times and $2 - \alpha$ times of that of an optimal solution respectively. By extending the technique in [7], the approximation ratio can be further improved to $(1, \ln n)$. To the best of our knowledge, this is the first approximation algorithm with a polylogarithmic ratio for the kRSP problem. Note that the extension of the technique in [7] is non-trivial, since we do not know the bound of the cost sum. Due to the length limitation, this paper will omit the details of the extension.

Like other rounding algorithms, the tricky task is to show that the round-up solutions are feasible for the kRSP problem, as the major results of this paper. The basic idea is to show that the fractional edges of a basic solution to the given LP formula will compose a graph, in which the degree of every vertex is exactly 2. Based on this observation, we show that the round-up edges in the first algorithm could collectively k-connect s and t. The correctness proof of the second algorithm follows a similar line to the first one, but requires a more sophisticated ratio proof because of its more complicated rounding method.

The following paragraphs are organized as follows: Section 2 gives the LP-rounding algorithm and shows that the resulting solution is with a bifactor ratio$(2, 2)$; Section 3 gives the proof of the correctness of the algorithm; Section 4 improves the ratio $(2, 2)$ to a pseudo ratio $(\alpha, 2 - \alpha)$ for $0 \leq \alpha \leq 2$ and then to $(1, \ln n)$; Section 5 concludes this paper.

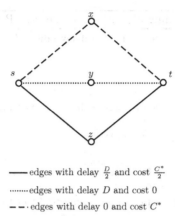

—— edges with delay $\frac{D}{2}$ and cost $\frac{C^*}{2}$

······ edges with delay D and cost 0

— — · edges with delay 0 and cost C^*

Fig. 1. $\{x_e = \frac{1}{2} | e \in E(G) \setminus \{e(s, z), e(z, t)\}\}$ is a basic optimum solution to LP (1) wrt the given graph G and $k = 1$. It is worth to note that this solution is not integral. $\{x_e = \frac{1}{3} | e \in E(G)\}$ is an optimum solution to LP (1) over this instance, but not a basic optimum solution.

2   An $(2, 2)$-Approximation Algorithm for the kRSP Problem

The linear programming (LP) formula for the kRSP problem is formally as in the following:

$$\min \sum c(e)x_e \tag{1}$$

subject to

$$\sum_{e \in \delta^+(v)} x_e - \sum_{e \in \delta^-(v)} x_e = \begin{cases} k & \text{for } v = s \\ 0 & \text{for } v \in V \setminus \{s, t\} \end{cases} \tag{2}$$

$$\sum_{e \in E} x_e d(e) \leq D \tag{3}$$

$$\forall e \in E(G): \quad 0 \leq x_e \leq 1 \tag{4}$$

where $\delta^+(v)$ and $\delta^-(v)$ denotes the set of edges leaving and entering v in G respectively. Before the technique paragraphs, we would like first do some discussion on this LP formula. If $x_e \in \{0, 1\}$, the above will be exactly the integral linear programming (ILP) formula for the kRSP problem. Moreover, with the relaxation over x_e (i.e. x_e is required to satisfy only Inequality (4) instead of $x_e \in \{0, 1\}$), a basic solution to the LP formula remains integral (i.e. , x_e remains integral for each e) for some special cases. When the delay constraint (i.e. Inequality (3)) is removed, any basic optimum solution to LP (1) would be exactly a set of st-paths with minimum cost. The reason is that,

---

Algorithm 1. A LP-rounding algorithm for the k-RSP problem.

Input: Graph G, distinct vertices s and t, a delay bound $D \in \mathbb{R}_0^+$, a cost function $c(e)$ and a delay function $d(e)$;
Output: k disjoint st-paths.

1. $E_{SOL} \leftarrow \emptyset$;
2. Solve LP (1) and get a basic optimum solution χ by Karmarkar's algorithm [12];
3. **For each** x_e in χ **do**
   (a) **if** $x_e = 1$ **then** $E_{SOL} \leftarrow \{e\} \cup E_{SOL}$
   (b) **if** $\frac{1}{2} \leq x_e < 1$ **then**
       i. Round the value of x_e to 1;
       ii. $E_{SOL} \leftarrow \{e\} \cup E_{SOL}$.
   /* As shown later in Theorem 2, s is k-connected to t by edges of E_{SOL} after the execution of Step 2.*/
4. **For each** e in E_{SOL} **do**
       if s is k connected to t in $E_{SOL} \setminus e$ **then**
       $E_{SOL} \leftarrow E_{SOL} \setminus \{e\}$;
   /* Remove redundant edges of E_{SOL}, such that E_{SOL} exactly compose k-disjoint paths. */
5. Return E_{SOL} as the k-disjoint paths.

---

in this case, LP (1) is totally unimodular [1,12], so any basic optimum solution to LP (1) must be integral. But with the delay constraint, LP (1) is no longer totally unimodular, and hence a basic optimum solution to LP (1) is no longer integral even when $k = 1$ (as depicted in Figure 1). As the first main result of this paper, we show that a basic optimum solution to LP (1) still acquires an interesting property as stated below:

Lemma 2. *The set of edges with $x_e \geq \frac{1}{2}$ in a basic optimum solution to LP (1) can collectively provide k-connectivity between s and t.*

Since Lemma 2 is one of our major results and its proof is the most tricky part of this paper, its proof is deferred to the next section. Based on this lemma, the key idea of our algorithm is simply as below: compute a basic optimum solution to LP (1), and round x_e to 1 for every edge e with $x_e \geq \frac{1}{2}$ in the solution. Then the algorithm outputs a set of edges with $x_e = 1$ as a solution to the kRSP problem. The detailed algorithm is formally as in Algorithm 1.

Step 2 of Algorithm 1 takes $O(n^{3.5} \log T)$ time to run Karmarkar's algorithm [12], where T is the maximum absolute value of the input numbers. Step 3 and Step 4 take $O(m)$ time to round the edges of E_{SOL} and remove the redundant edges of E_{SOL} respectively. Hence, the time complexity of Algorithm 1 is $O(n^{3.5} \log T)$ in the worst case.

For the approximation ratio of Algorithm 1, obviously E_{SOL} contains only edges with $x_e \geq \frac{1}{2}$ in the basic optimum solution χ, so we have

$$c(E_{SOL}) = \sum_{e \in E(G), x_e \geq \frac{1}{2}} c(e) \leq \sum_{e \in E(G), x_e \geq \frac{1}{2}} 2x_e c(e) \leq 2 \sum_{e \in E(G)} x_e c(e). \qquad (5)$$

Then because $\sum_{e \in E(G)} x_e c(e)$ is the cost of an optimum solution to LP (1), it is not larger than that of an optimum solution to the kRSP problem. This yields

$$\sum_{e \in E(G)} x_e c(e) \leq \sum_{e \in OPT} c(e), \tag{6}$$

where OPT is an optimum solution to the kRSP problem. Combining Inequality (5) and Inequality (6), we have $c(E_{SOL}) \leq 2c(OPT)$. Similarly, we have $d(E_{SOL}) \leq 2d(OPT)$. Therefore, the time complexity and approximation ratio of Algorithm 1 are as in the following theorem:

Theorem 3. *Algorithm 1 outputs E_{SOL}, a solution to the kRSP problem, in $O(n^{3.5} \log T)$ time. The cost and delay of E_{SOL} are at most two times of that of an optimum solution to the kRSP problem, where T is the maximum absolute value of the input numbers.*

3   Proof of Lemma 2

Before the technique paragraphs, we would like first to give some definitions. Let χ be a basic optimum solution to LP (1), and G_χ be the graph composed by the edges with $x_e > 0$ in χ. We say $e \in G_\chi$ is a full edge if and only if $x_e = 1$. Let Z be the set of full edges of G_χ and $E_{res} = G_\chi \setminus Z$ be the set of edges with $0 < x_e < 1$ in the solution to LP (1). The notation E_{res} or E_{SOL} also denotes the graph composed by the edges of E_{res} or E_{SOL} while no confusion arises.

To prove Lemma 2, the key idea is first to show that each vertex in graph E_{res} is exactly incident with two edges of E_{res} (as Lemma 4). Based on this property, we then show that the edges with $1 > x_e \geq \frac{1}{2}$, together with the edges with $x_e = 1$, collectively k connect s and t. Therefore, we shall first focus on the properties of only the edges in E_{res}, leaving the edges with $x_e = 0$ and $x_e = 1$ for a moment. To make the proof brief, we consider the following residual LP formula instead of LP (1), which is LP (1) except that the edges with $x_0 = 0$ or $x_0 = 1$ of G_χ are removed.

$$\min \sum_{e \in E_{res}} c_e x_e \tag{7}$$

subject to

$$\sum_{e \in \delta^+_{E_{res}}(v)} x_e - \sum_{e \in \delta^-_{E_{res}}(v)} x_e - \sum_{e \in \delta^+_Z(v)} 1 + \sum_{e \in \delta^-_Z(v)} 1 = \begin{cases} k & \text{for } v = s \\ 0 & \text{for } v \in V_{res} \setminus \{s,t\} \end{cases} \tag{8}$$

$$\sum_{e \in E_{res}} x_e d(e) \leq kD - \sum_{e \in Z} d(e) \tag{9}$$

$$\forall e \in E_{res} : \quad 0 < x_e < 1 \tag{10}$$

It is easy to see that $\{x_e | e \in E_{res}\}$ is a solution to the above LP formula. Let χ_{res} be a basic solution to LP (7). Then χ_{res} is χ except that the x_es of value 0 or 1 are removed. Let $\mathcal{A}$ be the whole constraint matrix of LP (7), $\mathcal{A}_G(v)$ be the row corresponding to v, and $\mathcal{A}_G(D)$ be the row corresponding to the delay constraint, i.e. Inequality (9). Assuming that V_{res} is the set of the vertices of E_{res}, we denote the vector space spanned by the vectors $\mathcal{A}_G(v)$, $v \in V_{res}$, by $Span(V_{res})$, and the space spanned by vectors $\mathcal{A}_G(D)$ together with $\mathcal{A}_G(v)$, $v \in V_{res}$, by $Span(D \cup V_{res})$.

Lemma 4. *Every vertex of V_{res} is incident to exactly two edges in graph E_{res}.*

Proof. Firstly, we shall show that every vertex of $v \in V_{res}$ is incident to at least two edges in E_{res}. Clearly, v must be incident with at least one edge of E_{res}, such that it can belong to V_{res}. Suppose only one edge of E_{res} joins v. Then on one hand, because $0 < x_e < 1$ for every $e \in E_{res}$, the degree of v is not integral. On the other hand, following the LP formula (7), every vertex of V_{res}, including $v \in V_{res}$, must be with an integral degree in the basic optimum solution χ_{res}. Hence a contradiction arises. Therefore, every vertex of V_{res} is incident to at least 2 edges.

Secondly, we shall show that every vertex in V_{res} is incident to at most two edges in E_{res}. Suppose there exists a vertex $v_0 \in V_{res}$ with a degree of at least 3, i.e. $|\delta^-(v_0)| + |\delta^+(v_0)| \geq 3$. Then, on one hand, since every vertex of V_{res} must be incident to at least 2 edges, we have:

$$|E_{res}| = \frac{1}{2} \sum_{v \in V_{res}} |\delta^-(v)| + |\delta^+(v)| \geq \frac{1}{2}\Big(\sum_{v \in V_{res} \setminus \{v_0\}} 2 + \sum_{v = v_0} 3\Big) > |V_{res}|.$$

On the other hand, we can show that $|V_{res}| \geq |E_{res}|$ must hold and hence obtain a contradiction. The proof is as below. Since χ_{res} is a basic optimum solution and there are $|E_{res}|$ edges with $x_e > 0$ in χ_{res}, the dimension of $Span(D \cup V_{res})$ is $|E_{res}|$. Then, since $\sum_{e \in E_{res}} x_e d(e) = D - \sum_{e \in Z} d(e)$ may holds for Inequality (9), the dimension of $Span(V_{res})$ is at least $|E_{res}| - 1$. Then, because the $|V_{res}|$ rows of the constraint matrix for $Span(V_{res})$ contains at most $|V_{res}| - 1$ linear independent vector, $|V_{res}| - 1 \geq |E_{res}| - 1$ holds, and hence $|V_{res}| \geq |E_{res}|$. This completes the proof.

Let V_{full} be the set of vertices of $V(E_{res}) \cap V(E_{full})$. Then, V_{full} are with degree 2 or -2 in E_{res}. Following Lemma 4, the degree of $v \in V_{res} \setminus V_{full}$ is 0. So for each $v \in V_{res} \setminus V_{full}$, E_{res} contains exactly an edge entering and the other edge leaving v. Therefore, E_{res} is actually a set of paths between the vertices of V_{full}. Further, we have the following lemma:

Lemma 5. *The edges of E_{res} compose a set of paths P_{res} between the vertices of V_{full}. For every $v \in V_{full}$, P_{res} contains exactly two paths leaving v or entering v. Moreover, any two paths of P_{res} share no common edge.*

Proof. Following Lemma 4, for every $v \in V_{full}$, E_{res} contains exactly two edges which either enter or leave v, while for $v \in V_{res} \setminus V_{full}$, E_{res} contains exactly one edge entering v and one another edge leaving v. So following flow theory, for each $v \in V_{full}$, E_{res} contains exactly two paths either entering or leaving v. In addition, E_{res} contains no cycles, since E_{res} is a subgraph of G_χ and G_χ contains no cycles because G_χ is a minimum cost fractional flow from s to t. Therefore, the edges of E_{res} compose exactly the paths of P_{res}.

It remains to show that the paths of P_{res} are edge-disjoint. Since every $v \in V_{res} \setminus V_{full}$ is with degree 2, v can appear on only one path of P_{res}. That is, any two distinct paths of P_{res} cannot go through a common vertex of $V_{res} \setminus V_{full}$. So the paths of P_{res} are internal vertex-disjoint, and hence edge-disjoint. This completes the proof.

Now we are to show that the edges with $x_e \geq \frac{1}{2}$ in G_χ could provide k connectivity for s and t. Remind that the output of Algorithm 1, E_{SOL}, is exactly the set of edges with $x_e \geq \frac{1}{2}$ in G_χ. That is, E_{SOL} is equivalently the edges with $x_e \geq \frac{1}{2}$ in E_{res} together with the full edges of Z. Suppose Lemma 2 is not true, then there must exist $k - 1$ edges, say $e_1, \dots, e_{k-1}$, which separate s and t in E_{SOL}. Then according to Lemma 6 as below, the flow between s and t wrt the solution χ to LP (1) is at most of value $k - 1$. This contradicts with the fact that the flow wrt χ is of value k, and completes the proof of Lemma 2.

Lemma 6. *Assume $e_1, \dots, e_{k-1}$ separate s and t in E_{SOL}, then the flow between s and t of the solution to LP (1) is at most of a value $k - 1$.*

Proof. Let $G_s \supset \{s\}$ be the component of $E_{SOL} \setminus \{e_1, \dots, e_{k-1}\}$ which contains s. Then $\{e_1, \dots, e_{k-1}\}$ would separate G_s and $E_{SOL} \setminus G_S$ in graph E_{SOL}. W.l.o.g., assume $e_1, \dots, e_h$ are with $x_e = 1$, while $e_{h+1}, \dots, e_{k-1}$ are with $\frac{1}{2} \leq x_e < 1$. Let $p_{h+1}^1, \dots, p_{k-1}^1$ be the paths between vertices of V_{full} in $E_{res} \subseteq G_\chi$. W.l.o.g., assume that $e_i \in p_i^1$ for each i. Following Lemma 5, there exists exactly one another path, say p_j^2, which leaves the same vertex as p_j^1 in E_{res}. Then the set of p_j^1s and p_j^2s, i.e. the set of paths $\{p_j^i | i \in \{1, 2\}, j \in \{h+1, \dots, k-1\}\}$, can only provide a flow of value at most $k - 1$ together with $\{e_1, \dots, e_h\}$.

It remains to show that there exists neither a full edge outside $\{e_1, \dots, e_h\}$, nor a path of P_{res} outside $\{p_j^i | i \in \{1, 2\}, j \in \{h+1, \dots, k\}\}$ that leaves G_s in graph G_χ. Suppose otherwise, as the two cases analyzed below, such a full edge or a path must belong to E_{SOL}, and hence it would connect G_s and $E_{SOL} \setminus G_s$ in $E_{SOL} \setminus \{e_1, \dots, e_{k-1}\}$. This contradicts with the assumption that edges of$\{e_1, \dots, e_k\}$ separate G_s and $E_{SOL} \setminus G_s$ in E_{SOL}, and completes the proof.

1. Suppose there exists a full edge $e \notin \{e_1, \dots, e_h\}$ in G_χ that leaves G_s. Since e is a full edge, then $x_e = 1$ and $e \in E_{SOL}$ holds. Hence, $e_1, \dots, e_k$ can not separate G_s and $E_{SOL} \setminus G_s$, because e connects them.
2. Suppose there exists a path $p \in P_{res} \setminus \{p_j^i | i \in \{1, 2\}, j \in \{h+1, \dots, k\}\}$ that leaves G_s at v. Following Lemma 5, there must be exactly one another path p' that leaves v in G_χ. Then either flow p or p' is with value at least $\frac{1}{2}$. That is, the edges of either p or p' would belong to E_{SOL} and connect

G_s and $E_{SOL} \setminus G_s$. So removal of $\{e_1, \ldots, e_k\}$ can not disconnect G_s and $E_{SOL} \setminus G_s$ in E_{SOL}.

4 An $(\alpha, 2 - \alpha)$-Approximation Algorithm for the kRSP Problem

In Section 2, Algorithm 1 adds every edge with $x_e \geq \frac{1}{2}$ to the solution. However, not all the edges with $x_e \in [\frac{1}{2}, 1)$ are good choices for constructing a solution to the kRSP problem. This section will give an improved rounding algorithm which selects the edges of the solution more carefully, such that the algorithm is with a pseudo approximation ratio $(\alpha, 2 - \alpha)$. Thus, for any output solution of the algorithm, there always exists $0 \leq \alpha \leq 2$, such that the delay and cost of the solution are bounded by α times and $2 - \alpha$ times of that of an optimal solution respectively.

Assume that χ is a basic optimum solution to LP (1). The main idea of the improved rounding algorithm is to select the edges with less cost and delay, rather than to select the edges with $\frac{1}{2} \leq x_e < 1$. To do this, the algorithm combines the cost and delay as one new cost and then selects a set of edges, which are with the new cost sum minimized and provide k connectivity between s and t together with the edges of $x_e = 1$ in χ. Let $c(\chi)$ and $d(\chi)$ be the cost and delay of the basic optimum solution to LP (1). The new mixed cost for every edge is $b(e) = \frac{c(e)}{c(\chi)} + \frac{d(e)}{d(\chi)}$.

According to Lemma 5, the edges of E_{res} compose exactly a set of internal vertex disjoint paths, say $P_{res} = \{p_j^i | i \in \{1, 2\}, j \in \{1, \ldots, h\}\}$, where p_j^1 and p_j^2 leaves a same vertex of V_{full}. Then the task is now to choose h paths $\{p_j^{i_j} | j \in \{1, \ldots, h\}\}$ from P_{res} to provide the k connectivity between s and t. According to Lemma 5, we could divide P_{res} into two path sets $\mathcal{P}_1, \mathcal{P}_2$, such that every two paths in $\mathcal{P}_i$ shares no common vertex. Then following the same line of the proof of Lemma2, it can be shown that $Z \cup E(\mathcal{P}_i)$ provides k-connectivity between s and t for either $i = 1$ or $i = 2$. Therefore, the main idea of our algorithm is to divide P_{res} into two path sets $\mathcal{P}_1, \mathcal{P}_2$, and then select $\mathcal{P}_i$ with smaller $\sum_{e \in E(\mathcal{P}_i)} b(e)$ for $i = 1, 2$. Formally, the algorithm is as below:

It remains to show the approximation ratio of the algorithm, which is stated as follows:

Theorem 7. *There exists a real number $0 \leq \alpha \leq 2$, such that the delay and cost of E_{SOL} are bounded by α and $2 - \alpha$ times of that of the optimum solution of the $kRSP$ problem.*

Proof. Clearly, $b(E_{SOL}) = b(Z) + \beta b(E(\mathcal{P}_i)) + (1 - \beta)b(E(\mathcal{P}_i))$ holds. Then because $\sum_{e \in E(\mathcal{P}_i)} b(e) \leq \sum_{e \in E(\mathcal{P}_{3-i})} b(e)$, we have

$$b(E_{SOL}) \leq b(Z) + \beta \sum_{e \in E(\mathcal{P}_i)} b(e) + (1 - \beta) \sum_{e \in E(\mathcal{P}_{3-i})} b(e) \leq \sum_{e \in G_\chi} x_e b(e) \leq 2. \quad (11)$$

Algorithm 2. A LP-rounding algorithm for the k-RSP problem.

Input: E_{res} with new cost $b(e)$, a basic optimum solution χ to LP (1);
Output: k disjoint st-paths.

1. $E_{SOL} \leftarrow Z$; /*E_{SOL} is initially the set of edges with $x_e = 1$ in G_χ. */
2. Divide P_{res} into two path sets $\mathcal{P}_1$, $\mathcal{P}_2$, such that every two paths in $\mathcal{P}_i$ shares no common vertex for $i = 1, 2$;
3. Select i, $i \in \{1, 2\}$, such that $\mathcal{P}_i$ is with smaller new cost sum, i.e., $\sum_{e \in E(\mathcal{P}_i)} b(e) \leq \sum_{e \in E(\mathcal{P}_{3-i})} b(e)$, $i \in \{1, 2\}$;
4. Return $E_{SOL} \leftarrow E(\mathcal{P}_i) \cup E_{SOL}$.

Assume that the delay-sum of E_{SOL} is α times of D, where α is a real number and $0 \leq \alpha \leq 2$. Then $b(E_{SOL}) = \sum_{e \in E_{SOL}} b(e) = \alpha + \frac{c(E_{SOL})}{c(\chi)}$ holds. From Inequality (11), $\alpha + \frac{c(E_{SOL})}{c(\chi)} \leq 2$ holds. That is, $c(E_{SOL}) \leq (2 - \alpha)c(\chi) \leq (2 - \alpha)C^*$. This completes the proof.

By extending the technique of combining cycle cancelation and layer graph as in [7], the approximation ratio can be improved to $(1, \ln n)$. However, we do not know the value of $c(OPT)$, so we can only construct an auxiliary graph based on the delay of edges instead of cost. The auxiliary graph constructed is no longer a layer graph, but the costs of the edges therein are nonnegative. Therefore, we can compute minimum delay-to-cost cycle in the residual graph of G by computing shortest paths in the constructed auxiliary graph, and improve the output of Algorithm 2 by using the cycle cancelation method against the minimum delay-to-cost cycle repeatedly, until the solution satisfies the delay constraint strictly. To the best of our knowledge, this is the first non-trivial approximation algorithm with single factor polylogarithmic ratio for the kRSP problem.

5   Conclusion

This paper investigated approximation algorithms for the k-restricted shortest paths (kRSP) problem. As the main contribution, this paper first developed an improved approximation algorithm with bifactor ratio $(2, 2)$ by rounding a basic optimum solution to the proposed LP formula of the kRSP problem. The algorithm was then improved by choosing the round-up edges more carefully, such that for any output solution there exists $0 \leq \alpha \leq 2$ that the delay and the cost of the solution are bounded, respectively, by α and $2 - \alpha$ times of that of the optimum solution. This ratio can be further improved to $(1, \ln n)$ by extending the technique of [7]. To the best of our knowledge, this is the first approximation with single factor polylogarithm ratio for the kRSP problem.

References

1. Ahuja, R.K., Magnanti, T.L., Orlin, J.B.: Network flows: theory, algorithms, and applications (1993)
2. Bhatia, R., Kodialam, M.: TV Lakshman. Finding disjoint paths with related path costs. Journal of Combinatorial Optimization 12(1), 83–96 (2006)
3. Chao, P., Hong, S.: A new approximation algorithm for computing 2-restricted disjoint paths. IEICE Transactions on Information and Systems 90(2), 465–472 (2007)
4. Garey, M.R., Johnson, D.S.: Computers and intractability. Freeman, San Francisco (1979)
5. Guo, L., Shen, H.: On the complexity of the edge-disjoint min-min problem in planar digraphs. Theoretical Computer Science 432, 58–63 (2012)
6. Guo, L., Shen, H.: On finding min-min disjoint paths. Algorithmica 66(3), 641–653 (2013)
7. Guo, L., Shen, H., Liao, K.: Improved approximation algorithms for computing k disjoint paths subject to two constraints. In: Du, D.-Z., Zhang, G. (eds.) COCOON 2013. LNCS, vol. 7936, pp. 325–336. Springer, Heidelberg (2013)
8. Li, C.L., McCormick, T.S., Simich-Levi, D.: The complexity of finding two disjoint paths with min-max objective function. Discrete Applied Mathematics 26(1), 105–115 (1989)
9. Lorenz, D.H., Raz, D.: A simple efficient approximation scheme for the restricted shortest path problem. Operations Research Letters 28(5), 213–219 (2001)
10. Misra, S., Xue, G., Yang, D.: Polynomial time approximations for multi-path routing with bandwidth and delay constraints. In: INFOCOM 2009, pp. 558–566. IEEE (2009)
11. Orda, A., Sprintson, A.: Efficient algorithms for computing disjoint QoS paths. In: INFOCOM 2004, vol. 1, pp. 727–738. IEEE (2004)
12. Schrijver, A.: Theory of linear and integer programming. John Wiley & Sons Inc. (1998)
13. Suurballe, J.W.: Disjoint paths in a network. Networks 4(2) (1974)
14. Suurballe, J.W., Tarjan, R.E.: A quick method for finding shortest pairs of disjoint paths. Networks 14(2) (1984)
15. Xu, D., Chen, Y., Xiong, Y., Qiao, C., He, X.: On the complexity of and algorithms for finding the shortest path with a disjoint counterpart. IEEE/ACM Transactions on Networking 14(1), 147–158 (2006)
16. Xue, G., Zhang, W., Tang, J., Thulasiraman, K.: Polynomial time approximation algorithms for multi-constrained qos routing. IEEE/ACM Transactions on Networking 16(3), 656–669 (2008)

Monotone Grid Drawings of Planar Graphs

Md. Iqbal Hossain and Md. Saidur Rahman

Graph Drawing and Information Visualization Laboratory,
Department of Computer Science and Engineering,
Bangladesh University of Engineering and Technology (BUET),
Dhaka-1000, Bangladesh
{mdiqbalhossain,saidurrahman}@cse.buet.ac.bd

Abstract. A monotone drawing of a planar graph G is a planar straight-line drawing of G where a monotone path exists between every pair of vertices of G in some direction. Recently monotone drawings of planar graphs have been proposed as a new standard for visualizing graphs. A monotone drawing of a planar graph is a monotone grid drawing if every vertex in the drawing is drawn on a grid point. In this paper we study monotone grid drawings of planar graphs in a variable embedding setting. We show that every connected planar graph of n vertices has a monotone grid drawing on a grid of size $O(n) \times O(n^2)$, and such a drawing can be found in $O(n)$ time. Our result immediately resolves two open problems on monotone drawings of planar graphs posted by Angelini *et al.*

1 Introduction

A *straight-line drawing* of a planar graph G is a drawing of G in which each vertex is drawn as a point and each edge is drawn as a straight-line segment without any edge crossing. A path P in a straight-line drawing of a planar graph is *monotone* if there exists a line l such that the orthogonal projections of the vertices of P on l appear along l in the order induced by P. A straight-line drawing Γ of a planar graph G is a *monotone drawing* of G if Γ contains at least one monotone path between every pair of vertices. In the drawing of a graph in Fig. 1, the path between the vertices s and t drawn as a thick line is a monotone path with respect to the direction d, whereas no monotone path exists with respect to any direction between the vertices s' and t'. We call a monotone drawing of a planar graph a *monotone grid drawing* if every vertex is drawn on a grid point.

Monotone drawings of graphs are well motivated by human subject experiments by Huang *et al.* [8], who showed that the "geodesic tendency" (paths following a given direction) is important in comprehending the underlying graph. *Upward drawings* [4,5,9,10] are related to monotone drawings where every directed path is monotone with respect to the vertical line, while in a monotone drawing each monotone path, in general, is monotone with respect to a different line. Arkin *et al.* [3] showed that any strictly convex drawing of a planar graph is monotone and they gave an $O(n \log n)$ time algorithm for finding

J. Chen, J.E. Hopcroft, and J. Wang (Eds.): FAW 2014, LNCS 8497, pp. 105–116, 2014.
© Springer International Publishing Switzerland 2014

Fig. 1. The path between vertices s and t (as shown by thick line) is monotone with respect to direction d

such a path between a pair of vertices in a strictly convex drawing of a planar graph of n vertices. Angelini *et al.* [1] showed that every biconnected planar graph of n vertices has a monotone drawing in real coordinate space. They also showed that every tree of n vertices admits a monotone grid drawing on a grid of $O(n) \times O(n^2)$ or $O(n^{1.6}) \times O(n^{1.6})$ area, and posed an open problem "Is it possible to construct planar monotone drawings of biconnected planar graphs in polynomial area?". Addressing the problem, Hossain and Rahman [7] showed that every series-parallel graph of n vertices admits a monotone grid drawing on an $O(n) \times O(n^2)$ grid, and such a drawing can be found in $O(n \log n)$ time. However, the problem mentioned above remains open.

It is known that not every plane graph (with fixed embedding) admits a monotone drawing [1]. However, every outerplane graph of n vertices admits a monotone grid drawing on a grid of area $O(n) \times O(n^2)$ [2]. Angelini *et al.* showed that every connected plane graph admits a "polyline" monotone grid drawing on $O(n) \times O(n^2)$ grid using at most two bends per edges [2], and posed an open problem "Given a graph G and an integer $k \in \{0, 1\}$, what is the complexity of deciding whether there exists an embedding G_ϕ such that G_ϕ admits a monotone drawing with curve complexity k?"

In this paper we investigate whether every connected planar graph has a monotone drawing and what are the area requirements for such a drawing on a grid. We show that every connected planar graph of n vertices has a monotone grid drawing on an $O(n) \times O(n^2)$ grid, and such a drawing can be computed in $O(n)$ time. We thus solve the pair of open problems mentioned above by showing that every planar graph has an embedding G_ϕ such that G_ϕ admits a monotone grid drawing with curve complexity 0 on an $O(n) \times O(n^2)$ grid, and such an embedding and also a drawing can be found in linear time. As a byproduct, we introduce a spanning tree of a plane graph with some interesting properties. Such a spanning tree may find applications in other areas of graph algorithms as well.

We now give an outline of our algorithm for constructing a monotone grid drawing of a planar graph G. We first construct a "good spanning tree" T of G and find a monotone drawing of T by the method given in [1]. We then draw each non-tree edge by a straight-line segment by shifting the drawing of some subtree of T, if necessary. Figure 2 illustrates the steps of our algorithm. The input planar graph G is shown in Fig 2(a). We first find a planar embedding of G containing a good spanning tree as illustrated in Fig. 2(b), where the edges

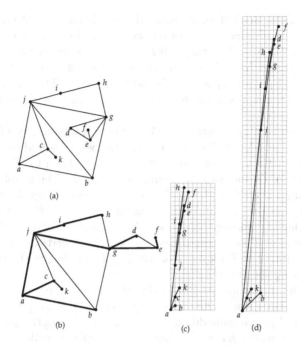

Fig. 2. Illustration for an outline of our algorithm

of the spanning tree are drawn by thick lines. We then find a monotone drawing of T on $O(n) \times O(n^2)$ grid using the algorithm in [1] as illustrated in Fig. 2(c). Finally we elongate the drawing of some edges and draw the non-tree edges of G using straight-line segments as illustrated in Fig. 2(d).

The rest of the paper is organized as follows. Section 2 describes some of the definitions that we have used in our paper. Section 3 deals with monotone drawings of connected planar graphs. Finally, Section 4 concludes the paper with discussions.

2   Preliminaries

In this section we give some definitions and present a known result. For the graph theoretic terminologies not given here, see [9].

Let $G = (V, E)$ be a connected graph with vertex set V and edge set E. A *subgraph* of G is a graph $G' = (V', E')$ such that $V' \subseteq V$ and $E' \subseteq E$. The degree of a vertex v in G is denoted by $d(v)$. We denote an edge joining vertices u and v of G by (u, v). A pair $\{u, v\}$ of vertices in G is a split pair if there exist two subgraphs $G_1 = (V_1, E_1)$ and $G_2 = (V_2, E_2)$ satisfying the following two conditions: 1. $V = V_1 \cup V_2, V_1 \cap V_2 = \{u, v\}$; and 2. $E = E_1 \cup E_2, E_1 \cap E_2 = \emptyset, |E_1| \geq 1, |E_2| \geq 1$. Thus every pair of adjacent vertices is a split pair. A $\{u, v\}$-*split component* of a split pair u, v in G is either an edge (u, v) or a maximal connected subgraph H of G such that $\{u, v\}$ is not a split pair of H. If v is a vertex in G, then $G - v$ is the subgraph of G obtained by deleting the vertex

v and all the edges incident to v. Similarly, if e is an edge of G, then $G - e$ is a subgraph of G obtained by deleting the edge e. Let v be a cut-vertex in a connected graph G. We call a subgraph H of G a *v-component* if H consists of a connected component H' of $G - v$ and all edges joining v to the vertices of H'.

Let $G_1 = (V_1, E_1)$ and $G_2 = (V_2, E_2)$ be two graphs. The *union* of G_1 and G_2, denoted by $G_1 \cup G_2$, is a graph $G_3 = (V_3, E_3)$ such that $V_3 = (V_1 \cup V_2)$ and $E_3 = (E_1 \cup E_2)$.

Let $G = (V, E)$ be a graph and $T = (V, E')$ be a spanning tree of G. An edge $e \in E$ is called a *tree edge* of G for T if $e \in E'$ otherwise e is said to be a *non-tree edge* of G for T. Let e be a non-tree edge of G for T. Then by $T \cup e$ we denote the subgraph G' of G obtained by adding edge e to T. The graph $G' = T \cup e$ always has a single cycle C and we call C the *cycle induced by the non-tree edge* e. If X is a set of edges of G and T is a spanning tree of G then $T \cup X$ denotes the graph obtained by adding the edges in X to T and replacing each multi-edge by a single edge. Let T be a rooted tree and let u be a vertex of T. Then by T_u we denote the subtree of T rooted at u. By $T - T_u$ we denote the tree obtained from T by deleting the subtree T_u.

A graph is *planar* if it can be embedded in the plane without edge intersections except at the vertices where the edges are incident. A *plane graph* is a planar graph with a fixed planar embedding. A plane graph divides the plane into some connected regions called *faces*. The unbounded region is called the *outer face* and each of the other faces is called an *inner face*. Let G be a plane graph. The boundary of the outer face of G is called the *outer boundary* of G. We call a simple cycle induced by the outer boundary of G an *outer cycle* of G. We call a vertex v of G an *outer vertex* of G if v is on the outer boundary of G, otherwise v is an *inner vertex* of G.

Let p be a point in the plane and l be a half-line with an end at p. The slope of l, denoted by *slope(l)*, is the angle spanned by a counterclockwise rotation that brings a horizontal half-line started at p and directed towards increasing x-coordinates to coincide with l. Let Γ be a drawing of a graph G and let (u, v) be an edge of G. We denote the direction of a half-line by $d(u, v)$ which is started at u and passed through v. The direction of a drawing of an edge e is denoted by $d(e)$ and the slope of the drawing of e is denoted by *slope(e)*.

Let T be a tree rooted at a node r. Denote by $T(u)$ the subtree of T rooted at a node u. A *slope-disjoint* drawing of T satisfies the following conditions [1] (see Fig. 3):

- For every node $u \in T$, there exist two angles $\alpha_1(u)$ and $\alpha_2(u)$, with $0 < \alpha_1(u) < \alpha_2(u) < \pi$, such that, for every edge e that is either in $T(u)$ or that connects u with its parent, it holds that $\alpha_1(u) < slope(e) < \alpha_2(u)$;
- for every two nodes $u, v \in T$ with v child of u, it holds that $\alpha_1(u) < \alpha_1(v) < \alpha_2(v) < \alpha_2(u)$;
- for every two nodes v_1, v_2 with the same parent, it holds that either $\alpha_1(v_1) < \alpha_2(v_1) < \alpha_1(v_2) < \alpha_2(v_2)$ or $\alpha_1(v_2) < \alpha_2(v_2) < \alpha_1(v_1) < \alpha_2(v_1)$.

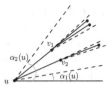

Fig. 3. A slope-disjoint drawing of a tree

Let G be a planar graph and Γ be a straight-line drawing of G. A path $u = u_1, \ldots, u_k = v$ between vertices u and v in G is denoted by $P(u, v)$. The drawing of the path $P(u, v)$ in Γ is *monotone* with respect to a direction d if the orthogonal projections of vertices $u_1, \ldots, u_k$ on d appear in the same order as the vertices appear on the path. The drawing Γ is a *monotone drawing* of G if there exists a direction d for every pair of vertices u and v such that at least a path $P(u, v)$ is monotone with respect to d. A monotone drawing is a *monotone grid drawing* if every vertex is drawn on a grid point. The following lemma is known from [1].

Lemma 1. *Let T be a tree of n vertices. Then T admits a monotone grid drawing on a grid of area $O(n) \times O(n^2)$, and such a drawing can be found in $O(n)$ time.*

In this paper we use a modified version of the algorithm for monotone grid drawing of a tree in [1], which we call Algorithm **Draw-Monotone-Tree** throughout this paper. Algorithm **Draw-Monotone-Tree** first assigns a slope to each vertex of a planar embedded tree then obtains a slope-disjoint drawing of the tree which is monotone. The algorithm draws each edge $(u, p(u))$ as a straight-line segment by using the assigned slope to u where $p(u)$ is the parent of u. A brief description of the algorithm is given in the rest of this section. Let T be an embedded rooted tree of n vertices. (Note that in [1] T is not embedded, but here we use T as an embedded tree for the sake of our algorithm.) Let $S = \{s_1, s_2, \ldots, s_{n-1}\} = \{1/1, 2/1, 3/1, \ldots, (n-1)/1\}$ be an ordered set of $n-1$ slopes in increasing order, where each slope is represented by the ratio y/x. Let $v_1, v_2, \ldots, v_n$ be an ordering of vertices in T in a counterclockwise postorder traversal. (In a counterclockwise postorder traversal of a rooted ordered tree, subtrees rooted at the children of the root are recursively traversed in counterclockwise order and then the root is visited.) Then we assign the slope s_i to vertex v_i ($i \neq n$). Let $u_1, u_2, \ldots, u_k$ be the children of v in T. Then the subtree T_{u_i} gets $|T_{u_i}|$ consecutive elements of S from the $(1 + \sum_{j=1}^{i-1} |T_{u_j}|)$-th to the $(\sum_{j=1}^{i} |T_{u_j}|)$-th. Let v' be the parent of v. If v is not the root of T then the drawing of the edge $e = (v', v)$ will be a straight-line with slope s_i.

We now describe how to find a monotone grid drawing of T using the slope assigned to each vertex of T. We first draw the root vertex r at $(0,0)$, and then use a counter clockwise preorder traversal for drawing each vertex of T. (In a counterclockwise preorder traversal of a rooted ordered tree, first the root is visited and then the subtrees rooted at the children of the root are visited recursively in counterclockwise order.) We fix the position of a vertex u when we traverse u. Note that when we traverse u, the position of the parent $p(u)$ has already been fixed. Let $(p_x(u), p_y(u))$ be the position of $p(u)$. Then we place u at grid point $(p_x(u) + x_b, p_y(u) + y_b)$, where $s_b = y_b/x_b$. Figure 4(b) illustrates a monotone grid drawing of the tree as shown in the Fig. 4(a). Algorithm **Draw-Monotone-Tree** computes a slope-disjoint monotone drawing of a tree on an $O(n) \times O(n^2)$ grid in linear time [1]. The following lemma is from [1].

Lemma 2. *Let Γ be a slope-disjoint monotone drawing of a rooted tree T. Let Γ' be a straight-line drawing of T obtained from Γ by elongation of the drawing of an edge e of T preserving the slope of e. Then Γ' is also a slope-disjoint monotone drawing. (See Fig. 4(c), where the edge (j, g) is elongated.)*

Proof. Let $e = (u, v)$ be an edge of T where u is the parent of v in T. The elongation of the drawing of e does not change $slope(e)$ and the drawing of T_v in Γ is shifted outwards preserving its drawing in Γ. The drawing of $T - T_v$ is remained same in Γ'. Since the elongation does not change the slope of the drawing of any edge, the new drawing Γ' preserves slope-disjoint monotone drawing of T. □

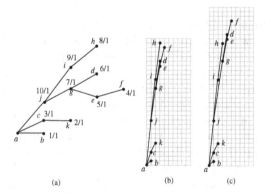

Fig. 4. (a) A tree T with assigned slope to each vertex, (b) a monotone drawing Γ of T and (c) a monotone drawing Γ' of T with elongation of the edge (j, g)

3   Monotone Grid Drawings

In this section we show that every connected planar graph of n vertices has a monotone grid drawing on an $O(n) \times O(n^2)$ grid.

Let G be a planar graph and let G_ϕ be a plane embedding of G. Let T be an ordered rooted spanning tree of G_ϕ such that the root r of T is an outer vertex of G_ϕ, and the ordering of the children of each vertex v in T is consistent with the ordering of the neighbors of v in G_ϕ. Let $P(r, v) = (r = u_1), u_2, \ldots, (v = u_k)$ be the path in T from the root r to a vertex $v \neq r$. The path $P(r, v)$ divides the children of u_i, $(1 \leq i < k)$, except u_{i+1}, into two groups; the left group L and the right group R. A child x of u_i is in group L and denoted by u_i^L if the edge (u_i, x) appears before the edge (u_i, u_{i+1}) in clockwise ordering of the edges incident to u_i when the ordering is started from the edge (u_i, u_{i+1}), as illustrated in the Fig. 5(a). Similarly, a child x of u_i is in the group R and denoted by u_i^R if the edge (u_i, x) appears after the edge (u_i, u_{i+1}) in clockwise order of the edges incident to u_i when the ordering is started from the edge (u_i, u_{i+1}). We call T a *good spanning* tree of G_ϕ if every vertex v ($v \neq r$) of G satisfies the following two conditions with respect to $P(r, v)$.

(Cond1) G does not have a non-tree edge (v, u_i), $i < k$; and

(Cond2) the edges of G incident to the vertex v excluding (u_{k-1}, v) can be partitioned into three disjoint (possibly empty) sets X_v, Y_v and Z_v satisfying the following conditions (a)-(c) (see Fig. 5(b)):

(a) Each of X_v and Z_v is a set of consecutive non-tree edges and Y_v is a set of consecutive tree edges.

(b) Edges of set X_v, Y_v and Z_v appear clockwise in this order from the edge (u_{k-1}, v).

(c) For each edge $(v, v') \in X_v$, v' is contained in $T_{u_i^L}$, $i < k$, and for each edge $(v, v') \in Z_v$, v' is contained in $T_{u_i^R}$, $i < k$.

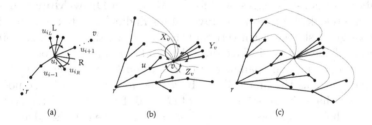

(a)   (b)   (c)

Fig. 5. (a) An illustration for $P(r, v)$, L and R groups, (b) an illustration for X_v, Y_v and Z_v sets of edges, and (c) an illustration for a good spanning tree T on G_ϕ where bold edges are tree edges

Figure 5(c) illustrates a good spanning tree T in a plane graph.

We now have the following lemma based on the properties of a good spanning tree and Lemma 1.

Lemma 3. *Let T be a good spanning tree of G_ϕ. Let v be a vertex in G, and let u be the parent of v in T. Let $X \subseteq X_v$ and $Z \subseteq Z_v$. Assume that Γ is the monotone drawing of $T \cup X \cup Z$ where a monotone path exists between every pair of vertices in the drawing of T in Γ. If a straight-line drawing Γ' of $T \cup X \cup Z$*

is obtained from Γ by elongation of the drawing of the edge (u, v) preserving the slope of (u, v), then Γ' is a monotone drawing of $T \cup X \cup Z$ where a monotone path exists between every pair of vertices in the drawing of T in Γ'.

Proof. Let r be the root of T. Let m be the slope assigned to the vertex v in T.

Let M_X and M_Z be the sets of slopes assigned to $T_{u_i^L}$ and $T_{u_i^R}$, respectively. According to assignment of slopes, for any $m_x \in M_X$ and $m_z \in M_Z$ the relation $m_x > m > m_z$ holds.

Since each vertex in X and Z are visible from the vertex v in Γ and $m_x < m < m_z$, v must be visible from each vertex in X and Z even after elongation of edge (u, v) without changing the slope of (u, v). Note that the elongation only changes the slopes of the drawings of non-tree edges in X and Z. The drawing of T_v is shifted outwards preserving the slopes of the edges in T_v and the drawing of $T - T_v$ is remained same. Let Γ' be the new drawing of $T \cup X \cup Z$. Then obviously the edges in X and in Z do not produce any edge crossing in Γ'. By Lemma 2, the elongation of the edge (u, v) does not break the monotone property in the drawing of T in Γ'. Thus a monotone path exists between every pair of vertices in the drawing of T in Γ'. $\square$

We now have the following lemma on monotone grid drawings of a plane graph with a good spanning tree.

Lemma 4. *Let G be a planar graph of n vertices and let G_ϕ be a plane embedding of G. Assume that G_ϕ has a good spanning tree T. Then G_ϕ admits a monotone grid drawing on an $O(n) \times O(n^2)$ grid.*

Proof. Let T be a good spanning tree of G_ϕ. We prove the claim by induction on the number z of non-tree edges of G for T. Algorithm **Draw-Monotone-Tree** uses a counterclockwise postorder traversal for finding a vertex ordering in the tree and assigns a slope to each vertex of the tree using that ordering. Note that the ordering of the vertices is fixed once the child of r that has to be visited first is fixed. Let T be a good spanning tree of G_ϕ and let r be the root of T. We take a child s of r as the first child to be visited in counterclockwise postorder traversal such that s is an outer vertex of G_ϕ and if s is on an outer cycle C of G_ϕ then s is the counterclockwise neighbor of r on C. We call the edge (r, s) the *reference edge* of T. (Later in Corollary 1 we show that such a reference edge always exists.) By induction on the number z of non-tree edges of G_ϕ, we now prove the claim that G_ϕ admits a monotone grid drawing on an $O(n) \times O(n^2)$ grid and a monotone path exists between every pair of vertices of G_ϕ through the edges of T in the drawing.

We first assume that $z = 0$. In this case $G_\phi = T$. We then find monotone drawing Γ of T on an $O(n) \times O(n^2)$ grid using Algorithm **Draw-Monotone-Tree** taking the reference edge (r, s) as the starting edge for traversals (counterclockwise postorder traversal for slope assignment and counterclockwise preorder traversal for drawing vertices). Since $G_\phi = T$, Γ is a monotone drawing of G_ϕ. That is, a monotone path exists between every pair of vertices of T in Γ.

We thus assume that $z > 0$ and the claim holds for any plane graph G_ϕ with number of non-tree edges z', where $z' < z$.

Let G_ϕ have z non-tree edges for T and let $e = (u, v)$ be a non-tree edge of the outer boundary of G_ϕ.

Let m_u and m_v be the slopes assigned to the vertices u and v, respectively in T by Algorithm **Draw-Monotone-Tree**. Without loss of generality let us assume $m_u > m_v$. Let w be the lowest common ancestor of u and v in T and let u' and v' be the parents of u an v in T. According to (Cond1) u does not lie on the path $P(v, r)$ and v does not lie on the path $P(u, r)$. Let $C = \{P(u, w) \cup P(v, w) \cup (u, v)\}$ and $G'_\phi = \{P(r, w) \cup C\}$. Clearly $G'_\phi - (u, v)$ has l_1 non-tree edges where $l_1 < z$. By induction hypothesis, $G'_\phi - (u, v)$ has a straight-line monotone drawing on $O(n) \times O(n^2)$ grid where the edges in T are drawn with the slope assigned to them and a monotone path exists between every pair of vertices through the edges in T. Let Γ' be the drawing of $G'_\phi - (u, v)$ in Γ. Let p_x be the largest x-coordinate used for the drawing of Γ'. We now shift the drawing of T_u and T_v such that u and v lie on the line $x = p_x + 1$ by preserving slopes of the drawings of the edges (u', u) and (v', v). Since the slopes are integer numbers, it guarantees that all vertices remain on grid points after the shifting operation. According to Lemma 2 elongations of (u', u) and (v', v) do not produce any edge crossing in the drawing. According to (Cond2) in the good spanning tree T, e belongs to set Z_u and set X_v. Then no tree edge incident to u exists between the edge (u', u) and (u, v) in counterclockwise from the edge (u', u), and no tree edge incident to v exists between the edge (v', v) and (v, u) in clockwise from the edge (v', v). (Remember that we have used counterclockwise postorder traversal starting from a reference edge for ordering the vertices in algorithm **Draw-Monotone-Tree**.) Hence we can draw the edge e on the line $x = p_x + 1$ by a straight-line segment without any edge crossings. In worst case, y-coordinate can be at most $O(n^2) + O(n^2)$. Hence the drawing takes a grid of size $O(n) \times O(n^2)$.   □

We now prove that every connected planar graph G has an embedding G_ϕ where a good spanning tree T of G_ϕ exists. We give a constructive proof for our claim. Before giving our formal proof we give an outline of our construction using an illustrative example in Fig. 6. We take an arbitrary plane embedding G_γ of G and start a breath-first-search (BFS) visit from an arbitrary outer vertex v of G_γ and regard r as the root of our desired spanning tree. In Fig. 6(a) BFS is started from vertex a, and vertex b, c and d are visited from a in this order by BFS, as illustrated in Fig. 6(b). Next we visit e from b by BFS, as illustrated in Fig. 6(c). When we visit a new vertex x then we check whether there is an edge (x, y) such that y is already visited and there is an (x, y)-split component or an x-component or a y-component inside the cycle induced by the edge (x, y) which does not contain the root r. The $\{e, d\}$-split component H_1 induced by the vertices $\{d, h, i, j, e\}$ is such a split component in Fig. 6(c) and the subgraph H_2 induced by vertices d, m, n is such a y-component for $y = d$ which are inside the cycle induced by the edge (e, d). We move the subgraphs H_1 and H_2 out of the cycle induced by the non-tree edge (e, d), as illustrated in Fig. 6(d). Since (b, e) is a tree edge and (e, d) is a non-tree edge, according to the definition of a good spanning tree, the edges (e, f) and (e, k) must be non-tree edges. Similarly since (a, d) is a tree edge and (e, d) a non-tree edge, then the edge (d, l) must be

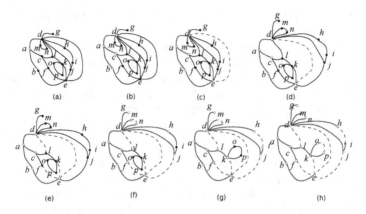

Fig. 6. Illustration for an outline of construction of a good spanning tree T. White vertices are visited vertices. Black vertices are not visited. Solid edges are tree edges. Dashed edges are non-tree edges.

non-tree edge. We mark (e, f), (e, k) and (d, l) non-tree edges as shown in the Fig. 6(e). We then visit vertices f, l, m, n and g, as illustrated in Fig. 6(f). When we visit k, we find a k-component H induced by vertices $\{k, p, o\}$ and we move H out of the cycle induced by (e, k) as shown in Fig. 6(g). Finally, at the end of BFS we find an embedding G_ϕ of G and a good spanning tree T as illustrated in Fig. 6(h), where the good spanning tree T is drawn by solid lines, and non-tree edges are drawn by dashed lines.

We now formally present our claim as in the following lemma whose proof is omitted due to page limit.

Lemma 5. *Let G be a planar graph of n vertices. Then G has a plane embedding G_ϕ that contains a good spanning tree.*

We have the following corollary based on the proof of Lemma 5.

Corollary 1. *Let T be a good spanning tree in the embedding G_ϕ of G obtained by the construction given in the proof of Lemma 5. Then T always has an edge with the property of the reference edge, mentioned in the proof of Lemma 4.*

Proof. The DFS-Start edge can be taken as the reference edge, mentioned in the proof of Lemma 4.   □

The following theorem is the main result of this paper.

Theorem 1. *Every connected planar graph of n vertices admits a monotone grid drawing on a grid of area $O(n) \times O(n^2)$, and such a drawing can be found in $O(n)$ time.*

Proof. Let G be a connected planar graph. By Lemma 5 G has a plane embedding G_ϕ such that G_ϕ contains a good spanning tree T. By Lemma 4 G_ϕ admits a monotone grid drawing on a grid of area $O(n) \times O(n^2)$.

We can find a good spanning tree G_ϕ of G using the construction in the proof of Lemma 5. After visiting each vertex during construction we need to identify v-components and $\{u, v\}$-split components for a non-tree edge (u, v) then we need to check whether these components are inside of $G_j(C)$ in the intermediate step. If any component is found then we need to move out the component.

A v-component is introduced by a cut vertex. All cut vertices can be found in $O(n + m)$ time using DFS. v-components and $\{u, v\}$-split components can also be found in linear time [6]. We maintain a data structure to store each cut vertex or every pair of vertices with split components. We then use this record in the intermediate steps for finding G_ϕ in Lemma 5. Let us assume we are traversing (u, v) non-tree edge in an intermediate step j of our algorithm. We can check whether any u-components, v-components and $\{u, v\}$-split components of G exist in $G_j(C)$ of Lemma 5 by checking each edges incident to u and v. This checking costs $O(d(u) + d(v))$ time. Throughout the algorithm it needs $O(m)$ time. For moving a component outside of cycle C we need to change at most four pointers in the adjacency list of u and v, which takes $O(1)$ time. Hence the required time is $O(m)$. Since G is a planar graph, G_ϕ can be constructed in $O(n)$ time.

After constructing G_ϕ we can construct a monotone drawing of G using a recursive algorithm based on the inducting proof of Lemma 4. It is not difficult to implement the recursive algorithm in $O(n)$ time.   □

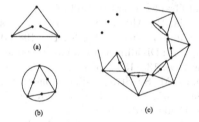

Fig. 7. Some examples of plane graph that has not good spanning tree. a) smallest connected plane graph, b) smallest biconnected plane graph, and c) a graph with many vertices that has no good spanning tree.

4   Conclusion

In this paper we have studied monotone grid drawings of planar graphs. We have shown that a connected planar graph of n vertices has an embedding that admits a straight-line monotone grid drawing on a grid of size $O(n) \times O(n^2)$, and such a drawing can be found in $O(n)$ time. Finding a straight-line monotone grid drawing of a planar graph on smaller grid is remained as a future work. In this research we have newly defined a good spanning tree of a planar graph with interesting properties which may have applications in other areas of graph algorithms. Since every plane graph does not have a good spanning tree as illustrated in Fig. 7, a natural question is "how to determine whether a given plane graph has a good spanning tree or not?"

Acknowledgment. This work is done in CSE Department of BUET as a part of a PhD research work. The authors gratefully acknowledge the support received from BUET. We also thank Bangladesh Academy of Sciences for a partial travel support to the second author.

References

1. Angelini, P., Colasante, E., Di Battista, G., Frati, F., Patrignani, M.: Monotone drawings of graphs. Journal of Graph Algorithms and Applications 16(1), 5–35 (2012)
2. Angelini, P., Didimo, W., Kobourov, S., Mchedlidze, T., Roselli, V., Symvonis, A., Wismath, S.: Monotone drawings of graphs with fixed embedding. Algorithmica, 1–25 (2013)
3. Arkin, E.M., Connelly, R., Mitchell, J.B.: On monotone paths among obstacles with applications to planning assemblies. In: Proceedings of the Fifth Annual Symposium on Computational Geometry, SCG 1989, pp. 334–343. ACM, New York (1989)
4. Di Battista, G., Eades, P., Tamassia, R., Tollis, I.G.: Graph Drawing: Algorithms for the Visualization of Graphs. Prentice-Hall (1999)
5. Di Battista, G., Tamassia, R.: Algorithms for plane representations of acyclic digraphs. Theoretical Computer Science 61(2–3), 175–198 (1988)
6. Hopcroft, J.E., Tarjan, R.E.: Dividing a graph into triconnected components. SIAM Journal on Computing 2(3), 135–158 (1973)
7. Hossain, M. I., Rahman, M. S.: Straight-line monotone grid drawings of series-parallel graphs. In: Du, D.-Z., Zhang, G. (eds.) COCOON 2013. LNCS, vol. 7936, pp. 672–679. Springer, Heidelberg (2013)
8. Huang, W., Eades, P., Hong, S.-H.: A graph reading behavior: Geodesic-path tendency. In: Proceedings of the 2009, IEEE Pacific Visualization Symposium, PACIFICVIS 2009, pp. 137–144. IEEE Computer Society, Washington, DC (2009)
9. Nishizeki, T., Rahman, M. S.: Planar Graph Drawing. Lecture notes series on computing. World Scientific, Singapore (2004)
10. Hassan Samee, M.A., Rahman, M.S.: Upward planar drawings of series-parallel digraphs with maximum degree three. In: WALCOM 2007, pp. 28–45 (2007)

Spanning Distribution Forests of Graphs

(Extended Abstract)

Keisuke Inoue and Takao Nishizeki

School of Science and Technology, Kwansei Gakuin University,
2-1 Gakuen, Sanda, 669-1337 Japan
keikei1338@gmail.com, nishi@kwansei.ac.jp

Abstract. Assume that a graph G has l sources, each assigned a non-negative integer called a supply, that all the vertices other than the sources are sinks, each assigned a non-negative integer called a demand, and that each edge of G is assigned a non-negative integer, called a capacity. Then one wishes to find a spanning forest F of G such that F consists of l trees, each tree T in F contains a source w, and the flow through each edge of T does not exceed the edge-capacity when a flow of an amount equal to a demand is sent from w to each sink in T along the path in T. Such a forest F is called a spanning distribution forest of G. In the paper, we first present a pseudo-polynomial time algorithm to find a spanning distribution forest of a given series-parallel graph, and then extend the algorithm for graphs with bounded tree-width.

Keywords: spanning distribution forest, series-parallel graph, network flow, supply, demand, bounded tree-width.

1 Introduction

Assume that a graph $G = (V, E)$ has l vertices $w_1, w_2, \cdots, w_l$ called *sources* and each source $w_i, 1 \leq i \leq l$, is assigned a non-negative integer $\sup(w_i)$, called the *supply* of w_i. Every other vertex is called a *sink*, and is assigned a non-negative integer $\mathrm{dem}(v)$, called the *demand* of v. Each edge $e \in E$ is assigned a non-negative integer $\mathrm{cap}(e)$, called the *capacity* of e. Figure 1 illustrates such a graph, in which the sources w_1 and w_2 are drawn as squares, each sink is drawn as a circle, the integer in a square or circle is a supply or demand, and the integer attached to an edge is the capacity.

As in an ordinary network flow problem, each source $w_i, 1 \leq i \leq l$, can send at most an amount $\sup(w_i)$ of flow to sinks through edges in G. However, in our flow problem, every sink v must receive an amount $\mathrm{dem}(v)$ of flow from exactly *one* of the l sources, and the flows from each source w_i must be sent along paths in a tree of G, say T_i. Of course, the *capacity constraint* must be satisfied for every edge e of T_i: the amount $\psi(e)$ of flow through e should not exceed the capacity $\mathrm{cap}(e)$ of e. Regard T_i as a tree rooted at the source w_i as illustrated in Fig. 1, and assume that an end u of edge $e = (u, v)$ is the parent of the other end v in the rooted tree T_i, then $\psi(e) = \sum_x \mathrm{dem}(x)$ where x runs over v and all the descendants of

J. Chen, J.E. Hopcroft, and J. Wang (Eds.): FAW 2014, LNCS 8497, pp. 117–127, 2014.
© Springer International Publishing Switzerland 2014

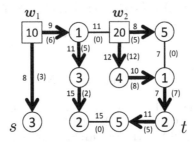

Fig. 1. A series-parallel graph G and a spanning distribution forest F

v in T_i. Furthermore, these l trees $T_1, T_2, \cdots, T_l$ must be pairwise vertex-disjoint. The forest F consisting of these l trees is called a *spanning distribution forest* of G. The *spanning distribution forest problem* asks whether a given graph has a spanning distribution forest. The problem has some applications to the power supply problem for power delivery networks [1, 4–7, 11, 13], the server-client problem in computer networks [9], etc. In Fig. 1, a spanning distribution forest is drawn by thick lines, the integer in parentheses attached to an edge e is the amount $\psi(e)$ of flow through e and the arrow of e indicates the direction of the flow.

The spanning distribution forest problem is strongly NP-Complete, and hence it is unlikely that there is a polynomial or pseudo-polynomial time algorithm for the problem. Furthermore, the problem is NP-Complete even for series-parallel graphs [8], although many combinatorial problems can be solved in polynomial time or even in linear time for series-parallel graphs [12]. Thus, it is unlikely the problem can be solved in polynomial time for series-parallel graphs. However, Kawabata and Nishizeki [8] showed that it can be solved in pseudo-polynomial time for a series-parallel graph G if $l = 1$, that is, G has exactly one source w; the computation time is $O((\sup(w))^4 n)$, where n is the number of vertices in G.

In this paper, we first present a pseudo-polynomial time algorithm for the spanning distribution forest problem on series-parallel graphs. It takes time $O(l^2 S^4 n)$, where $S = \max\{\sup(w_i) \mid 1 \leq i \leq l\}$. We then extend the algorithm for graphs with bounded tree-width.

2   Series-Parallel Graphs

A *series-parallel graph* is recursively defined as follows:

1. A graph G of a single edge (s, t), depicted in Fig. 2(a), is a series-parallel graph. The vertices s and t are called the *terminals* of G.
2. Let G_1 be a series-parallel graph with terminals s_1 and t_1, and let G_2 be a series-parallel graph with terminals s_2 and t_2, as depicted in Fig. 2(b).
   (a) A graph G obtained from G_1 and G_2 by identifying vertex t_1 with s_2, as illustrated in Fig. 2(c), is a series-parallel graph, whose terminals are $s = s_1$ and $t = t_2$. Such a connection is called a *series connection*.

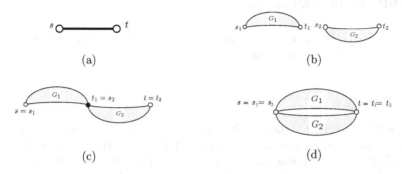

Fig. 2. Illustration for the definition of series-parallel graphs

(b) A graph G obtained from G_1 and G_2 by identifying vertex s_1 with s_2
    and t_1 with t_2, as illustrated in Fig. 2(d), is a series-parallel graph, whose
    terminals are $s = s_1 = s_2$ and $t = t_1 = t_2$. Such a connection is called a
    parallel connection.

The graph in Fig. 1 is a series-parallel graph.

A series-parallel graph G can be represented by a *binary decomposition tree*
T_{BD}, as illustrated in Fig. 3. Every node u of T_{BD} corresponds to a subgraph G_u
of G, which is drawn in a square in Fig. 3. If u is a leaf of T_{BD}, then G_u consists
of a single edge. Every inner node u of T_{BD} is labeled by **s** or **p**, which represents
a series or parallel connection. Let u_1 and u_2 be the children of u in T_{BD}, let G_1
be the graph corresponding to u_1, and let G_2 be the graph corresponding to u_2.
Then G_u is a series (or parallel) connection of G_1 and G_2 if u is labeled by **s** (or
p). If u is the root r of T_{BD}, then $G = G_u$. T_{BD} can be obtained from G in linear
time [12].

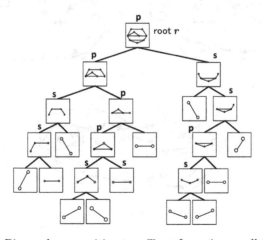

Fig. 3. Binary decomposition tree T_{BD} of a series-parallel graph

3   Outline of Algorithm

In this section, we outline the algorithm for series-parallel graphs and define functions $f_{\alpha,\beta}$ used by the algorithm.

3.1   Outline

Suppose that a series-parallel graph G has two sources w_1 and w_2, as illustrated in Fig. 4. Then a spanning distribution forest $F = T_1 \cup T_2$ of G consists of two vertex-disjoint trees T_1 and T_2. Such a forest is often called a 2-*tree* of G. Suppose further that, for some inner nodes u_1, u_2 and u_3 of a binary decomposition tree T_{BD} of G, G is a parallel connection of G_{u_3} and a graph which is a series connection of G_{u_1} and G_{u_2}, as illustrated in Fig. 4. The source w_1 is in G_{u_3} and w_2 in G_{u_2}. Three spanning distribution forests $F = T_1 \cup T_2$ of G are depicted in Figs. 4 (a), (b), and (c), where T_1 is drawn by thick lines, and T_2 by dotted lines. Although T_1 and T_2 are trees in G, a subgraph of G_{u_j}, $1 \leq j \leq 3$, induced by T_i, $i = 1$ or 2, is not necessarily a tree, but may be a 2-tree. Therefore, a subgraph of G_{u_j} induced by a forest F is not necessarily a 2-tree, but may be a tree, 2-tree or 3-tree. The forest F in Fig. 4(a) induces a tree for G_{u_3} and a 2-tree for G_{u_1}, while the forest F in Fig. 4(b) induces a 3-tree for G_{u_2}. For the case of $l = 2$, our algorithm finds trees, 2-trees and 3-trees of G_u for each node u of T_{BD}. For general, we find l'-trees of G_u for each integer l', $1 \leq l' \leq l+1$. It should be noted that the tree containing a terminal s of G_u and the tree containing a terminal t would be merged to a single tree in G although every other tree of G_u remains in G as it is.

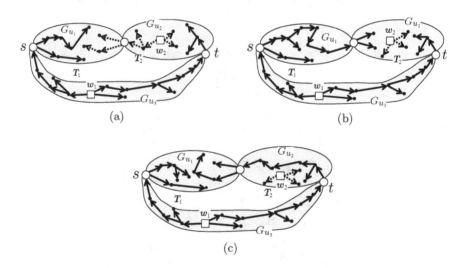

(a)

(b)

(c)

Fig. 4. Three spanning distribution forests of a series-parallel graph

We indeed give an algorithm to compute the following function f for series-parallel graphs G:

$$f(G) = \begin{cases} 1 & \text{if } G \text{ has a spanning distribution forest;} \\ 0 & \text{otherwise.} \end{cases}$$

Slightly modifying the algorithm, one can actually find a spanning distribution forest F of G if there exists.

For each node u of T_{BD}, we compute functions $f_{\alpha,\beta}(G_u, x, y)$. Let $S = \max\{\sup(w_i) \mid 1 \le i \le l\}$, then for every edge e of G_u the flow $\psi(e)$ through e is an integer and $0 \le \psi(e) \le S$. We denote by $\mathbb{Z}_S$ the set of all integers z with $0 \le z \le S$. The domain of variables x and y is $\mathbb{Z}_S$, and the range of $f_{\alpha,\beta}$ is $\{0,1\}$. The function $f_{\alpha,\beta}(G_u, x, y)$ takes value 1 if and only if G_u has a spanning distribution forest specified by the parameters α, β and the variables x, y. Let W be the set of all sources $w_1, w_2, \cdots, w_l$ in G, and let U be the set of all sources in G_u. Then $U \subseteq W$. The two parameters α and β run over all elements in set $U \cup \{\text{in}, \text{out}\}$, that is, $\alpha, \beta \in U \cup \{\text{in}, \text{out}\}$. The parameter α (or β) indicates the source of flow either entering to a terminal s (or t) or emanating from s (or t). If $\alpha \in U$, then an amount x of flow is sent to s from source α; for example, $\alpha = w_1$ for $f_{\alpha,\beta}(G_{u_3}, x, y)$ representing the forest (tree in this case) of G_{u_3} in Fig. 4(a). If $\alpha = \text{in}$, then an amount x of flow from a source in W/U enters in G_u through terminal s; for example, $\alpha = \text{in}$ for $f_{\alpha,\beta}(G_{u_1}, x, y)$ representing the forest of G_{u_1} in Fig. 4(a). On the other hand, if $\alpha = \text{out}$, then $\beta = \text{in}$ must hold and an amount x of flow entering in G_u through t goes out G_u from s; for example, $\alpha = \text{out}$ for $f_{\alpha,\beta}(G_{u_2}, x, y)$ representing the forest of G_{u_2} in Fig. 4(c). Similarly, the parameter β indicates the flow either entering to a terminal t or emanating from t. The functions $f_{\alpha,\beta}$ will be formally defined in Section 3.2. We recursively computes $f_{\alpha,\beta}(G_u, x, y)$ for each node u of T_{BD} from leaves to the root r.

3.2 Definition of $f_{\alpha,\beta}$

Since $\alpha, \beta \in U \cup \{\text{in}, \text{out}\}$, there are nine combinations of α and β, but $f_{\alpha,\beta}$ is defined only for six of them, as follows.

Case (i): $\alpha, \beta \in U$.

Let s and t be the terminals of a series-parallel graph G_u. Add to G_u dummy sinks s' with $\text{dem}(s') = x$ and t' with $\text{dem}(t') = y$, and add dummy edges (s, s') and (t, t') with infinite capacities, as in Fig. 5(a). Let G'_u be the resulting graph. Then $f_{\alpha,\beta}$ is defined as follows: $f_{\alpha,\beta}(G_u, x, y) = 1$ if G'_u has a spanning distribution forest F such that the tree in F containing source α contains s' and the tree containing source β contains t'. Intuitively speaking, $f_{\alpha,\beta}(G_u, x, y) = 1$ means that G_u has a spanning distribution forest which can output x units of flow from source α through terminal s and y units of flow from source β through t. For the sake of convenience, we assume that $f_{\alpha,\beta}(G_u, x, y) = 0$ for all integers x and y such that $x \notin \mathbb{Z}_S$ or $y \notin \mathbb{Z}_S$. The same assumption applies to the other five cases.

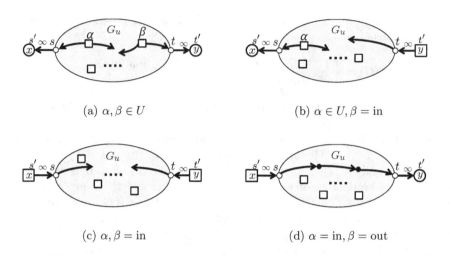

(a) $\alpha, \beta \in U$ (b) $\alpha \in U, \beta = \text{in}$

(c) $\alpha, \beta = \text{in}$ (d) $\alpha = \text{in}, \beta = \text{out}$

Fig. 5. Graphs G'_u and their forests

Case (ii): $\alpha \in U$ and $\beta = \text{in}$.

Add to G_u a dummy sink s' with $\text{dem}(s') = x$ and a dummy source t' with $\text{sup}(t') = y$, and add two dummy edges (s, s') and (t, t') with infinite capacities, as in Fig. 5(b). Let G'_u be the resulting graph. Then the function $f_{\alpha,\text{in}}$ is defined as follows: $f_{\alpha,\text{in}}(G_u, x, y) = 1$ if G'_u has a spanning distribution forest such that the tree containing α contains s' and the tree containing source t' contains t. Intuitively speaking, $f_{\alpha,\text{in}}(G_u, x, y) = 1$ means that G_u has a spanning distribution forest which can output x units of flow from source α through s if y units is inputted to G_u through t.

Case (iii): $\alpha = \text{in}$ and $\beta \in U$.

The function $f_{\text{in},\beta}$ is similarly defined as $f_{\alpha,\text{in}}$ although the roles of s and t are interchanged.

Case (iv): $\alpha, \beta = \text{in}$.

Add to G_u dummy sources s' with $\text{sup}(s') = x$ and t' with $\text{sup}(t') = y$, and add dummy edges (s, s') and (t, t') with infinite capacities, as in Fig. 5(c). Let G'_u be the resulting graph. Then $f_{\text{in},\text{in}}(G_u, x, y) = 1$ if G'_u has a spanning distribution forest such that the tree containing s' contains s and the tree containing t' contains t.

Case (v): $\alpha = \text{in}$ and $\beta = \text{out}$.

Add to G a dummy source s' with $\text{sup}(s') = x$ and a dummy sink t' with $\text{dem}(t') = y$, and add dummy edges (s', s) and (t', t) with infinite capacities, as in Fig. 5(d). Let G'_u be the resulting graph. Then $f_{\text{in},\text{out}}(G_u, x, y) = 1$ if G'_u has a spanning distribution forest such that the tree containing s' contains t'.

Case (vi): $\alpha = \text{out}$ and $\beta = \text{in}$.

Define $f_{\text{out},\text{in}}$ similarly as $f_{\text{in},\text{out}}$ although the roles of s and t are interchanged.

4 Algorithm

The terminal s or t of G_u may be a source or a sink of positive demand. However, we consider a virtual graph G_u^* in which both s and t are regarded as sinks of zero demand, and compute the functions $f_{\alpha,\beta}$ for G_u^*. Such a convention makes the description of our algorithm simple. One can easily compute the functions $f_{\alpha,\beta}$ for G_u from those for G_u^*. Let U be the set of all sources in G_u^*. We compute $f_{\alpha,\beta}(G_u^*, x, y)$ for each node u of T_{BD} from leaves to the root r.

(a) How to decide the existence of a spanning distribution forest in G

Suppose that the functions $f_{\alpha,\beta}$ have been computed for $G^* = G_r^*$ where r is the root of T_{BD}. We decide the existence of a spanning distribution forest in G, as follows. When both of the terminals s and t of G are sinks, G has a spanning distribution forest if and only if $f_{\alpha,\beta}(G^*, \mathrm{dem}(s), \mathrm{dem}(t)) = 1$ for some sources $\alpha, \beta \in U$. When either s or t, say s, is a source in G, G has a spanning distribution forest if and only if $f_{\mathrm{in},\beta}(G^*, \sup(s), \mathrm{dem}(t)) = 1$ for some $\beta \in U \cup \{\mathrm{out}\}$. When both of s and t are sources in G, G has a spanning distribution forest if and only if $f_{\mathrm{in},\mathrm{in}}(G^*, \sup(s), \sup(t)) = 1$.

(b) Computation at a leaf u of T_{BD}

Let u be a leaf of T_{BD}, then G_u consists of a single edge $e = (s,t)$ as illustrated in Fig. 2(a). The virtual graph G_u^* contains no source, and hence $U = \emptyset$. Since $\mathrm{dem}(s) = \mathrm{dem}(t) = 0$ in G_u^*, we can compute $f_{\alpha,\beta}(G_u, x, y)$ for $x, y \in \mathbb{Z}_S$ as follows:

$$f_{\mathrm{in},\mathrm{in}}(G_u^*, x, y) = 1;$$

$$f_{\mathrm{in},\mathrm{out}}(G_u^*, x, y) = \begin{cases} 1 & \text{if } y \le x \text{ and } y \le \mathrm{cap}(e); \\ 0 & \text{otherwise;} \end{cases}$$

and

$$f_{\mathrm{out},\mathrm{in}}(G_u^*, x, y) = \begin{cases} 1 & \text{if } x \le y \text{ and } x \le \mathrm{cap}(e); \\ 0 & \text{otherwise.} \end{cases}$$

(c) Computation at an inner node u of T_{BD}

Let u be an inner node of T_{BD}, let u_1 and u_2 be the children of u, and let G_{u_1} and G_{u_2} be the subgraphs of G corresponding to u_1 and u_2, respectively. Then G_u is a series or parallel connection of G_{u_1} and G_{u_2} as illustrated in Figs. 2(c) and (d). One can compute the functions $f_{\alpha,\beta}(G_u^*, x, y)$ from those of G_{u_1} and G_{u_2}. Assume that G_u is a parallel connection of G_{u_1} and G_{u_2}, $\alpha, \beta \in U, \alpha \ne \beta$, and both sources α and β are contained in G_{u_1}, as illustrated in Fig. 6. (The proof for the other cases is similar, and is omitted in this extended abstract.) The equation $f_{\alpha,\beta}(G_u^*, x, y) = 1$ holds if and only if a certain amount x_1 of flow goes from source

α to terminal s through G_1^* and the flow is split to an amount x of flow going to s' and the remaining amount $x_1 - x$ of flow entering in G_2^*, and similarly a certain amount y_1 of flow goes from source β to terminal t through G_1^* and the flow is split to an amount y of flow going to t' and the remaining amount $y_1 - y$ of flow entering in G_2^*. Therefore, $f_{\alpha,\beta}(G_u^*, x, y) = 1$ if only if there exist $x_1, y_1 \in \mathbb{Z}_S$ such that $f_{\alpha,\beta}(G_1^*, x_1, y_1) \wedge f_{\text{in,in}}(G_2^*, x_1 - x, y_1 - y) = 1$.

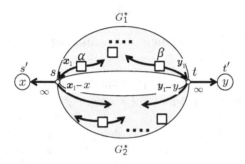

Fig. 6. Graph $G_u^{*\prime}$ when G_u is a parallel connection of G_{u_1} and G_{u_2}

The functions $f_{\alpha,\beta}$ can be computed for a leaf u of T_{BD} in time $O(|\mathbb{Z}_S|^2) = O(S^2)$ as in (b) above. Let n be the number of vertices in a given series-parallel graph G. One may assume that G is a simple graph and has no multiple edges. Then G has at most $2n - 3$ edges [12], and hence T_{BD} has at most $2n - 3$ leaves. Thus, the computation at all the leaves of T_{BD} takes time $O(S^2 n)$.

For an inner node u of T_{BD}, let u_1 and u_2 be the children of u in T_{BD}, let U_1 and U_2 be the sets of all sources in $G_{u_1}^*$ and $G_{u_2}^*$, respectively, and let $l_1 = |U_1|$ and $l_2 = |U_2|$. Then $f_{\alpha,\beta}(G_u^*, x, y)$ can be computed from $f_{\alpha_1,\beta_1}(G_{u_1}^*, x_1, y_1)$ and $f_{\alpha_2,\beta_2}(G_{u_2}^*, x_2, y_2)$ as in (c) above, where $\alpha, \beta \in U \cup \{\text{in, out}\}$, $\alpha_1, \beta_1 \in U_1 \cup \{\text{in, out}\}$, and $\alpha_2, \beta_2 \in U_2 \cup \{\text{in, out}\}$. Clearly, $|U_1 \cup \{\text{in, out}\}| \leq l_1 + 2$, $|U_2 \cup \{\text{in, out}\}| \leq l_2 + 2$, $|U \cup \{\text{in, out}\}| \leq l_1 + l_2 + 3$, and $l_1 + l_2 \leq l$. Furthermore $x, y, x_1, y_1, x_2, y_2 \in \mathbb{Z}_S$ and $|\mathbb{Z}_S| = S + 1$. Therefore, the computation time for node u is $(l_1 + 2)^2 (S + 1)^2 \times (l_2 + 2)^2 (S + 1)^2 = O(l^2 S^4)$.

Since T_{BD} has at most $2n - 3$ leaves, T_{BD} has at most $2n - 4$ inner nodes. Therefore, the computation at all inner nodes take time $O(l^2 S^4 n)$.

As shown in (a) above, one can decide from $f_{\alpha,\beta}(G_r^*, x, y)$ in time $O(1)$ whether $G = G_r$ has a spanning distribution forest.

We thus have the following theorem.

Theorem 1. *The spanning distribution forest problem can be solved in time* $O(l^2 S^4 n)$ *for series-parallel graphs.*

5 Graphs with Bounded Tree-Width

In this section, we outline an algorithm to solve the spanning distribution forest problem for graphs whose tree-widths are bounded by a fixed constant k.

Such a graph $G = (V, E)$ can be decomposed to several "pieces," each contains at most $k+1$ vertices of G, and these pieces form a tree structure, called a *binary decomposition tree* T_{BD} [5]. Every node u of T_{BD} corresponds to a subset $V(u)$ of V with $|V(u)| \leq k + 1$. Node u corresponds to a subgraph G_u of G, whose tree-width is bounded by k, too. Let W be the set of all the l sources in G, let U be the set of all sources in G_u, and let $l_u = |U|$. It sufficed to compute only the functions $f_{\alpha,\beta}$, $\alpha, \beta \in U \cup \{\mathrm{in}, \mathrm{out}\}$, for a series-parallel graph, which has two terminals s and t. In a sense, G_u has at most $k + 1$ "terminals" contained in $V(u)$. For simplicity, we assume that

$$V(u) = \{v_1, v_2, \cdots, v_{k+1}\}.$$

Then we compute functions $f_{\alpha}(G_u, x)$, where

$$x = (x_1, x_2, \cdots, x_{k+1}) \in \mathbb{Z}_S^{k+1},$$

$$\alpha = (\alpha_1, \alpha_2, \cdots, \alpha_{k+1}),$$

and

$$\alpha_i \in U \cup \{\mathrm{in}\} \cup V(u), 1 \leq i \leq k + 1.$$

The function f_{α} satisfies $f_{\alpha}(G_u, x) = 1$ if and only if G_u has a spanning distribution forest such that

(a) if $\alpha_i \in U$, then an amount $x_i \in \mathbb{Z}_S$ of flow from source α_i in G_u can be outputted from terminal $v_i \in V(u)$;
(b) if $\alpha_i = \mathrm{in}$, then an amount $x_i \in \mathbb{Z}_S$ of flow from a source in W/U is inputted to G_u through terminal $v_i \in V(u)$; and
(c) if $\alpha_i = v_j \in V(u)$, then $\alpha_j = \mathrm{in}$ must hold and an amount $x_i \in \mathbb{Z}_S$ of flow entering to G_u through terminal v_j is outputted from G_u through terminal v_i.

For example,

$$\alpha = (w_1, w_1, w_2, \mathrm{in}, v_4, v_4, \mathrm{in}, v_7, \cdots, v_7)$$

for the forest drawn by thick lines in Fig. 7. The number of parameters α is at most

$$(l_u + k + 2)^{k+1} = O(l_u^{k+1}),$$

and

$$|\mathbb{Z}_S^{k+1}| = O(S^{k+1}).$$

Therefore, the functions $f_{\alpha}(G_u, x)$ can be computed from $f_{\alpha}(G_{u_1}, x)$ and $f_{\alpha}(G_{u_2}, x)$ in time $O(l_u^{k+1} S^{2k+2})$ where u_1 and u_2 are the children of an inner node u of T. Since T_{BD} contains $O(n)$ nodes, the algorithm takes time $O(l^{k+1} S^{2k+2} n)$.

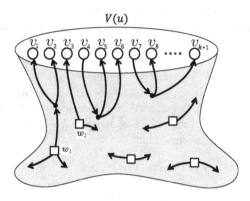

Fig. 7. Forest in G_u

6   Concluding Remarks

In the paper, we first gave a pseudo-polynomial time algorithm to solve the spanning distribution forest problem for a series-parallel graph G in time $O(l^2 S^4 n)$, where n is the number of vertices in G, l is the number of sources and S is the maximum supply. If S is bounded by a polynomial in n, then the algorithm takes polynomial time. The algorithm can be modified so that it actually finds a spanning distribution forest whenever G has one. We then showed that our algorithm can be extended for graphs of bounded tree-width.

If a given graph G has no spanning distribution forest, then one wishes to find the *maximum distribution forest*, that is, a distribution forest with the maximum sum of demands, which is not necessarily a spanning forest of G. Such a *maximum distribution forest problem* is, of course, NP-hard for series-parallel graphs. Extending our algorithm, one can find the maximum distribution forest in pseudo-polynomial time for series-parallel graphs or graphs with bounded tree-width. There are fully polynomial-time approximation schemes (FPTASs) for the maximum distribution forest problem when a given graph G is either a tree [6, 7] or a series-parallel graph with exactly one source [4]. Thus, it is desired to obtain an FPTAS for the maximum distribution forest problem on series-parallel graphs with two or more sources.

References

1. Boulaxis, N.G., Papadopoulos, M.P.: Optimal feeder routing in distribution system planning using dynamic programming technique and GIS facilities. IEEE Trans. on Power Delivery 17(1), 242–247 (2002)
2. Chekuri, C., Ene, A., Korula, N.: Unsplittable flow in paths and trees and column-restricted packing integer programs. In: Dinur, I., Jansen, K., Naor, J., Rolim, J. (eds.) APPROX and RANDOM 2009. LNCS, vol. 5687, pp. 42–55. Springer, Heidelberg (2009)

3. Dinitz, Y., Garg, N., Goemans, M.X.: On the single-source unsplittable flow problem. Combinatorica 19(1), 17–41 (1999)
4. Ito, T., Demaine, E.D., Zhou, X., Nishizeki, T.: Approximability of partitioning graphs with supply and demand. Journal of Discrete Algorithms 6, 627–650 (2008)
5. Ito, T., Zhou, X., Nishizeki, T.: Partitioning graphs of supply and demand. Discrete Applied Math. 157, 2620–2633 (2009)
6. Ito, T., Zhou, X., Nishizeki, T.: Partitioning trees of supply and demand. IJFCS 16(4), 803–827 (2005)
7. Kawabata, M., Nishizeki, T.: Partitioning trees with supply, demand and edge-capacity. IEICE Trans. on Fundamentals of Electronics, Communications and Computer Science 96-A(6), 1036–1043 (2013)
8. Kawabata, M., Nishizeki, T.: Spanning distribution trees of graphs. In: Fellows, M., Tan, X., Zhu, B. (eds.) FAW-AAIM 2013. LNCS, vol. 7924, pp. 153–162. Springer, Heidelberg (2013); also IEICE Trans. E97-D(3) (2014)
9. Kim, M.S., Lam, S.S., Lee, D.-Y.: Optimal distribution tree for internet streaming media. In: Proc. 23rd Int. Conf. on Distributed Computing System (ICDCS 2003), pp. 116–125 (2003)
10. Kleinberg, J.M.: Single-source unsplittable flow. In: Proc. of 37th FOCS, pp. 68–77 (1996)
11. Morton, A.B., Mareels, I.M.Y.: An efficient brute-force solution to the network reconfiguration problem. IEEE Trans. on Power Delivery 15(3), 996–1000 (2000)
12. Takamizawa, K., Nishizeki, T., Saito, N.: Linear-time computability of combinatorial problems on series-parallel graphs. J. Assoc. Comput. Mach. 29, 623–641 (1982)
13. Teng, J.-H., Lu, C.-N.: Feeder-switch relocation for customer interruption cost minimization. IEEE Trans. on Power Delivery 17(1), 254–259 (2002)

A (1.408+ε)-Approximation Algorithm for Sorting Unsigned Genomes by Reciprocal Translocations

Haitao Jiang[1], Lusheng Wang[2], Binhai Zhu[3], and Daming Zhu[1]

[1] School of Computer Science and Technology, Shandong University, Jinan, China
htjiang@mail.sdu.edu.cn, dmzhu@sdu.edu.cn
[2] Department of Computer Science, City University of Hong Kong, Kowloon, Hongkong
cswangl@cityu.edu.hk
[3] Department of Computer Science, Montana State University,
Bozeman, MT 59717-3880, USA
bhz@cs.montana.edu

Abstract. Sorting genomes by translocations is a classic combinatorial problem in genome rearrangements. The translocation distance for signed genomes can be computed exactly in polynomial time, but for unsigned genomes the problem becomes NP-Hard and the current best approximation ratio is 1.5+ε. In this paper, we investigate the problem of sorting unsigned genomes by translocations. Firstly, we propose a tighter lower bound of the optimal solution by analyzing some special sub-permutations; then, by exploiting the two well-known algorithms for approximating the maximum independent set on graphs with a bounded degree and for set packing with sets of bounded size, we devise a new polynomial-time approximation algorithm, improving the approximation ratio to 1.408+ε, where $\varepsilon = O(1/\log n)$.

1 Introduction

Genome rearrangement is an important area in computational biology, and some combinatorial problems in this area have attracted more and more attention during the last two decades. Genome rearrangement mainly deals with sorting genomes by a series of rearrangement operations. Sankoff *et al.* proposed three basic rearrangement operations: reversal, transposition and translocation [1]. Subsequently, sorting single-chromosome genome by reversals or transpositions and sorting multi-chromosome genome by translocations have become three famous problems and researchers devoted a lot of effort on these problems, both on determining their computational complexity and on designing efficient algorithms.

Generally speaking, a translocation is an operation swapping the tails of two chromosomes. For sorting signed genomes by translocations, the problem was first studied by Kececioglu and Ravi [2]. Hannenhalli devised a polynomial time exact algorithm which runs in $O(n^3)$ [3]. Wang *et al.* improved the running time to $O(n^2)$ [4], which was further improved to $O(n^{\frac{3}{2}}\sqrt{\log n})$ by Ozery-Flato and Shamir [5]. Wang *et al.* and Bergeron *et al.* showed that the translocation distance can be computed by a formula in linear time [6,7].

J. Chen, J.E. Hopcroft, and J. Wang (Eds.): FAW 2014, LNCS 8497, pp. 128–140, 2014.

Zhu and Wang proved that the problem for sorting unsigned genomes by translocations is NP-hard and APX-hard [8]. In [2], Kececioglu and Ravi also gave a ratio-2 approximation algorithm for computing the translocation distance beween unsigned genomes. Cui *et al.* presented a 1.75-approximation algorithm [9], and further improved the approximation ratio to $1.5 + \varepsilon$ [10].

The problem of sorting unsigned genome by translocations has a close relationship with the maximum cycle decomposition of the breakpoint graph, which is also NP-hard [11]. Finding a good cycle decomposition of the breakpoint graph is critical to approximating the translocation distance [9,10] and some other genome rearrangement distances, such as sorting by reversals [12], sorting by double cut and join [13]. About ten years ago, Caprara and Rizzi proposed a better cycle decomposition method and improved the approximation ratio for sorting unsigned genomes by reversals [14]. Basically, this method exploits the approximation algorithm for computing a maximum independent set on bounded degree graphs [15] and computing a maximum set packing with each set containing bounded number of elements [16]. Recently, Chen *et al.* used this method to improve the approximation ratio of the sorting multi-chromosome genomes by double cut and join problem to $\frac{13}{9} + \varepsilon$ [17].

Our Contribution. In this paper, we devised a new approximation algorithm for sorting unsigned genomes by translocations with an approximation ratio of $1.408 + \varepsilon$. We present a new lower bound for the optimal solution by identifying a special local structure in the breakpoint graph, then devise a cycle decomposition algorithm by combining the approximation algorithm for computing a maximum independent set on bounded degree graphs and for computing a maximum set packing with bounded set size, and enumerate the cycle decomposition of sub-permutations with no more than $O(\log n)$ genes.

The paper is organized as follows. In the preliminaries section, we review some notions about the sorting unsigned genomes by translocations problem. In section 3, we review the relation between sorting signed genomes with sorting unsigned genomes. In section 4, we introduce the new lower bound and then our new algorithm. In section 5, we analyze the approximation ratio. We conclude the paper in section 6.

2   Preliminaries

In this section, we present the precise definition of a translocation operation and review some useful notions.

Gene, Chromosome and Genome. We use a signed integer to represent a directed gene and use an unsigned integer to represent a gene with unknown direction. A chromosome, denoted by a permutation, is a sequence of genes and a genome is a set of chromosomes. A gene which lies at the end of some linear chromosome is called an *ending* gene. Throughout this paper, without loss of generality, we assume that the target genome Y is *perfect*, i.e., each chromosome in Y is a segment of consecutive positive integers, for example, $i, i + 1, i + 2, \ldots, j - 1, j$, and, the concatenation of these chromosomes forms the identity permutation (i.e., no duplication of any gene is allowed). In the context of sorting genomes, the comparative order of the genes in the same chromosome matters, but the order of chromosomes and the direction of a whole chromosome does not matter, which implies that each chromosome can be viewed in both directions. In the case of signed genomes, a chromosome $\langle a_i, a_{i+1}, \cdots, a_j \rangle$ is equivalent to $\langle -a_j, \cdots, -a_{i+1}, -a_i \rangle$; and in the case of unsigned genomes, a chromosome $\langle b_i, b_{i+1}, \cdots, b_j \rangle$ is equivalent to $\langle b_j, \cdots, b_{i+1}, b_i \rangle$.

Signed and Unsigned Translocations. For two signed chromosomes $S = [x_1, x_2, \cdots, x_n]$ and $T = [y_1, y_2, \cdots, y_m]$ in a genome, a translocation swaps the segments in the chromosomes and generates two new chromosomes. If we cut the adjacency (x_i, x_{i+1}) $(1 < i < n)$ in S and (y_j, y_{j+1}) $(1 < j < m)$ in T. There are two cases of two new resulting chromosomes (See Figure 1),

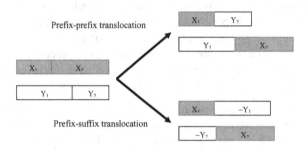

Fig. 1. Examples of translocation

- prefix-prefix translocation: $S' = [x_1, \cdots, x_i, y_{j+1}, \cdots, y_m]$, and $T' = [y_1, \cdots, y_j, x_{i+1}, \cdots, x_n]$.
- prefix-suffix translocation: $S' = [x_1, \cdots, x_i, -y_j, \cdots, -y_1]$, and $T' = [-y_m, \cdots, -y_{j+1}, x_{i+1}, \cdots, x_n]$.

Note that each signed chromosome can be viewed in both directions, such as $T = [y_1, y_2, \cdots, y_m]$ or $T = [-y_m, -y_{m-1}, \cdots, -y_1]$.

If $S = [x_1, x_2, \cdots, x_n]$ and $T = [y_1, y_2, \cdots, y_m]$ are two unsigned genomes, then the resulting new chromosomes in each case are shown as follows,

- prefix-prefix translocation: $S' = [x_1, \cdots, x_i, y_{j+1}, \cdots, y_m]$, and $T' = [y_1, \cdots, y_j, x_{i+1}, \cdots, x_n]$.
- prefix-suffix translocation: $S' = [x_1, \cdots, x_i, y_j, \cdots, y_1]$, and $T' = [y_m, \cdots, y_{j+1}, x_{i+1}, \cdots, x_n]$.

Again, an unsigned chromosome T can also be treated as $T = [y_1, y_2, \cdots, y_m]$ or $T = [y_m, y_{m-1}, \cdots, y_1]$. It can be seen that to transform a genome X into Y with unsigned translocations alone, the set of genes, the number of chromosomes and the set of ending genes in the chromosomes in X and Y must all be the same. This can be checked easily, so we just assume that the input genomes satisfy these conditions. Now, we formally put forward the problem we investigate in this paper.

Problem Description: Sorting Unsigned Genomes by Translocations
Input: Two multi-chromosome genomes X and Y, where Y is perfect.
Question: Is there a sequence of unsigned translocations $\rho_1, \rho_2, \ldots, \rho_t$ to transform X into Y, and t is minimized? The minimum t is called the *unsigned translocation distance* between X and Y.

Recall that if the input genomes are signed and the operation is signed translocations, then the problem becomes sorting signed genomes by translocations.

3 Sorting Signed Genomes

First, let us review the computation method for signed genomes that was originally devised by Hannenhalli [3]. The breakpoint graph plays a critical role in computing the signed translocation distance.

3.1 The Breakpoint Graph of Signed Genomes

Given signed genomes A and B, the breakpoint graph $G_s(A, B)$ can be obtained as follows: For every chromosome $S = [x_1, x_2, \ldots, x_{n_i}]$ of A, replace each x_i with an ordered pair $(l(x_i), r(x_i))$ of vertices. If x_i is positive, then $(l(x_i), r(x_i)) = (x_i^t, x_i^h)$; and if x_i is negative, then $(l(x_i), r(x_i)) = (x_i^h, x_i^t)$. If the genes x_i and x_{i+1} are adjacent in A, then we connect $r(x_i)$ and $l(x_{i+1})$ by a black edge in $G_s(A, B)$. If the genes x_i and x_{i+1} are adjacent in B, then we connect $r(x_i)$ and $l(x_{i+1})$ by a gray edge in $G_s(A, B)$. Every vertex (except the ones at the two ends of a chromosome) in $G_s(A, B)$ is incident to one black and one gray edge. Therefore, $G_s(A, B)$ can be uniquely decomposed into cycles, on which the black edges and gray edges appears alternatively. A cycle containing exactly i black (gray) edges is called an i-cycle.

3.2 The Signed Translocation Distance Formula

Let $S = [x_1, x_2, \ldots, x_{n_i}]$ be a chromosome in A. A *sub-permutation*, denoted by SP, is a segment $[x_i, x_{i+1}, \ldots, x_{i+l}]$ of at least three genes in S such that there is another segment $[y_j, y_{j+1}, \ldots, y_{j+l}]$ of the same length in some chromosome T of B satisfying that

- $\{|x_i|, |x_{i+1}|, \ldots, |x_{i+l}|\} = \{|y_j|, |y_{j+1}|, \ldots, |y_{i+l}|\}$;
- $|x_i| = |y_j|$, $|x_{i+l}| = |y_{j+l}|$;
- $[x_i, x_{i+1}, \ldots, x_{i+l}] \neq [y_j, y_{j+1}, \ldots, y_{j+l}]$.

Here, x_i and x_{i+l} are the two ending genes of the SP, $(r(x_i), l(x_{i+1}))$ and $(r(x_{i+l-1}), l(x_{i+l}))$ are the two boundary black edges of the SP. A $minSP$ is a SP not containing any other SP. If all $minSP$s in $G_s(A, B)$ are in a SP, say, I, and the total number of $minSP$s is even, then I is an *even isolation*.

Let b (resp. c, s) be the number of black edges (resp. cycles, $minSP$s) in $G_s(A, B)$. Let f be a remaining index, which is defined as follows: 1) $f=1$ if s is odd, 2) $f=2$ if there is an even isolation, and 3) $f=0$ otherwise. Hannenhalli showed the following formula to compute the translocation distance $d_s(A, B)$ between A and B [3].

Lemma 1. $d_s(A, B) = b - c + s + f$.

As we only focus on designing an approximation algorithm for unsigned genomes (by converting them to some signed ones), we ignore f throughout the paper. But we certainly need to focus on approximating s.

3.3   The Breakpoint Graph of Unsigned Genomes

The breakpoint graph for two unsigned genomes is a bit different from that of signed genomes. In the breakpoint graph $G(X, Y)$ of two unsigned genomes X and Y, each vertex corresponds to a gene in X; two genes are connected by a black edge if they appear consecutively in some chromosome of X and two genes are connected by a gray edge if they appear consecutively in some chromosome of Y. So the breakpoint graph $G(X, Y)$ can be decomposed into a set of edge-disjoint cycles, denoted by **D**, and on each cycle, the black edge and gray edge appears alternatively. Each ending gene of X and Y is incident to only one black edge and one gray edge; and each of the rest genes is incident to exactly two blacks edge and two gray edges. Thus, the ways to decompose $G(X, Y)$ into cycles might not be unique.

If we split each non-ending vertex x_i in $G(X, Y)$ into two vertices $l(x_i), r(x_i)$, make $l(x_i)$ incident to one black edge and one gray edge which is incident to x_i in $G(X, Y)$, and make $r(x_i)$ incident to the other black edge and gray edge which is incident to x_i in $G(X, Y)$, we obtain a the new graph **D**, which is called a *cycle decomposition* of $G(X, Y)$. In the next subsection, we will show that each cycle decomposition of $G(X, Y)$ is identical to the breakpoint graph $G_s(\bar{X}, \bar{Y})$, where $\bar{X}$ and $\bar{Y}$ is obtained by assigning a proper sign to each gene in X and Y.

In unsigned genomes, a *candidate SP* (abbreviated as CSP), is made of a substring of at least three genes in X such that another substring exists in Y with the same gene content, but with a different gene order. A *candidate $minSP$* (abbreviated as $CminSP$) is a candidate SP not containing any other candidate SP.

3.4   Converting Unsigned Genomes into Signed Ones

Once we have a cycle decomposition **D** of $G(X, Y)$, we can obtain two signed genomes $\bar{X}$ and $\bar{Y}$ by assigning a sign to each gene in X and Y, so that $G_s(\bar{X}, \bar{Y}) = \mathbf{D}$. (We can view $\bar{X} = A$ and $\bar{Y} = B$ in Lemma 1.) As aforementioned, all genes in Y are positive. Therefore, all gray edges in $G_s(\bar{X}, \bar{Y})$ have the form $((x_i)^h, (x_i + 1)^t)$.

Next, we show how to assign a proper sign to each gene in X (to obtain $\bar{X}$). An ending gene is positive if it lies at the same (i.e., both left or both right) ends of some chromosome in X and some chromosome in Y; otherwise, it is negative in X. For a non-ending gene x_i, according to the two gray edges, $((x_i)^h, (x_i + 1)^t)$ and $((x_i - 1)^h, (x_i)^t)$ in the cycle decomposition, we assign x_i positive if $((x_i - 1)^h, l(x_i))$ is a gray edge in the given cycle decomposition; if $((x_i - 1)^h, r(x_i))$ is a gray edge in the given cycle decomposition, then x_i is assigned a negative sign. See Figure 2 for an example.

As the translocation distance between signed genomes can be computed in polynomial time, a natural and efficient method to solve the problem of sorting unsigned

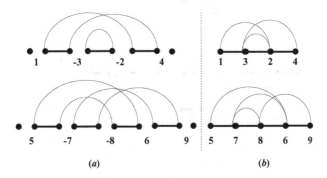

Fig. 2. (a) The breakpoint graph $G_s(\bar{X}, \bar{Y})$ for the signed case, where $\bar{X}$={[1,-3,-2,4], [5,-7,-8,6,9]} and $\bar{Y}$={[1,2,3,4], [5,6,7,8,9]}. (b) The breakpoint graph $G(X,Y)$ for the unsigned case, where X={[1,3,2,4], [5,7,8,6,9]} and Y={[1,2,3,4], [5,6,7,8,9]}. (a) is a cycle decomposition of (b).

genomes by translocations is to assign proper signs to the unsigned genome, or equivalently, to find a proper cycle decomposition of the breakpoint graph. The following lemma shows that the influence by changing the sign of some gene is local.

Lemma 2. *Let Δs and Δc be the difference of $minSPs$ and cycles by changing the sign of some gene respectively. Δs, $\Delta c \in \{-1,0,1\}$*

4 Our Algorithm

To solve the problem of sorting unsigned genomes by translocations, we aim at finding a proper cycle decomposition of the breakpoint graph. But the maximum cycle decomposition of the breakpoint graph between two unsigned genomes is NP-hard, an intuitive method is to decompose as many cycles as possible. Cui [10] showed that all the possible 1-cycles could be kept in the cycle decomposition by the following lemma.

Lemma 3. *There exists some optimal cycle decompositions containing all the existing 1-cycles.*

4.1 A New Lower Bound

To design a better approximation algorithm, we need a lower bound better than this lemma. We achieve this by looking at some special CSP's which could be included in some optimal solution.

Let I be a CSP and let $I_s = [x_i, x_{i+1}, \ldots, x_j]$ be a cycle decomposition of I in $G_s(\bar{X}, \bar{Y})$, if there exists a cycle C in I_s that contains the two black edges (possibly some other black edges), $(r(x_i), l(x_{i+1}))$ and $(r(x_{j-1}), l(x_j))$, then C is called the *boundary cycle* of I_s. Based on the traversal order of the four vertices $(r(x_i), l(x_{i+1})), (r(x_{j-1})$ and $l(x_j))$ in cycle C, there are two kinds of boundary cycles. If

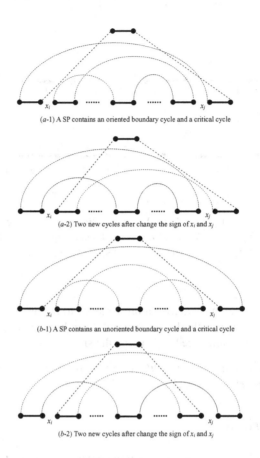

(a-1) A SP contains an oriented boundary cycle and a critical cycle

(a-2) Two new cycles after change the sign of x_i and x_j

(b-1) A SP contains an unoriented boundary cycle and a critical cycle

(b-2) Two new cycles after change the sign of x_i and x_j

Fig. 3. Candidate removable SP and boundary cycle: (a-1) A $RCSP$ containing a positive boundary cycle. (b-1) A $RCSP$ containing a negative boundary cycle.

the boundary cycle C has the form $[r(x_i), l(x_{i+1}), \ldots, r(x_{j-1}), l(x_j), \ldots, r(x_i)]$, C is *positive*; if C has the form $[r(x_i), l(x_{i+1}), \ldots, l(x_j), r(x_{j-1}), \ldots, r(x_i)]$, C is *negative*. If there exists another cycle C' containing the two black edges $(r(x_{i-1}), l(x_i))$ and $(r(x_j), l(x_{j+1}))$ such that if we change the signs of x_i and x_j, C' and C become two other cycles C'' and C''', then we say that C is a *removable boundary cycle*, C' is a *critical cycle* for C, and I_s is a *removable candidate SP*, abbreviated as $RCSP$. An example on candidate removable CSP is given in Figure 3. We comment that the two boundary black edges might be in two different cycles, but we will mostly focus on the boundary cycles in this paper.

A CSP that contains no less than $O(\log n)$ genes is called a *long CSP*, otherwise it is *short*. A CSP is *bad* if it is short and there is no other short CSP containing it. Note that by scanning each chromosome from left to right, we could find out the (disjoint) bad CSPs one by one.

Let I_s be a CSP, we say that a cycle decomposition D_s of the breakpoint graph of I_s is *feasible* if it maximize c-s, where c is the number of cycles in D_s, s is the number of $minSP$s in D_s. A feasible cycle decomposition D_s is *efficient*, if I_s is a $minSP$.

The overall idea of our algorithm is as follows. We first identify all bad CSPs. For each bad CSP (with length $O(\log n)$), we enumerate all its possible cycle decompositions in polynomial time. Then, depending on the properties of these cycle decompositions (shown in Lemma 5, Lemma 6 and Lemma 7), we pick certain cycle decompositions. After that, we select disjoint 2-cycles and 3-cycles which have not been picked yet (using approximate subroutines for the maximum independent set and maximum set packing). Finally, we process the remaining bad CSPs optimally while at least keeping the number of cycles computed by the independent set and set packing (shown in Lemma 1). Due to space constraint, we omit the proofs of the following lemmas.

Lemma 4. *There exists some optimal cycle decomposition in which the two ending genes of a CSP are either both positive or both negative.*

Lemma 5. *Let $I = [x_i, x_{i+1}, \ldots, x_j]$ be a CSP, if there is no efficient cycle decomposition of the breakpoint graph of I, then any feasible cycle decomposition could appear as a component in some optimal cycle decomposition.*

Lemma 6. *Let $I = [x_i, x_{i+1}, \ldots, x_j]$ be a CSP, if there is no efficient cycle decomposition of the breakpoint graph of I containing a boundary cycle, then any efficient cycle decomposition could appear as a component in some optimal cycle decomposition.*

Lemma 7. *Let $I = [x_i, x_{i+1}, \ldots, x_j]$ be a CSP, the boundary cycles in any efficient cycle decomposition are of the same type, then for any efficient cycle decomposition with a boundary cycle, all the cycles except the boundary cycle could appear in some optimal cycle decomposition.*

Based on Lemma 7, we immediately have the following corollary.

Corollary 1. *Let $I=[x_i, x_{i+1}, \ldots, x_j]$ be a CSP, whose breakpoint graph has efficient cycle decompositions and the boundary cycles in these efficient cycle decompositions are of two types. Then there exists at least one type of boundary cycle such that for any efficient cycle decomposition with a boundary cycle of this type, all the cycles except the boundary cycle could appear in some optimal cycle decomposition.*

The difficulty here to handle the case in Corollary 1 is that if the boundary cycles in these efficient cycle decompositions are of two types, we do not know which type we should use. We will carefully deal with this case when designing the algorithm. The rough idea is to decide the signs of some 2- and 3-cycle which contains x_i or x_j, using approximate solution for Maximum Independent Set (for graphs of degree at most 4) or Maximum Set Packing (for sets of size at most 3). Then we can decide the type of the boundary cycle, as the cycle decomposition is known from the outside of this cycle.

4.2   The Algorithm

Now, we are ready to describe our approximation algorithm. Here we give an approximate cycle decomposition and then use the exact algorithm for the signed case to obtain the solution. The basic idea of our cycle decomposition algorithm is as follows: (I) We identify all the bad CSPs, and fix the signs of each gene in each bad CSP by a feasible cycle decomposition. Since the size of a bad CSP is at most $O(\log n)$, we can obtain all the feasible cycle decompositions in polynomial time. Note that these bad CSPs are disjoint. (II) We try to obtain a large number of 2-cycles for the genes outside of the bad CSPs (possibly including the ending genes of bad CSPs) using an approximate algorithm for Maximum Independent Set (for graphs with degree at most 4). (III) We try to obtain a large number of 2-cycles and 3-cycles for the genes outside of the bad CSPs (possibly including the ending genes of bad CSPs) using an approximate algorithm for the Maximum Set Packing (for sets of size at most 3). (IV) We choose the cycle decomposition with a bigger number of cycles obtained at step (II) and (III). (V) Arbitrarily fix the sign for the rest of genes to get a signed version of the problem and use the exact algorithm for the signed version to get a solution. Let us look at the details.

Identifying all bad CSPs: From Lemma 3, we know that each chromosome is split into segments by the 1-cycles. So we deal with each segment respectively. Firstly, it is easy to see that by scanning each segment from left to right, we can identify all CSPs. Some CSPs may (completely) overlap, i.e., some CSP is entirely contained in another CSP. A CSP containing no more than $O(\log n)$ elements (i.e., is short) and not contained in any other short CSP is a bad CSP. Secondly, by scanning a CSP with no more than $O(\log n)$ elements from the left to right, we identify the longest CSP with no more than $O(\log n)$ elements (if any) which is a bad CSP, and then identify other bad CSPs from the right end of the newly found bad CSP. This can be done as follows: Start from the leftmost gene t_1 of a segment T, identify t_q such that $q = O(\log n)$. Then set $i \leftarrow 1, j \leftarrow q$ and check whether $T[i..j]$ is a CSP by sorting all the elements in $T[i..j]$, and if there is no gap between two neighbors in the sorted list, then $T[i..j]$ is a CSP. If the test is negative, then decrease j by one and repeat the process. When $T[i..j]$ is found to be a bad CSP, then set $i \leftarrow j + 1, j \leftarrow i + q - 1$ and repeat the process. If no $T[i..j]$ is found to be a bad CSP, then set $i \leftarrow 2, j \leftarrow q + 1$ and repeat. So, the above procedure to consecutively identify the bad CSPs can be completed in $n \times \log n \times O(\log n \log \log n)$ $= O(n \log^2 n \log \log n)$ time.

Fix the sign of genes in a bad CSP I: For each bad CSP, we can enumerate all possible cycle decompositions in $O(2^{\log n})$ time. We have to consider the following cases:

Case 1: No efficient cycle decomposition of I can be found. We arbitrarily select any feasible cycle decomposition (by Lemma 5).

Case 2: No efficient cycle decomposition exists and no efficient cycle decomposition containing a boundary cycle of I can be found. We can arbitrarily select any efficient cycle decomposition (by Lemma 6).

Case 3: The boundary cycles in all efficient cycle decompositions of I are of the same type. We arbitrarily select any efficient cycle decomposition with a boundary cycle and keep all the cycles except the boundary cycle (by Lemma 7).

Case 4: There are two types of boundary cycles in efficient cycle decompositions. According to Corollary 1, we can use one type of efficient cycle decomposition. However, at present we do not know which type to choose. Thus, we will decide which type of boundary cycle we have to use after steps (II), (III) and (IV).

Finding a large number of 2-cycles: In order to get a large number of 2-cycles, we construct a *2-cycle graph* G' as follows: For each possible 2-cycle satisfies: (a) it is not a boundary cycle from a bad CSP, and (b) it is not contained entirely in any bad CSP, we create a vertex (corresponding to such a 2-cycle) in G'. There is an edge between two vertices if and only if their corresponding 2-cycles share edges. It should be emphasized that for a bad CSP I (including Case 4), we do not consider any 2-cycle entirely contained in I when constructing G'.

Lemma 8. *There exists an approximation algorithm with ratio $\frac{5}{7} - \varepsilon$ for any $\varepsilon > 0$ for the maximum independent set problem on a graph with maximum degree 4.*

Proof. Refer to reference [15]. □

It is known that the constructed graph G' has degree at most 6, but can be converted to a graph of degree at most 4 while searching for the independent set [14]. Thus, we can use the ratio $\frac{5}{7} - \varepsilon$ approximate algorithm here.

Finding a large number of 2-cycles and 3-cycles: Another strategy is to get a large number of 2-cycles and 3-cycles together. Based on the two input genomes, we construct an instance of the maximum set packing problem. Here each set contains the edges from a possible 2-cycle or a possible 3-cycle which again satisfies (a) and (b).

Lemma 9. *There is a ratio $\frac{3}{7} - \varepsilon$ approximate algorithm for any $\varepsilon > 0$ for the maximum set packing problem with set size at most 3.*

Proof. Refer to reference [16]. □

Algorithm *Cycle-Decomposition$(G(X,Y))$*
Input: $G(X,Y)$
Output: $G(\bar{X}, \bar{Y})$, which is a cycle decomposition of $G(X,Y)$.
1   Keep all the 1-cycles and assign each gene involved in them a proper sign.
2   Identify all the bad CSPs in $G(X,Y)$.
3   For each bad CSP: I,
3.1   Enumerate all possible cycle decompositions of the breakpoint graph of I.
3.2   For a bad CSP I of Cases 1-3, fix the signs of genes (except possibly the ending genes) in I using Lemmas 5-7.
4   Construct the 2-cycle graph G' and compute an approximate independent set I_1 of G' as the selected 2-cycles.
5   Construct all the sets for 2- and 3-cycles and compute an approximate set packing S_1.
6   Choose the greater one of I_1 and S_1. Assign signs to each gene in the selected 2-cycles or 2,3-cycles.
7   For each bad CSP of Case 4, choose the type of the boundary cycle according to S' (if possible).
8   For each of the remaining bad CSP, arbitrarily choose an efficient cycle decomposition for it.
9   Arbitrarily assign signs to the rest of genes so that every gene has a sign.

We can use the ratio $\frac{3}{7} - \varepsilon$ approximate algorithm here. After we obtain the ratio $\frac{5}{7} - \varepsilon$ solution S_1 for G' and the ratio $\frac{3}{7} - \varepsilon$ solution S_2 for the constructed instance of the maximum set packing problem, we choose the solution S' which is one of S_1 and S_2 with bigger number of cycles. We refer S^* as the *approximate partial cycle decomposition*. Now the sign of some of the ending genes for bad CSPs of Case 4 might be determined by S' and we can determine the type of boundary cycles accordingly and select all the cycles in the CSPs based on Corollary 2. Finally, for the genes that have not been assigned signs, we can arbitrarily fix the sign and thus obtain an instance of the signed translocation problem. The complete algorithm is given separately.

5   Proof of the Approximation Ratio

To analyze the performance of our algorithm, we need to compare our algorithm to some optimal cycle decomposition. So we assume that the optimal cycle decomposition S^* fulfils Lemma 4, 5, 6, 7, and Corollary 1, i.e., in the optimal cycle decomposition, we have

1. The ending genes of any CSP are either both negative or both positive;
2. If the ending genes of a CSP are both positive, then the cycle decomposition for this CSP is either efficient or feasible;
3. If the ending genes of a CSP are both negative, then by changing them to positive, the CSP becomes a $minSP$.

Lemma 10. *Let c'_2 and c'_3 be the number of 2-cycles and 3-cycles in S_2 obtained by our algorithm, c^*_2 and c^*_3 be the number of 2-cycles and 3-cycles in the optimal cycle decomposition, c^*_a be the number of boundary 2-cycles and 3-cycles in the optimal cycle decomposition, then we have $c'_2 + c'_3 \geq (3/7 - \varepsilon)(c^*_2 + c^*_3 - c^*_a)$.*

Proof. Let $\bar{c}_2$ and $\bar{c}_3$ be the number of 2-cycles and 3-cycles entirely contained in the bad CSPs, from Lemma 7 and Corollary 1, we know that, for any bad CSP, all cycles except the boundary cycle in our cycle decomposition also appears in the optimal cycle decomposition. Let $\tilde{c}_2$ and $\tilde{c}_3$ be the number of 2-cycles and 3-cycles not entirely contained in any bad CSP, which are figured out at Step 5. So $c'_2 = \bar{c}_2 + \tilde{c}_2$ and $c'_3 = \bar{c}_3 + \tilde{c}_3$. Let $\tilde{c}^*_2$ and $\tilde{c}^*_3$ be the number of 2-cycles and 3-cycles not entirely contained in any bad CSP in the optimal cycle decomposition. So $c^*_2 = \bar{c}^*_2 + \tilde{c}^*_2$ and $c^*_3 = \bar{c}^*_3 + \tilde{c}^*_3$. Since we do not consider any boundary cycle at Step 5, from Lemma 9, we have $\tilde{c}_2 + \tilde{c}_3 \geq (3/7 - \varepsilon)(\tilde{c}^*_2 + \tilde{c}^*_3 - c^*_a)$. By adding $\bar{c}_2$ and $\bar{c}_3$ to both side, we have $c'_2 + c'_3 \geq (3/7 - \varepsilon)(c^*_2 + c^*_3 - c^*_a)$.   □

Corollary 2. *Let c''_2 be the number of 2-cycles returned by our algorithm, c^*_2 be the number of 2-cycles in the optimal cycle decomposition, c^*_b be the number of boundary 2-cycles in the optimal cycle decomposition, then we have $c''_2 \geq (5/7 - \varepsilon)(c^*_2 - c^*_b)$.*

Lemma 11. *If the algorithm Cycle-Decomposition($\bullet$) converts a long CSP into a SP, then either it also appears in some optimal cycle decomposition or it contains at least one cycle of length at least 4 or $O(\log n)$ number of 2,3-cycles.*

Proof. Since the CSP is long, i.e., it contains at least $O(\log n)$ genes, then either it contains a cycle of length at least 4 or it contains $O(\log n)$ number of 2,3-cycles.   □

Lemma 12. *If the algorithm Cycle-Decomposition($\bullet$) converts a bad CSP into a SP, either this SP also appears in some optimal cycle decomposition or the boundary cycle is not a selected cycle.*

Proof. If the CSP does not appear in any optimal cycle decomposition, it is either the third case shown in Lemma 7 or the case shown Corollary 1, from Step 4 and 5, in either case, the boundary cycle is not a selected cycle. $\square$

Lemma 13. *Let s be the number of $minSPs$ by the algorithm Cycle-Decomposition($\bullet$), s^* be the number of $minSPs$ in some optimal cycle decomposition, c_i be the number of i-cycles computed by the algorithm Cycle-Decomposition($\bullet$), c_a be the number of boundary 2,3-cycles returned by Step 8. We have $s \leq s^* + \sum_{i \geq 4} c_i + (c_2 + c_3)/O(\log n) + c_a$.*

Proof. We use an amortized analysis. For each $minSP$ by the algorithm Cycle-Decomposition($\bullet$), there are three cases.

1. It is also a $minSP$ of some optimal cycle decomposition, then we are done.
2. It is long, from Lemma 11, there are $O(\log n)$ selected 2,3-cycles or at least one i-cycle ($i \geq 4$) in it.
3. It is bad, from Lemma 12, the boundary cycle is not a selected cycle but a cycle counted in c_a,

$\square$

Corollary 3. *Let s be the number of $minSPs$ by the algorithm Cycle-Decomposition($\bullet$), s^* be the number of $minSPs$ in some optimal cycle decomposition, c_i be the number of i-cycles by the algorithm Cycle-Decomposition($\bullet$), c_b be the number of boundary 2-cycles returned by Step 8. We have $s \leq s^* + \sum_{i \geq 3} c_i + (c_2)/O(\log n) + c_b$.*

Theorem 1. *The algorithm Cycle-Decomposition($\bullet$) approximates the translocation distance within a ratio $\alpha = 1.408 + \varepsilon$.*

Proof. Omitted due to space constraint. $\square$

6   Concluding Remarks

Sorting unsigned genomes by translocations is a famous NP-hard problem in genome arrangements. It is a great challenge to obtain smaller approximation ratio. In this paper, we improve the approximation ratio from $1.5 + \varepsilon$ to $1.408 + \varepsilon$, based on a new lower bound and some cycle decomposition methods. What is more, the description of our algorithm and the analysis is much simpler than what is in [10]. Devising algorithms with smaller approximation ratio is a valuable future work.

Acknowledgments. This research is supported by NSF of China under grant 60928006, 61070019 and 61202014, by Doctoral Fund of Chinese Ministry of Education under grant 20090131110009, and by China Postdoctoral Science Foundation funded project under grant 2011M501133 and 2012T50614. LW is supported by a grant from the Research Grants Council of the Hong Kong SAR, China (CityU 123013).

References

1. Sankoff, D., Leduc, G., Antoine, N., Paquin, B., Lang, B.F., Cedergren, R.: Gene order comparisons for phylogenetic inference: Evolution of the mitochondrial genome. Proc. Nat. Acad. Sci. USA 89, 6575–6579 (1992)
2. Kececioglu, J., Ravi, R.: Of Mice and Men: Algorithms for Evolutionary Distances between Genomes with Translocation. In: Proceedings of the 6th Annual ACM-SIAM Symposium on Discrete Algorithms (SODA 1995), pp. 604–613 (1995)
3. Hannenhalli, S.: Polynomial-time Algorithm for Computing Translocation Distance Between Genomes. Discrete Applied Mathematics 71(1-3), 137–151 (1996)
4. Wang, L., Zhu, D., Liu, X., Ma, S.: An $O(n^2)$ algorithm for signed translocation. J. Comput. Syst. Sci. 70(3), 284–299 (2005)
5. Ozery-Flato, M., Shamir, R.: An $O(n^{\frac{3}{2}}\sqrt{logn})$ algorithm for sorting by reciprocal translocations. J. Discrete Algorithms 9(4), 344–357 (2011)
6. Li, G., Qi, X., Wang, X., Zhu, B.: A linear-time algorithm for computing translocation distance between signed genomes. In: Sahinalp, S.C., Muthukrishnan, S.M., Dogrusoz, U. (eds.) CPM 2004. LNCS, vol. 3109, pp. 323–332. Springer, Heidelberg (2004)
7. Bergeron, A., Mixtacki, J., Stoye, J.: On Sorting by Translocations. Journal of Computational Biology 13(2), 567–578 (2006)
8. Zhu, D., Wang, L.: On the complexity of unsigned translocation distance. Theor. Comput. Sci. 352(1-3), 322–328 (2006)
9. Cui, Y., Wang, L., Zhu, D.: A 1.75-approximation algorithm for unsigned translocation distance. J. Comput. Syst. Sci. 73(7), 1045–1059 (2007)
10. Cui, Y., Wang, L., Zhu, D., Liu, X.: A $(1.5 + \epsilon)$-Approximation Algorithm for Unsigned Translocation Distance. IEEE/ACM Trans. Comput. Biology Bioinform. 5(1), 56–66 (2008)
11. Caprara, A.: Sorting Permutations by Reversals and Eulerian Cycle Decompositions. SIAM J. Discrete Math. 12, 91–110 (1999)
12. Berman, P., Hannenhalli, S., Karpinski, M.: 1.375-Approximation Algorithm for Sorting by Reversals. In: Möhring, R.H., Raman, R. (eds.) ESA 2002. LNCS, vol. 2461, pp. 200–210. Springer, Heidelberg (2002)
13. Jiang, H., Zhu, B., Zhu, D.: Algorithms for sorting unsigned linear genomes by the DCJ operations. Bioinformatics 27(3), 311–316 (2011)
14. Caprara, A., Rizzi, R.: Improved Approximation for Breakpoint Graph Decomposition and Sorting by Reversals. J. Comb. Optim. 6(2), 157–182 (2002)
15. Berman, P., Fürer, M.: Approximating maximum independent set in bounded degree graphs. In: Proceedings of the 5th Annual ACM-SIAM Symposium on Discrete Algorithms (SODA 1994), pp. 365–371 (1994)
16. Halldórsson, M.M.: Approximating discrete collections via local improvements. In: Proceedings of the 6th Annual ACM-SIAM Symposium on Discrete Algorithms (SODA 1995), pp. 160–169 (1995)
17. Chen, X., Sun, R., Yu, J.: Approximating the double-cut-and-join distance between unsigned genomes. BMC Bioinformatics 12(suppl. 9), S17 (2011)

Space-Efficient Approximate String Matching Allowing Inversions in Fast Average Time[*],[**]

Hwee Kim and Yo-Sub Han

Department of Computer Science, Yonsei University
50, Yonsei-Ro, Seodaemun-Gu, Seoul 120-749, Korea
{kimhwee,emmous}@cs.yonsei.ac.kr

Abstract. An inversion is one of the important operations in bio sequence analysis and the sequence alignment problem is well-studied for
efficient bio sequence comparisons. We investigate the string matching
problem allowing inversions: Given a pattern P and a text T, find all
indices of matching substrings of T when non-overlapping inversions are
allowed. We design an $O(nm)$ algorithm using $O(m)$ space, where n is
the size of T and m is the size of P. The proposed algorithm improves
the space complexity of the best-known algorithm, which runs in $O(nm)$
time with $O(m^2)$ space. We, furthermore, improve the algorithm and
achieve $O(\max\{n, \min\{nm, nm^{\frac{5-t}{2}}\}\})$ average runtime for an alphabet
of size t, which is faster than $O(nm)$ when $t \geq 4$.

1 Introduction

In modern biology, it is important to determine exact orders of DNA sequences,
retrieve relevant information of DNA sequences and align these sequences [1,
7, 10, 11]. For a DNA sequence, a *chromosomal translocation* is to relocate a
piece of the DNA sequence from one place to another and, thus, rearrange the
sequence [8]. A *chromosomal inversion* is a reversal and complement of a DNA
sequence. Chromosomal inversion occurs when a single chromosome undergoes
breakage and rearrangement within itself [9]. For instance, consider a string $w =
w_1 w_2 \cdots w_{i-1} w_i w_{i+1} \cdots w_k$.

$$w' = w_1 \cdots \overbrace{w_{i+2} w_{i+3}}^{\text{translocation}} \overbrace{w_{i-1} w_i w_{i+1}}^{} \cdots w_k, \qquad w'' = w_1 \cdots \overbrace{w_{i+2} w_{i+1} w w_{i-1}}^{\text{inversion}} \cdots w_k.$$

Here w' is a translocation of w and w'' is an inversion of w.

Based on these biological events, there is a well-defined string matching problem: given two strings P and T with two operations—translocation or inversion—
the string matching problem is to find all matching substrings of T that match
P under the operations.

---

[*] This research was supported by the Basic Science Research Program through NRF
funded by MEST (2012R1A1A2044562).

[**] Kim was supported by NRF (National Research Foundation of Korea) Grant funded
by the Korean Government (NRF-2013-Global Ph.D. Fellowship Program).

J. Chen, J.E. Hopcroft, and J. Wang (Eds.): FAW 2014, LNCS 8497, pp. 141–150, 2014.

Many researchers investigated efficient algorithms for this problem [1–6, 11, 12]. Note that inversions in strings are not automatically detected by the traditional alignment algorithms [12]. Schöniger and Waterman [11] introduced the alignment problem with non-overlapping inversions and showed a dynamic programming algorithm that computes local alignments with inversions between two strings of length n and m in $O(n^6)$, where $n \geq m$. Vellozo et al. [12] presented an improved $O(n^2 m)$ algorithm. They built a matrix for one string and partially inverted the other string using table filling method in the extended edit graph. Recently, Cantone et al. [1] introduced an $O(nm)$ algorithm using $O(m^2)$ space for the string matching problem, which is to find all locations of a pattern P of length m with respect to a text T of length n based on non-overlapping inversions. Cho et al. [4] introduced an $O(n^3)$ algorithm using $O(n^2)$ space that solves the alignment problem allowing inversions to two strings of length n.

We consider the same problem that Cantone et al. [1] examined and tackle the problem by taking a different approach based on the relationship between substrings generated by inversions. We start from designing an algorithm for two strings of the same length. Based on the algorithm, we design an $O(m^2)$ algorithm using $O(m^2)$ space that checks the existence of the matching at a given index of T with respect to P. We, furthermore, modify the algorithm to process m indices simultaneously and present a new algorithm that finds all matching substrings of T in $O(nm)$ time using $O(m)$ space, which uses less amount of space compared with the best-known algorithm—$O(nm)$ time and $O(m^2)$ space [1]. Finally, based on the frequency of character appearances in a random string, we prove that our algorithm can achieve $O(\max\{n, \min\{nm, nm^{\frac{5-t}{2}}\}\})$ average runtime for an alphabet of size t by adding a permutation match filter suggested by Grabowski et al. [5]. For a DNA pattern, our algorithm runs in $O(n\sqrt{(m)})$, which outperforms the previous algorithm. Moreover, for $t \geq 5$, our algorithm runs in $O(n)$ time.

2   Preliminaries

Let $A[a_1][a_2] \cdots [a_n]$ be an n-dimensional array, where the size of each dimension is a_i for $1 \leq i \leq n$. Let $A[i_1][i_2] \cdots [i_n]$ be the element of A at an index $(i_1, i_2, \ldots, i_n)$. Given a finite set Σ of characters and a string w over Σ, let $|w|$ be the length of w and $w[i]$ be the symbol of w at position i, for $1 \leq i \leq |w|$. We use $w[i : j]$ to denote a substring $w[i]w[i+1] \cdots w[j]$, where $1 \leq i \leq j \leq |w|$.

Let the symbol θ denote inversion[1]. Then, $\theta(w)$ is the inversion of w. Given a range (i, j), we define the inversion operation $\theta_{(i,j)}(w)$ to be $\theta(w[i : j])$, the inversion of the substring $w[i : j]$. We say that $\theta_{(i,j)}$ *yields* the string $\theta_{(i,j)}(w)$ from w. The *size* of $\theta_{(i,j)}$ is $|\theta_{(i,j)}| = j - i + 1$. When the context is clear, we denote $\theta_{(i,j)}$ as (i, j).

---

[1] In biology, inversion is a composition of a reversal operation and a complement operation. However, inversion is often regarded as reversal in the string matching and alignment literature for the simplicity of analysis. We also follow this convention.

We define a sequence $\Theta = ((p_1, q_1), (p_2, q_2), \ldots, (p_k, q_k))$ of inversions for a string w to be non-overlapping if it satisfies the following conditions: For $1 \leq i \leq k$, $p_1 = 1$, $q_k = |w|$, $p_i \leq q_i$ and $p_{i+1} > q_i$ for $1 \leq i \leq k - 1$. Since $\theta(w) = w$ if $|w| = 1$, we can assume that all positions of w are within one of the ranges in a sequence of non-overlapping inversions; For example, given a string of size 6, two sequences of non-overlapping inversions $((1, 2), (5, 6))$ and $((1, 2), (3, 3), (4, 4), (5, 6))$ result in the same string. Therefore, we further assume that $p_{i+1} = q_i + 1$ for $1 \leq i \leq k - 1$. When the context is clear, we call a sequence of non-overlapping inversions just a sequence of inversions.

Now, in summary, given a sequence $\Theta = ((p_1, q_1), (p_2, q_2), \ldots, (p_k, q_k))$ of inversions and a string w,

$$\Theta(w) = \Theta[1](w)\Theta[2](w) \cdots \Theta[k](w).$$

Definition 1 (Problem). *Given a text T and a pattern P, we define the approximate string matching problem allowing inversions to be finding all indices i of T that has a sequence Θ_i of inversion satisfying $\Theta_i(P) = T[i : i+m-1]$.*

3   The Algorithm

We tackle the approximate string matching problem allowing inversions using $\theta(P)$—the inversion of P. We start with a special case when $|P| = |T|$ and extend the idea to the general case when $|P| \leq |T|$. For example, suppose that $P = AGTCTAG$ and $T = TGACATG$. A sequence $\Theta = ((1, 3), (4, 4), (5, 6), (7, 7))$ of inversions makes $\Theta(P) = T$. The sequence of $\Theta[i](P)$ can be represented on $\theta(P)$ in the reverse order; the sequence (TGA, C, AT, G) of strings yielded by Θ appears as the reversed sequence (G, AT, C, TGA) in $\theta(P)$.

Observation 1 (Relationship between $\Theta(P)$ and $\theta(P)$). *Given a sequence Θ of inversions and a pattern P, $\theta(P) = \Theta[k](P)\Theta[k-1](P) \cdots \Theta[1](P)$, where $k = |\Theta|$.*

Definition 2 (Successfully comparable substring and paired string). *Based on Observation 1, we say that $T[i : i+l-1]$ is a successfully comparable substring (SCS in short) for the given index i, given length l, strings T and $\theta(P)$ of the same length m if $\theta(P)_{(m+2-i-l, m+1-i)} = T[i : i+l-1]$. Moreover, we call the left-handed side of the equality a paired string of a successfully comparable substring (PSCS in short).*

Definition 3 (Sequence of successfully comparable substrings). *We define a sequence $\mathcal{S}_{(a,b)}$ to be a sequence of successfully comparable substrings for the given range (a, b), strings T and $\theta(P)$, where $\mathcal{S}_{(a,b)}[j] = T[i_j : i_j+l_j-1]$ is an SCS for T and $\theta(P)$, $i_1 = a$, $i_{|\mathcal{S}_{(a,b)}|} + t_{|\mathcal{S}_{(a,b)}|} - 1 = b$ and $i_{j+1} = i_j + l_j$ for $1 \leq j < |\mathcal{S}_{(a,b)}|$.*

It is clear that if $\mathcal{S}_{(1,m)}$ exists for T and $\theta(P)$ such that $|T| = |P| = m$, then there exists Θ satisfying $\Theta(P) = T$. We design a simple algorithm that retrieves $\mathcal{S}_{(1,m)}$ from T and $\theta(P)$ as described in Algorithm 1. Fig. 1 is an example of the algorithm.

Algorithm 1.

Input: Strings T and $\theta(P)$ of length m
Output: Sequence $\mathcal{S}_{(1,m)}$ of SCSs
1 $i \leftarrow 1, l \leftarrow 1$
2 **while** $i \leq m$ **do**
3   **if** $T[i : i+l-1]$ *is an SCS* **then**
4   add $T[i : i+l-1]$ to $\mathcal{S}_{(1,m)}$
5   **if** $i+l-1 = m$ **then return** $\mathcal{S}_{(1,m)}$ $i \leftarrow i + l, l \leftarrow 1$
6   **else** $l \leftarrow l + 1$

7 **return** *"no $\mathcal{S}_{(1,m)}$ exists"*

Lemma 1. *It takes $O(m^2)$ time and $O(m)$ space to retrieve $\mathcal{S}_{(1,m)}$ from T and $\theta(P)$, where $|T| = |P| = m$.*

Note that the algorithm always compares substrings with increasing order of length and chooses the shortest SCS. It may seem possible that, at some index, the algorithm chooses a shorter SCS where there exists a longer SCS that should be chosen for $\mathcal{S}_{(1,m)}$. We prove that such a case is impossible.

Theorem 2. *For given strings T and $\theta(P)$ of length m, given an index i and given lengths l_1 and l_2 where $l_1 < l_2$, if $T[i : i+l_1-1]$ and $T[i : i+l_2-1]$ are SCSs, there always exists a sequence $\mathcal{S}_{(i,i+l_2-1)}$ of SCSs such that $\mathcal{S}_{(i,i+l_2-1)}[1] = T[i : i+l_1-1]$. Namely, if there exist two SCSs from an index i, then the longer string can always be expressed as a sequence of SCSs that starts with the shorter string.*

For all cases, at some index, if there exist both a shorter and a longer SCSs, then the longer string is always decomposed to a sequence of SCSs starting with the shorter string. Now we can adapt the algorithm for the approximate pattern matching problem.

As a running example for describing the algorithm for the approximate pattern matching, we use $P = GTTAG$ and $T = TGTGATTG$. For each index i, we start from building a table $R_P^i[m][m]$, where

$$R_P^i[j][k] = \begin{cases} R_P^i[j-1][k-1] + 1 & \text{if } T[i+k-1] = P[j], \\ 0 & \text{otherwise.} \end{cases} \tag{1}$$

The role of R_P^i is as follows: we regard each diagonal in R_P^i as $\theta(P)$, and calculate the length of the longest SCS that ends at each index of $\theta(P)$.

The process of finding $\mathcal{S}_{(1,m)}$ for each index of T is described in Algorithm 2.

We claim that Algorithm 2 returns a correct answer for the approximate string matching problem for an index i and establish the following result:

Lemma 2. *Algorithm 2 returns true if and only if there exists a sequence $\mathcal{S}_{(1,m)}$ of SCSs for strings P and $T[i : i+m-1]$.*

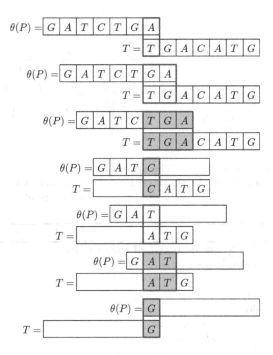

Fig. 1. Comparing $\theta(P)$ and T. Red boxes represent comparing lengths. SCSs (and their PSCSs in $\theta(P)$) are erased for better readability.

Algorithm 2 shows the existence of a sequence of SCSs for an index i and requires $O(m^2)$ space. Next, we show how to reduce the space requirement by relying on the properties of R_P^i.

Observation 3 (Properties of R_P^i). *For an index i, R_P^i has the following properties:*

1. $R_P^i[j][k] = R_P^{i+1}[j][k-1]$ *for all i, j, k.*
2. *The value of $R_P^i[j][k]$ is computed from $R_P^i[j-1][k-1]$ and $T[i+k-1]$ (from Equation (1)).*

The first property of Observation 3 ensures that there exists a table $R_P[m][n]$, where $R_P^i[j][k] = R_P[j][i+k-1]$ for all i, j, k. In other words, there exists a common table from which we can retrieve every $R_P^i[j][k]$ value. The second property of Observation 3 ensures that we need only the $(i+k-2)$th column of R_P and $T[i+k-1]$ to construct the $(i+k-1)$th column of R_P. Based on these properties, we design an improved algorithm that requires $O(m)$ space instead of $O(m^2)$.

In Algorithm 3, $R[m][2]$ is the table that keeps the $(i-1)$th and ith columns of R_P, and p_i is a pointer indicating the corresponding investigating cell for an index i. Note that line 8 to 9 of Algorithm 3 is same as running Algorithm 2

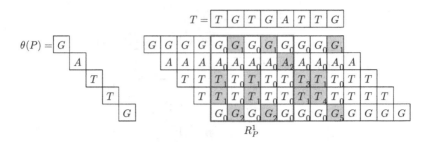

Fig. 2. An example of building R_P^1. The symbol in each cell represents the character in $\theta(P)$ which is compared to the character in T. The number at the bottom right of each cell represents the length of the longest SCS, which is the element in R_P^1.

Algorithm 2.

Input: Pattern P of length m, text T of length n, index i
Output: $\mathcal{S}_{(1,m)}$ for an index i

1 $j \leftarrow m$
2 construct $R_P^i[m][m]$
3 **for** $k \leftarrow 1$ **to** m **do**
4   **if** $R_P^i[j][k] \geq j + k - m$ **then**
5   add $T[i{+}m{-}j : i{+}k{-}1]$ to $\mathcal{S}_{(1,m)}$
6          $j \leftarrow m - k$

7 **if** $j = 0$ **then return** $\mathcal{S}_{(1,m)}$ **else return** *"no $\mathcal{S}_{(1,m)}$ exists"*

---

on m indices simultaneously to retrieve indices that match P. We show that Algorithm 3 runs in $O(nm)$ time using $O(m)$ space.

Lemma 3. *Algorithm 3 runs in $O(nm)$ time using $O(m)$ space, where $m = |P|$ and $n = |T|$.*

From Lemmas 2 and 3, we know that Algorithm 3 returns a correct answer for our problem in Definition 1. We establish the following statement for our problem:

Theorem 4. *We can solve an approximate string matching problem for inversions in Definition 1 in $O(nm)$ time using $O(m)$ space, where m is the size of the pattern and n is the size of the text.*

Note that Algorithm 3 improves the space complexity compared with the previous algorithm by Cantone et al. [1] while keeping the same runtime.

4 Average Runtime Analysis

Now we consider the frequency of character appearances in T and improve the average runtime when the alphabet size $|\Sigma| = t$ is larger than 3. The improved

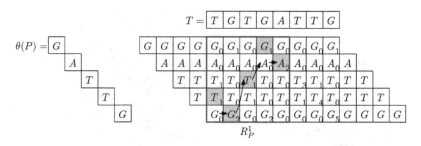

Fig. 3. Searching the existence of $\mathcal{S}_{(1,m)}$ for the first index. $\mathcal{S}_{(1,m)} = \{TG, T, GA\}$. The algorithm actually finds the ending cells of all PSCSs. Arrows in R_P^1 shows the track of inspection cells.

Algorithm 3.

Input: Pattern P of length m, text T of length n
Output: every index i where $\mathcal{S}_{(1,m)}$ exists

1   build $R[m][2]$.
2   initialize $R[j][1]$ as 0 for all j.
3   **for** $i \leftarrow 1$ **to** n **do**
4   **for** $j \leftarrow 1$ **to** m **do**
5        $\lfloor$ **if** $T[i] = P[j]$ **then** $R[j][2] \leftarrow R[j-1][1] + 1$ **else** $R[j][2] \leftarrow 0$
6      $p_i = m$
7   **for** $j \leftarrow 1$ **to** m **do**
8        $\lfloor$ **if** $R[p_{i-m+j}][2] \geq p_{i-m+j} - j + 1$ **then** $p_{i-m+j} \leftarrow j - 1$
9      $\lfloor$ **if** $p_{i-m+1} = 0$ **then return** $i - m + 1$ $R[j][1] = R[j][2]$ for all j.

---

algorithm runs in $O(n\sqrt{m})$ when $t = 4$, which is the case of a DNA or RNA pattern. Moreover, the algorithm runs in linear time when $t \geq 5$.

Observation 5. *If there exists a sequence Θ_i of inversions at an index i such that $\Theta_i(P) = T[i : i+m-1]$, then $T[i : i+m-1]$ is a permutation of P.*

Based on Observation 5, we add a filter module that checks whether or not $T[i : i+m-1]$ is a permutation of P at an index i before line 4 of Algorithm 3.

The following probability analysis is from Grabowski et al. [5]. Suppose $\Sigma = \{\sigma_1, \sigma_2, \ldots, \sigma_t\}$ and $|\Sigma| = t$. Without loss of generality, assume that m is divisible by t and $k = \frac{m}{t}$. It requires $O(n)$ time and $O(t)$ space to return all i such that $T[i : i + m - 1]$ is a permutation of P [5]. Assume that P and T are random strings in which each character has $\frac{1}{t}$ occurrence probability for each position. Let $Pr\{\pi\}$ be the probability that the substring $T[i : i+m-1]$ is a permutation of P. Then we have

$$Pr\{\pi\} \leq \frac{t^{\frac{t}{2}}}{m^{\frac{t-1}{2}}}. \qquad (2)$$

Based on Observation 5, instead of examining all substrings of size m of T, we only need to consider the substrings of size m of T that are permutations

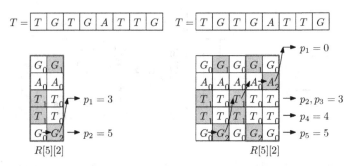

Fig. 4. An example of Algorithm 3 for $i = 2$ and $i = 5$. Since $p_1 = 0$ for $i = 5$, there exists a matching for $i = 1$.

of P. Suppose line 4 to 9 of Algorithm 3 takes cm number of calculations. Then, with $\Pr\{\pi\}$—the probability that $T[i : i+m-1]$ is a permutation of P—in Equation (2) and $(n-m+1)$ substrings of T, we have an upper bound of average runtime

$$AUB_1 = \frac{t^{\frac{t}{2}}}{m^{\frac{t-1}{2}}} \times (n - m + 1) \times cm^2 = O(nm^{\frac{5-t}{2}}).$$

AUB_1 is calculated under the assumption that we run the non-overlapping inversion matching algorithm to each permutation-matched index independently, similar to the average runtime analysis in [5]. Now we compute a smaller upper bound of the average runtime by tightly analyzing Algorithm 3.

In Algorithm 3, line 4 to 9, which takes cm, is repeated from index i to $i+m-1$ to check the non-overlapping inversion matching for i. Therefore, if we apply the permutation filter to Algorithm 3, cm is required for an index i if one of the indices from $i - m + 1$ to i has a permutation matching. The probability that one of the indices from $i - m + 1$ to i has such a matching is

$$1 - \left(1 - \frac{t^{\frac{t}{2}}}{m^{\frac{t-1}{2}}}\right)^m.$$

Therefore, we have another upper bound of average runtime

$$AUB_2 = \left(1 - \left(1 - \frac{t^{\frac{t}{2}}}{m^{\frac{t-1}{2}}}\right)^m\right) \times cm \times (n - m + 1) = O(nm^{\frac{5-t}{2}}). \quad (3)$$

Fig. 5 compares the upper bound calculations for AUB_1 and AUB_2. From the figure, we know that $AUB_2 \leq AUB_1$. Though the order of AUB_1 and AUB_2 are the same, we retrieved an upper bound of average runtime smaller than AUB_1.

Note that our average runtime in Equation (3) becomes faster than $O(nm)$ when $t \geq 4$ and becomes sublinear when $t \geq 6$. Combined with the runtime of the permutation filter [5], we establish the following statement for our problem:

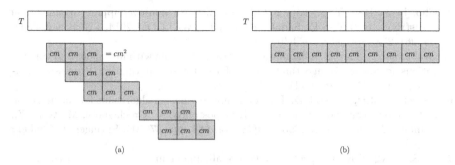

Fig. 5. Upper bound analysis of number of calculations for (a) AUB_1 and (b) AUB_2 for $m = 3$. Colored boxes in T represent the indices where permutation matching occurs.

Theorem 6. *We can solve an approximate string matching problem for inversions in Definition 1 in $O(\max\{n, \min\{nm, nm^{\frac{5-t}{2}}\}\})$ average runtime using $O(m)$ space, where m is the size of P, n is the size of T and t is the size of Σ.*

Theorem 6 guarantees a faster runtime—$O(n\sqrt{m})$—for a DNA string over $\Sigma = \{C, G, T, A\}$ and $t = 4$. Furthermore, for $t \geq 5$, the algorithm shows a linear runtime.

5   Conclusions

An inversion is an important operation for bio sequences such as DNA or RNA and is closely related to mutations. We have examined the string matching problem allowing inversions. Given a text T, a pattern P where $|P| = m \leq n = |T|$, our algorithm finds all indices i in which there is a matching alignment between P and $T[i : i+m-1]$ in $O(nm)$ time using $O(m)$ space. Moreover, we improve the algorithm and achieve $O(\max\{n, \min\{nm, nm^{\frac{5-t}{2}}\}\})$ average runtime using $O(m)$ space for an alphabet of size t. Compared with the previous algorithm, the new algorithm improves the space complexity, and shows a better average runtime for $t \geq 4$. A possible future direction is to consider multiple patterns instead of a single pattern for the pattern matching problem.

References

1. Cantone, D., Cristofaro, S., Faro, S.: Efficient string-matching allowing for non-overlapping inversions. Theoretical Computer Science 483, 85–95 (2013)
2. Cantone, D., Faro, S., Giaquinta, E.: Approximate string matching allowing for inversions and translocations. In: Proceedings of the Prague Stringology Conference 2010, pp. 37–51 (2010)
3. Chen, Z.-Z., Gao, Y., Lin, G., Niewiadomski, R., Wang, Y., Wu, J.: A space-efficient algorithm for sequence alignment with inversions and reversals. Theoretical Computer Science 325(3), 361–372 (2004)

4. Cho, D.-J., Han, Y.-S., Kim, H.: Alignment with non-overlapping inversions on two strings. In: Pal, S.P., Sadakane, K. (eds.) WALCOM 2014. LNCS, vol. 8344, pp. 261–272. Springer, Heidelberg (2014)
5. Grabowski, S., Faro, S., Giaquinta, E.: String matching with inversions and translocations in linear average time (most of the time). Information Processing Letters 111(11), 516–520 (2011)
6. Kececioglu, J.D., Sankoff, D.: Exact and approximation algorithms for the inversion distance between two chromosomes. In: Apostolico, A., Crochemore, M., Galil, Z., Manber, U. (eds.) CPM 1993. LNCS, vol. 684, pp. 87–105. Springer, Heidelberg (1993)
7. Li, S.C., Ng, Y.K.: On protein structure alignment under distance constraint. In: Dong, Y., Du, D.-Z., Ibarra, O. (eds.) ISAAC 2009. LNCS, vol. 5878, pp. 65–76. Springer, Heidelberg (2009)
8. Ogilvie, C.M., Scriven, P.N.: Meiotic outcomes in reciprocal translocation carriers ascertained in 3-day human embryos. European Journal of Human Genetics 10(12), 801–806 (2009)
9. Painter, T.S.: A New Method for the Study of Chromosome Rearrangements and the Plotting of Chromosome Maps. Science 78, 585–586 (1933)
10. Sakai, Y.: A new algorithm for the characteristic string problem under loose similarity criteria. In: Asano, T., Nakano, S.-i., Okamoto, Y., Watanabe, O. (eds.) ISAAC 2011. LNCS, vol. 7074, pp. 663–672. Springer, Heidelberg (2011)
11. Schniger, M., Waterman, M.S.: A local algorithm for DNA sequence alignment with inversions. Bulletin of Mathematical Biology 54(4), 521–536 (1992)
12. Vellozo, A.F., Alves, C.E.R., do Lago, A.P.: Alignment with non-overlapping inversions in $O(n^3)$-time. In: Bücher, P., Moret, B.M.E. (eds.) WABI 2006. LNCS (LNBI), vol. 4175, pp. 186–196. Springer, Heidelberg (2006)

Minimal Double Dominating Sets in Trees

Marcin Krzywkowski[*],[**]

marcin.krzywkowski@gmail.com

Abstract. We provide an algorithm for listing all minimal double dominating sets of a tree of order n in time $\mathcal{O}(1.3248^n)$. This implies that every tree has at most 1.3248^n minimal double dominating sets. We also show that this bound is tight.

Keywords: domination, double domination, minimal double dominating set, tree, combinatorial bound, exact exponential algorithm, listing algorithm.

1 Introduction

Let $G = (V, E)$ be a graph. The order of a graph is the number of its vertices. By the neighborhood of a vertex v of G we mean the set $N_G(v) = \{u \in V(G) : uv \in E(G)\}$. The degree of a vertex v, denoted by $d_G(v)$, is the cardinality of its neighborhood. By a leaf we mean a vertex of degree one, while a support vertex is a vertex adjacent to a leaf. We say that a support vertex is strong (weak, respectively) if it is adjacent to at least two leaves (exactly one leaf, respectively). The distance between two vertices of a graph is the number of edges in a shortest path connecting them. The eccentricity of a vertex is the greatest distance between it and any other vertex. The diameter of a graph G, denoted by $\mathrm{diam}(G)$, is the maximum eccentricity among all vertices of G. A path on n vertices we denote by P_n.

A vertex of a graph is said to dominate itself and all of its neighbors. A subset $D \subseteq V(G)$ is a dominating set of G if every vertex of G is dominated by at least one vertex of D, while it is a double dominating set of G if every vertex of G is dominated by at least two vertices of D. A dominating (double dominating, respectively) set D is minimal if no proper subset of D is a dominating (double dominating, respectively) set of G. A minimal double dominating set is abbreviated as mdds. Double domination in graphs was introduced by Harary and Haynes [6]. For a comprehensive survey of domination in graphs, see [7,8].

Observation 1. *Every leaf of a graph G is in every DDS of G.*

---

[*] Research fellow at the Department of Mathematics, University of Johannesburg, South Africa.

[**] Faculty of Electronics, Telecommunications and Informatics, Gdansk University of Technology, Poland. Research partially supported by the Polish National Science Centre grant 2011/02/A/ST6/00201.

J. Chen, J.E. Hopcroft, and J. Wang (Eds.): FAW 2014, LNCS 8497, pp. 151–157, 2014.

Observation 2. *Every support vertex of a graph G is in every DDS of G.*

One of the typical questions in graph theory is how many subgraphs of a given property a graph on n vertices can have. For example, the famous Moon and Moser theorem [12] says that every graph on n vertices has at most $3^{n/3}$ maximal independent sets.

Combinatorial bounds are of interest not only on their own, but also because they are used for algorithm design as well. Lawler [11] used the Moon-Moser bound on the number of maximal independent sets to construct an $(1 + \sqrt[3]{3})^n \cdot n^{O(1)}$ time graph coloring algorithm, which was the fastest one known for twenty-five years. For an overview of the field, see [5].

Fomin et al. [4] constructed an algorithm for listing all minimal dominating sets of a graph on n vertices in time $\mathcal{O}(1.7159^n)$. They also presented graphs ($n/6$ disjoint copies of the octahedron) having $15^{n/6} \approx 1.5704^n$ minimal dominating sets. This establishes a lower bound on the running time of an algorithm for listing all minimal dominating sets of a given graph.

The number of maximal independent sets in trees was investigated in [13]. Couturier et al. [3] considered minimal dominating sets in various classes of graphs. The authors of [9] investigated the enumeration of minimal dominating sets in graphs.

Bród and Skupień [1] gave bounds on the number of dominating sets of a tree. They also characterized the extremal trees. The authors of [2] investigated the number of minimal dominating sets in trees containing all leaves.

In [10] an algorithm was given for listing all minimal dominating sets of a tree of order n in time $\mathcal{O}(1.4656^n)$. This implies that every tree has at most 1.4656^n minimal dominating sets. An infinite family of trees for which the number of minimal dominating sets exceeds 1.4167^n was also given. This establishes a lower bound on the running time of an algorithm for listing all minimal dominating sets of a given tree.

We provide an algorithm for listing all minimal double dominating sets of a tree of order n in time $\mathcal{O}(1.3248^n)$. This implies that every tree has at most 1.3248^n minimal double dominating sets. We also show that this bound is tight.

2 Results

We describe a recursive algorithm which lists all minimal double dominating sets of a given input tree. We prove that the running time of this algorithm is $\mathcal{O}(1.3248^n)$, implying that every tree has at most 1.3248^n minimal double dominating sets.

Theorem 3. *Every tree T of order n has at most α^n minimal double dominating sets, where $\alpha \approx 1.32472$ is the positive solution of the equation $x^3 - x - 1 = 0$, and all those sets can be listed in time $\mathcal{O}(1.3248^n)$.*

Proof. The family of sets returned by our algorithm is denoted by $\mathcal{F}(T)$. To obtain the upper bound on the number of minimal double dominating sets of a tree, we prove that the algorithm lists these sets in time $\mathcal{O}(1.3248^n)$. If $\text{diam}(T) \leq 3$, then let $\mathcal{F}(T) = \{V(T)\}$. Every vertex of T is a leaf or a support vertex. Observations 1 and 2 imply that $V(T)$ is the only mdds of T. We have $n \geq 2$ and $|\mathcal{F}(T)| = 1$. Obviously, $1 < \alpha^n$.

Now assume that $\text{diam}(T) \geq 4$. Thus the order n of the tree T is at least five. The results we obtain by the induction on the number n. Assume that they are true for every tree T' of order $n' < n$.

First assume that some support vertex of T, say x, is strong. Let y and z be leaves adjacent to x. Let $T' = T - y$, and let

$$\mathcal{F}(T) = \{D' \cup \{y\}: D' \in \mathcal{F}(T')\}.$$

Let D' be an mdds of the tree T'. By Observation 2 we have $x \in D'$. It is easy to see that $D' \cup \{y\}$ is an mdds of T. Thus all elements of the family $\mathcal{F}(T)$ are minimal double dominating sets of the tree T. Now let D be any mdds of the tree T. By Observations 1 and 2 we have $x, y, z \in D$. Let us observe that $D \setminus \{y\}$ is an mdds of the tree T' as the vertex x is still dominated at least twice. By the inductive hypothesis we have $D \setminus \{y\} \in \mathcal{F}(T')$. Therefore the family $\mathcal{F}(T)$ contains all minimal double dominating sets of the tree T. We now get $|\mathcal{F}(T)| = |\mathcal{F}(T')| \leq \alpha^{n-1} < \alpha^n$. Henceforth, we can assume that every support vertex of T is weak.

We now root T at a vertex r of maximum eccentricity $\text{diam}(T)$. Let t be a leaf at maximum distance from r, v be the parent of t, u be the parent of v, and w be the parent of u in the rooted tree. If $\text{diam}(T) \geq 5$, then let d be the parent of w. By T_x we denote the subtree induced by a vertex x and its descendants in the rooted tree T.

Assume that u is adjacent to a leaf, say x. Let $T' = T - T_v$, and let

$$\mathcal{F}(T) = \{D' \cup \{v, t\}: D' \in \mathcal{F}(T')\}.$$

Let us observe that all elements of the family $\mathcal{F}(T)$ are minimal double dominating sets of the tree T. Now let D be any mdds of the tree T. By Observations 1 and 2 we have $t, x, v, u \in D$. It is easy to observe that $D \setminus \{v, t\}$ is an mdds of the tree T'. By the inductive hypothesis we have $D \setminus \{v, t\} \in \mathcal{F}(T')$. Therefore the family $\mathcal{F}(T)$ contains all minimal double dominating sets of the tree T. We now get $|\mathcal{F}(T)| = |\mathcal{F}(T')| \leq \alpha^{n-2} < \alpha^n$.

Now assume that all children of u are support vertices. Assume that $d_T(u) \geq 4$. Let $T' = T - T_v$, and let

$$\mathcal{F}(T) = \{D' \cup \{v, t\}: D' \in \mathcal{F}(T')\}.$$

Let us observe that all elements of the family $\mathcal{F}(T)$ are minimal double dominating sets of the tree T. Now let D be any mdds of the tree T. By Observations 1 and 2 we have $v, t \in D$. Let us observe that $D \setminus \{v, t\}$ is an mdds of the tree T' as the vertex u is still dominated at least twice. By the inductive hypothesis we have $D \setminus \{v, t\} \in \mathcal{F}(T')$. Therefore the family $\mathcal{F}(T)$ contains all minimal double dominating sets of the tree T. We now get $|\mathcal{F}(T)| = |\mathcal{F}(T')| \leq \alpha^{n-2} < \alpha^n$.

Now assume that $d_T(u) = 3$. Let x be the child of u other than v. The leaf adjacent to x we denote by y. Let $T' = T - T_u$ and $T'' = T - T_v - y$. Let $\mathcal{F}(T)$ be a family as follows,

$$\{D' \cup \{t, v, x, y\} : D' \in \mathcal{F}(T')\}$$
$$\cup \{D'' \cup \{v, t, y\} : D'' \in \mathcal{F}(T'') \text{ and } D'' \setminus \{u, x\} \notin \mathcal{F}(T')\}.$$

Let us observe that all elements of the family $\mathcal{F}(T)$ are minimal double dominating sets of the tree T. Now let D be any mdds of the tree T. By Observations 1 and 2 we have $v, t, x, y \in D$. If $u \notin D$, then observe that $D \setminus \{v, t, x, y\}$ is an mdds of the tree T'. By the inductive hypothesis we have $D \setminus \{v, t, x, y\} \in \mathcal{F}(T')$. Now assume that $u \in D$. It is easy to observe that $D \setminus \{v, t, y\}$ is an mdds of the tree T''. By the inductive hypothesis we have $D \setminus \{v, t, y\} \in \mathcal{F}(T'')$. Let us observe that $D \setminus \{u, v, t, x, y\}$ is not a double dominating set of the tree T', otherwise $D \setminus \{u\}$ is a double dominating set of the tree T, a contradiction to the minimality of D. Therefore the family $\mathcal{F}(T)$ contains all minimal double dominating sets of the tree T. We now get $|\mathcal{F}(T)| = |\mathcal{F}(T')| + |\{D'' \in \mathcal{F}(T'') : D'' \setminus \{u, x\} \notin \mathcal{F}(T')\}| \leq |\mathcal{F}(T')| + |\mathcal{F}(T'')| \leq \alpha^{n-5} + \alpha^{n-3} = \alpha^{n-5}(\alpha^2 + 1) < \alpha^{n-5} \cdot \alpha^5 = \alpha^n$.

Now assume that $d_T(u) = 2$. Assume that $d_T(w) \geq 3$. First assume that w is adjacent to a leaf, say k. Let $T' = T - T_u$, and let

$$\mathcal{F}(T) = \{D' \cup \{v, t\} : D' \in \mathcal{F}(T')\}.$$

Let us observe that all elements of the family $\mathcal{F}(T)$ are minimal double dominating sets of the tree T. Now let D be any mdds of the tree T. By Observations 1 and 2 we have $v, t, w, k \in D$. We have $u \notin D$ as the set D is minimal. Observe that $D \setminus \{v, t\}$ is an mdds of the tree T'. By the inductive hypothesis we have $D \setminus \{v, t\} \in \mathcal{F}(T')$. Therefore the family $\mathcal{F}(T)$ contains all minimal double dominating sets of the tree T. We now get $|\mathcal{F}(T)| = |\mathcal{F}(T')| \leq \alpha^{n-3} < \alpha^n$.

Now assume that there is a child of w, say k, such that the distance of w to the most distant vertex of T_k is two. Thus k is a support vertex of degree two. The leaf adjacent to k we denote by l. Let $T' = T - T_u - l$ and $T'' = T - T_w$. Let

$$\mathcal{F}(T) = \{D' \cup \{v, t, l\} : D' \in \mathcal{F}(T')\} \cup \{D'' \cup V(T_w) \setminus \{w\} : D'' \in \mathcal{F}(T'')\}.$$

Let us observe that all elements of the family $\mathcal{F}(T)$ are minimal double dominating sets of the tree T. Now let D be any mdds of the tree T. By Observations 1 and 2 we have $v, t, k, l \in D$. If $u \notin D$, then $w \in D$ as the vertex u has to be dominated twice. It is easy to observe that $D \setminus \{v, t, l\}$ is an mdds of the tree T'. By the inductive hypothesis we have $D \setminus \{v, t, l\} \in \mathcal{F}(T')$. Now assume that $u \in D$. We have $w \notin D$, otherwise $D \setminus \{u\}$ is a double dominating set of the tree T, a contradiction to the minimality of D. Observe that $D \cap V(T'')$ is an mdds of the tree T''. By the inductive hypothesis we have $D \cap V(T'') \in \mathcal{F}(T'')$. Therefore the family $\mathcal{F}(T)$ contains all minimal double dominating sets of the tree T. We now get $|\mathcal{F}(T)| = |\mathcal{F}(T')| + |\mathcal{F}(T'')| \leq \alpha^{n-4} + \alpha^{n-6} = \alpha^{n-6}(\alpha^2 + 1) < \alpha^{n-6} \cdot \alpha^6 = \alpha^n$.

Now assume that for every child of w, say k, the distance of w to the most distant vertex of T_k is three. Due to the earlier analysis of the degree of the vertex u, which is a child of w, it suffices to consider only the possibility when T_k is a path P_3. Let $T' = T - T_w$. Let T'' (T''', respectively) be a tree that differs from T' only in that it has the vertex w (the vertices w and u, respectively). Let $\mathcal{F}(T)$ be a family as follows,

$$\{D' \cup V(T_w) \setminus \{w\}: D' \in \mathcal{F}(T')\}$$
$$\cup \{D'' \cup V(T_w) \setminus (N_T(w) \setminus \{d\}): D'' \in \mathcal{F}(T'')\}$$
$$\cup \{D''' \cup V(T_w) \setminus (N_T(w) \setminus \{x\}): d \notin D''' \in \mathcal{F}(T'') \text{ and } x \in N_T(w) \setminus \{d\}\}.$$

Let us observe that all elements of the family $\mathcal{F}(T)$ are minimal double dominating sets of the tree T. Now let D be any mdds of the tree T. If $w \notin D$, then observe that $D \cap V(T')$ is an mdds of the tree T'. By the inductive hypothesis we have $D \cap V(T') \in \mathcal{F}(T')$. Now assume that $w \in D$. If no child of w belongs to the set D, then observe that $D \cap V(T'')$ is an mdds of the tree T''. By the inductive hypothesis we have $D \cap V(T'') \in \mathcal{F}(T'')$. Now assume that some child of w, say x, belongs to the set D. Let us observe that $(D \cup \{u\}) \cap V(T''')$ is an mdds of the tree T'''. By the inductive hypothesis we have $(D \cup \{u\}) \cap V(T''') \in \mathcal{F}(T''')$. Therefore the family $\mathcal{F}(T)$ contains all minimal double dominating sets of the tree T. We now get $|\mathcal{F}(T)| = |\mathcal{F}(T')| + |\mathcal{F}(T'')| + (d_T(w) - 1) \cdot |\{D''' \in \mathcal{F}(T'''): d \notin D'''\}|$ $\leq |\mathcal{F}(T')| + |\mathcal{F}(T'')| + (d_T(w) - 1) \cdot |\mathcal{F}(T''')| \leq \alpha^{n-3d_T(w)+2} + \alpha^{n-3d_T(w)+3}$ $+ (d_T(w) - 1) \cdot \alpha^{n-3d_T(w)+4}$. To show that $\alpha^{n-3d_T(w)+2} + \alpha^{n-3d_T(w)+3} + (d_T(w) - 1) \cdot \alpha^{n-3d_T(w)+4} < \alpha^n$, it suffices to show that $\alpha^2 + \alpha^3 + (d_T(w) - 1) \cdot \alpha^4 < \alpha^{3d_T(w)}$. We prove this by the induction on the degree of the vertex w. For $d_T(w) = 3$ we have $\alpha^2 + \alpha^3 + (d_T(w) - 1) \cdot \alpha^4 = 2\alpha^4 + \alpha^3 + \alpha^2 = 2\alpha^4 + \alpha^2(\alpha + 1)$ $= 2\alpha^4 + \alpha^5 = \alpha^4(\alpha + 1) + \alpha^4 = \alpha^7 + \alpha^4 = \alpha^6(\alpha^3 - 1) + \alpha^4 = \alpha^9 + \alpha^4 - \alpha^6$ $< \alpha^9 = \alpha^{3d_T(w)}$. We now prove that if the inequality $\alpha^2 + \alpha^3 + (k-1) \cdot \alpha^4 < \alpha^{3k}$ is satisfied for an integer $k = d_T(w) \geq 3$, then it is also satisfied for $k + 1$. We have $\alpha^2 + \alpha^3 + k\alpha^4 = \alpha^2 + \alpha^3 + (k-1) \cdot \alpha^4 + \alpha^4 < \alpha^{3k} + \alpha^4 < \alpha^{3k} + \alpha^{3k+1} = \alpha^{3k+3}$.

Now assume that $d_T(w) = 2$. If $d_T(d) = 1$, then let $\mathcal{F}(T) = \{\{d, w, v, t\}\}$. The tree T is a path P_5. It is easy to observe that $\{d, w, v, t\}$ is the only mdds of the tree T. We have $n = 5$ and $|\mathcal{F}(T)| = 1$. Obviously, $1 < \alpha^5$. Now assume that $d_T(d) \geq 2$. Due to the earlier analysis of the degrees of the vertices w and u, we may assume that for every child of d, say k, the tree T_k is a path on at most four vertices. Let $T' = T - T_u$, $T'' = T - T_w$ and $T''' = T - T_d$. If T''' is a single vertex, then let $\mathcal{F}(T) = \{\{r, d, w, v, t\}, \{r, d, u, v, t\}\}$. The tree T is a path P_6. Let us observe that $\{r, d, w, v, t\}$ and $\{r, d, u, v, t\}$ are the only two minimal double dominating sets of the tree T. We have $n = 6$ and $|\mathcal{F}(T)| = 2$. Obviously, $2 < \alpha^6$. Now assume that $|V(T''')| \geq 2$. Let $\mathcal{F}(T)$ be a family as follows,

$$\{D' \cup \{v, t\}: D' \in \mathcal{F}(T')\}$$
$$\cup \{D'' \cup \{u, v, t\}: d \in D'' \in \mathcal{F}(T'')\}$$
$$\cup \{D''' \cup V(T_d) \setminus \{d\}: D''' \in \mathcal{F}(T''')\},$$

where the third component is ignored if d is adjacent to a leaf. Let us observe that all elements of the family $\mathcal{F}(T)$ are minimal double dominating sets of the tree T. Now let D be any mdds of the tree T. By Observations 1 and 2 we have $v, t \in D$. If $u \notin D$, then observe that $D \setminus \{v, t\}$ is an mdds of the tree T'. By the inductive hypothesis we have $D \setminus \{v, t\} \in \mathcal{F}(T')$. Now assume that $u \in D$. If $w \notin D$, then observe that $D \setminus \{u, v, t\}$ is an mdds of the tree T''. By the inductive hypothesis we have $D \setminus \{u, v, t\} \in \mathcal{F}(T'')$. Now assume that $w \in D$. We have $d \notin D$, otherwise $D \setminus \{u\}$ is a double dominating set of the tree T, a contradiction to the minimality of D. Observe that $D \cap V(T''')$ is an mdds of the tree T'''. By the inductive hypothesis we have $D \cap V(T''') \in \mathcal{F}(T''')$. Therefore the family $\mathcal{F}(T)$ contains all minimal double dominating sets of the tree T. We now get $|\mathcal{F}(T)| = |\mathcal{F}(T')| + |\{D'' \in \mathcal{F}(T''): d \in D''\}| + |\mathcal{F}(T''')| \leq |\mathcal{F}(T')| + |\mathcal{F}(T'')| + |\mathcal{F}(T''')| \leq \alpha^{n-3} + \alpha^{n-4} + \alpha^{n-5} = \alpha^{n-5}(\alpha^2 + \alpha + 1) = \alpha^{n-5}(\alpha^2 + \alpha^3) = \alpha^{n-3}(\alpha + 1) = \alpha^{n-3} \cdot \alpha^3 = \alpha^n$.

We show that paths attain the bound from the previous theorem.

Proposition 4. *For positive integers n, let a_n denote the number of minimal double dominating sets of the path P_n. We have*

$$a_n = \begin{cases} 0 & \text{if } n = 1; \\ 1 & \text{if } n = 2, 3, 4, 5; \\ a_{n-5} + a_{n-4} + a_{n-3} & \text{if } n \geq 6. \end{cases}$$

Proof. Obviously, the one-vertex graph has no mdds. It is easy to see that a path on at most five vertices has exactly one mdds. Observe that the path P_6 has two minimal double dominating sets. Now assume that $n \geq 7$. Let $T' = T - v_n -v_{n-1} - v_{n-2}$, $T'' = T' - v_{n-3}$ and $T''' = T'' - v_{n-4}$. It follows from the last paragraph of the proof of Theorem 3 that $a_n = a_{n-5} + a_{n-4} + a_{n-3}$.

Solving the recurrence $a_n = a_{n-5} + a_{n-4} + a_{n-3}$, we get $\lim_{n \to \infty} \sqrt[n]{a_n} = \alpha$, where $\alpha \approx 1.3247$ is the positive solution of the equation $x^3 - x - 1 = 0$ (notice that $x^5 - x^2 - x - 1 = (x^2 + 1)(x^3 - x - 1)$). This implies that the bound from Theorem 3 is tight.

It is an open problem to prove the tightness of an upper bound on the number of minimal dominating sets of a tree. In [10] it has been proved that any tree of order n has less than 1.4656^n minimal dominating sets. A family of trees having more than 1.4167^n minimal dominating sets has also been given.

References

1. Bród, D., Skupień, Z.: Trees with extremal numbers of dominating sets. Australasian Journal of Combinatorics 35, 273–290 (2006)
2. Bród, D., Włoch, A., Włoch, I.: On the number of minimal dominating sets including the set of leaves in trees. International Journal of Contemporary Mathematical Sciences 4, 1739–1748 (2009)
3. Couturier, J.-F., Heggernes, P., van't Hof, P., Kratsch, D.: Minimal dominating sets in graph classes: combinatorial bounds and enumeration. In: Bieliková, M., Friedrich, G., Gottlob, G., Katzenbeisser, S., Turán, G. (eds.) SOFSEM 2012. LNCS, vol. 7147, pp. 202–213. Springer, Heidelberg (2012)
4. Fomin, F., Grandoni, F., Pyatkin, A., Stepanov, A.: Combinatorial bounds via measure and conquer: bounding minimal dominating sets and applications, ACM Transactions on Algorithms 5, article 9, 17 p. (2009)

5. Fomin, F., Kratsch, D.: Exact Exponential Algorithms. Springer, Berlin (2010)
6. Harary, F., Haynes, T.: Double domination in graphs. Ars Combinatoria 55, 201–213 (2000)
7. Haynes, T., Hedetniemi, S., Slater, P.: Fundamentals of Domination in Graphs. Marcel Dekker, New York (1998)
8. Haynes, T., Hedetniemi, S., Slater, P. (eds.): Domination in Graphs: Advanced Topics. Marcel Dekker, New York (1998)
9. Kanté, M., Limouzy, V., Mary, A., Nourine, L.: Enumeration of minimal dominating sets and variants. In: Owe, O., Steffen, M., Telle, J.A. (eds.) FCT 2011. LNCS, vol. 6914, pp. 298–309. Springer, Heidelberg (2011)
10. Krzywkowski, M.: Trees having many minimal dominating sets. Information Processing Letters 113, 276–279 (2013)
11. Lawler, E.: A note on the complexity of the chromatic number problem. Information Processing Letters 5, 66–67 (1976)
12. Moon, J., Moser, L.: On cliques in graphs. Israel Journal of Mathematics 3, 23–28 (1965)
13. Wilf, H.: The number of maximal independent sets in a tree. SIAM Journal on Algebraic and Discrete Methods 7, 125–130 (1986)

Parallel-Machine Scheduling Problem
under the Job Rejection Constraint
(Extended Abstract)

Weidong Li[1,*], Jianping Li[1], Xuejie Zhang[1], and Zhibin Chen[2]

[1] Yunnan University, Kunming 650091, P.R. China
{weidong,jianping,xjzhang}@ynu.edu.cn
[2] Kunming University of Science and Technology, Kunming 650500, P.R. China
chenzhibin11@gmail.com

Abstract. Given m identical machines and n independent jobs, each job J_j has a processing time (or size) p_j and a penalty e_j. A job can be either rejected, in which case its penalty is paid, or scheduled on one of the machines, in which case its processing time contributes to the load of that machine. The objective is to minimize the makespan of the schedule for accepted jobs under the constraint that the total penalty of the rejected jobs is no more than a given bound B. In this paper, we present a 2-approximation algorithm within strongly polynomial time and a polynomial time approximation scheme whose running time is $O(nm^{O(\frac{1}{\epsilon^2})} + mn^2)$ for the general case. Moreover, we present a fully polynomial time approximation scheme for the case where the number of machines is a fixed constant. This result improves previous best running time from $O(n^{m+2}/\epsilon^m)$ to $O(1/\epsilon^{2m+3} + mn^2)$.

Keywords: Scheduling, Rejection penalty, Polynomial time approximation scheme, Fully polynomial time approximation scheme.

1 Introduction

Given a set of machines and a set of jobs such that each job j has to be processed on one of the machines, the classical scheduling problem $P \parallel C_{\max}$ is to minimize the makespan. Since Graham [8] designed a classical *list scheduling* (LS for short) algorithm, which is to assign the next job in an arbitrary list to the first available machine, for the problem, this strongly NP-hard problem has been widely studied for more than four decades. In this paper we consider a generalized version of this problem in which a job can be rejected at a certain penalty : the so-called *parallel-machine scheduling under the job rejection constraint*, which was first introduced by Zhang et al [24]. Given are a set of identical machines $\mathcal{M} = \{M_1, M_2, \ldots, M_m\}$, a set of jobs $\mathcal{J} = \{J_1, J_2, \ldots, J_n\}$ with a processing time (or size) $p_j > 0$ and a penalty $e_j \geq 0$ for each J_j of $\mathcal{J}$, and a bound B. We wish to partition the set of jobs into two subsets, the subset

---

* Corresponding author.

J. Chen, J.E. Hopcroft, and J. Wang (Eds.): FAW 2014, LNCS 8497, pp. 158–169, 2014.

of accepted and the subset of rejected jobs, and to schedule the set of accepted jobs on the m machines such that the makespan is minimized under the constraint that the total penalty of the rejected jobs is no more than the bound B. Following the convention of Lawler et al. [13], we denote this problem and a special case of this problem where the number of machines is a fixed constant m by $P|\sum_{J_j \in \mathcal{R}} e_j \leq B|C_{\max}$ and $P_m|\sum_{J_j \in \mathcal{R}} e_j \leq B|C_{\max}$, respectively.

The problem $P|\sum_{J_j \in \mathcal{R}} e_j \leq B|C_{\max}$ is a generalized version of the classical scheduling problem $P \parallel C_{\max}$ as we just mentioned, but on the other hand, the problem $P|\sum_{J_j \in \mathcal{R}} e_j \leq B|C_{\max}$ can be viewed as a special case of the bicriteria problem on unrelated parallel machines as defined in [2], where we are given n independent jobs and m unrelated parallel machines with the requirement that when job J_j is processed on machine M_i, it requires $p_{ij} \geq 0$ time units and incurs a cost c_{ij}. The objective of the general bicriteria problem on unrelated parallel machines is to find a schedule obtaining a trade-off between the makespan and the total cost.

More attentions have been given to the general bicriteria problem on unrelated parallel machines as well as its many variants over the last two decades. For instance, readers are referred to [2, 14, 21, 12]. Given T, C and ϵ, assume that there exists a schedule with a makespan value of T and a cost of C. Lin and Vitter [14] presented a polynomial time algorithm, which finds a schedule with a makespan of at most $(2 + \frac{1}{\epsilon})T$ and a cost of at most $(1 + \epsilon)C$. This result was improved by Shmoys and Tardos [21], who proposed a polynomial time algorithm, which finds a solution with a makespan value of at most $2T$ and with a cost of at most C. It is worthwhile pointing out that the algorithm in [21] implied a 2-approximation algorithm for the problem $P|\sum_{J_j \in \mathcal{R}} e_j \leq B|C_{\max}$. However, their algorithm is not strongly polynomial, as it requires the solution of a linear programming formulation. One of objectives of this paper is devoted to providing a strongly polynomial 2-approximation algorithm with running time of $O(mn^2)$ for the problem $P|\sum_{J_j \in \mathcal{R}} e_j \leq B|C_{\max}$ in Section 2. Based on our 2-approximation algorithm, we also design a polynomial time approximation scheme (PTAS), which finds a solution of makespan at most $(1 + \epsilon)OPT$ for $P|\sum_{J_j \in \mathcal{R}} e_j \leq B|C_{\max}$ in Section 3.

When the number of machines m is a constant, Jansen and Porkolab [12] designed a fully polynomial time approximation scheme (FPTAS) for the bicriteria problem on unrelated parallel machines. The FPTAS computes a schedule in time $O(n(m/\epsilon)^{O(m)})$ with makespan at most $(1+\epsilon)T$ and cost at most $(1+\epsilon)C$, providing there exists a schedule with makespan T and cost C. This approximation ratio was improved by Angel et al. [2], who proposed a FPTAS, that finds a schedule with makespan at most $(1 + \epsilon)T$ and cost at most C, if there exists a schedule with makespan T and cost C, for the unrelated parallel machines scheduling problem with costs. However, the running time of the FPTAS in [2] is $O(n(n/\epsilon)^m)$, which is higher than that in [12]. As a consequence, the algorithm in [2] implies a FPTAS for the problem $P_m|\sum_{J_j \in \mathcal{R}} e_j \leq B|C_{\max}$. Recently, Zhang et al. [24] has presented a FPTAS with running time of $O(n^{m+2}/\epsilon^m)$ for the problem $P_m|\sum_{J_j \in \mathcal{R}} e_j \leq B|C_{\max}$. In this paper, we present a FPTAS with improved running time of $O(1/\epsilon^{2m+3} + mn^2)$ for $Pm|\sum_{J_j \in \mathcal{R}} e_j \leq B|C_{\max}$ in Section 3.

Recently, some variants of the parallel machine scheduling problem with rejection have received considerable attention. Engels et al. [7] studied the objective of minimizing the sum of the weighted completion times of the accepted jobs plus the sum of the penalties of the rejected jobs. The preemptive cases are considered in [10] and [18]. The batch cases are studied in [4, 15–17]. Cao and Zhang [5] presented a PTAS for scheduling with rejection and job release times, which generalized the result in [3]. For the single-machine scheduling with rejection, Cheng and Sun [6] designed some FPTAS for the case where the size of a job is a linear function of its starting time under different objectives. Zhang et al. [22] considered the single machine scheduling problem with release dates and rejection. The same authors [23] also considered the problems of minimizing some classic objectives under the job rejection constraint. Shabtay et al. [19] presented a bicriteria approach to scheduling a single machine with job rejection. More related results can be found in the recent survey [20].

We would like to make a remark that the problem $P|\sum_{J_j \in \mathcal{R}} e_j \leq B|C_{\max}$ we considered in this paper is closely related to the parallel machine scheduling problem with rejection introduced by Bartal et al. [3], in which we are given m identical parallel machines and a set of n jobs with each job J_j characterized by a size p_j and a penalty e_j, and the objective is to minimize the makespan of the schedule for accepted jobs plus the sum of the penalties of the rejected jobs. Bartal et al. [3] presented a PTAS for the general case and a FPTAS for the case where the number of machines is fixed. Unfortunately, their method can not be generalized to our problem directly, as it may violate the job rejection constraint.

The rest of the paper is organized as follows. In Section 2 we design a strongly polynomial 2-approximation algorithm for the problem $P|\sum_{J_j \in \mathcal{R}} e_j \leq B|C_{\max}$. Section 3 is devoted to presenting a polynomial time approximation scheme (PTAS) for $P|\sum_{J_j \in \mathcal{R}} e_j \leq B|C_{\max}$. As a consequence, Section 3 contains a fully polynomial time approximation scheme (FPTAS) for $Pm|\sum_{J_j \in \mathcal{R}} e_j \leq B|C_{\max}$. Finally, Section 4 contains the conclusion.

2   A Strongly Polynomial Time Approximation Algorithm for the Problem $P|\sum_{J_j \in \mathcal{R}} e_j \leq B|C_{\max}$

As mentioned in the last section, we can derive a 2-approximation algorithm from [21] for $P|\sum_{J_j \in \mathcal{R}} e_j \leq B|C_{\max}$. However, this kind of approximation algorithm is not strongly polynomial as it requires the solution of a linear programming formulation. So it would be desirable to have strongly polynomial 2-approximation algorithms for the problem $P|\sum_{J_j \in \mathcal{R}} e_j \leq B|C_{\max}$. In this section, we shall provide such an algorithm (ALGORITHM 2).

Let us introduce some notations and terminology before proceeding. An instance $\mathcal{I} = (\mathcal{M}, \mathcal{J}, p, e, B)$ of the problem $P|\sum_{J_j \in \mathcal{R}} e_j \leq B|C_{\max}$ consists of a number of machines m, a set of jobs J with $|J| = n$, a nonnegative pair (p_j, e_j)

for each job $J_j \in \mathcal{J}$ where p_j is the processing time and e_j is the penalty of the job J_j, and a given bound B. For a set of jobs $X \subseteq J$, $p(X) = \sum_{J_j \in \mathcal{R}} p_j$ is the total processing time of jobs in X, and $e(X) = \sum_{J_j \in \mathcal{R}} e_j$ is the total penalty of jobs in X. We call $(S_1, S_2, \ldots, S_m; R)$ a *schedule* for $\mathcal{I}$ if the followings hold:

(i) $\cup_{i=1}^m S_i \cup R = \mathcal{J}$; and
(ii) $S_i \cap S_j = \emptyset$, for distinct $i, j \in \{1, 2, \ldots, m\}$ and $S_k \cap R = \emptyset$ for each $k \in \{1, 2 \ldots, m\}$

We further call a schedule $(S_1, S_2, \ldots, S_m; R)$ *feasible* if $e(R) \leq B$. Obviously, for $i \in \{1, \ldots, m\}$, S_i stands for the set of jobs that have to be processed on machine M_i, and R stands for the set of rejected jobs in a feasible schedule $(S_1, S_2, \ldots, S_m; R)$. For $1 \leq i \leq m$, the completion time C_i of machine M_i in some fixed schedule equals the total processing time of the jobs that are assigned to M_i, and the makespan of the schedule equals $\max_{1 \leq i \leq m}\{C_i\}$. We call schedule $(S_1, S_2, \ldots, S_m; R)$ optimal if the schedule is feasible and the makespan of the schedule attains the minimum. As usual, when we say OPT is the optimal value for instance $\mathcal{I}$, we mean that there is an optimal schedule for instance $\mathcal{I}$ and OPT equals the makespan of this schedule. Note that we allow $OPT = +\infty$ in the case that there is no feasible schedule for instance $\mathcal{I}$.

Now let's consider an optimal schedule $(S_1^*, S_2^*, \ldots, S_m^*; R^*)$ and assume that we know the maximum size $p_{\max}$ of the accepted jobs in the schedule, where $p_{\max} = \max\{p_j \mid J_j \in \cup_i^m S_i^*\}$. To motivate the construction of a 2-approximation algorithm for the problem, we may have a simple idea that is to reject all jobs with size bigger than $p_{\max}$ and other some jobs, if any, as long as the total penalty of the rejected jobs is no more than the given bound B, and then greedily allocate the remaining jobs as LS algorithm does. Fortunately, this simple idea can indeed lead to a 2-approximation algorithm for the problem. Motivated by this, we actually do not need the prior knowledge about the maximum size $p_{\max}$ of the accepted jobs in an optimal schedule, as we can try at most n possibilities to meet our purpose of designing a 2-approximation algorithm.

Let us be more precise. Given an instance $\mathcal{I} = (\mathcal{M}, \mathcal{J}, p, e, B)$ for $P|\sum_{J_j \in \mathcal{R}} e_j \leq B|C_{\max}$, reindex, if there is a need, the jobs in $\mathcal{J}$ such that $p_1/e_1 \leq p_2/e_2 \leq \cdots \leq p_n/e_n$. Note that we assume that $e_j > 0$ for all jobs $J_j \in \mathcal{J}$ throughout the paper, as for otherwise, the jobs with penalty of 0 can be all rejected without reducing the given bound. Also we assume that $\sum_{j=1}^n e_j > B$, as for otherwise, all jobs should be rejected and $OPT = 0$, where OPT denotes the optimal value for instance $\mathcal{I}$. It is worth pointing out that any subset $\Lambda \subseteq \mathcal{J}$ enjoys a nice property under our sorting, i.e.,

$$\frac{p_i}{e_i} \leq \frac{p_j}{e_j} \quad if \quad i \leq j \tag{1}$$

for all jobs J_i, J_j in Λ. This property shall be used in the proof for the following Lemma 1.

For each index $k \in \{1, 2, \ldots, n\}$, consider the following restricted instance $\mathcal{I}_k = (\mathcal{M}, \mathcal{J}_k, p, e, B_k)$ of $\mathcal{I}$:

- The number of machines remains the same as in instance $\mathcal{I}$.
- The job set $\mathcal{J}_k$ consists of jobs in $\mathcal{J}$ with processing time no more than p_k, i.e., $\mathcal{J}_k = \{J_j \in \mathcal{J} \mid p_j \leq p_k\}$.
- The processing time (or size) p_j and penalty e_j of job J_j in $\mathcal{J}_k$ remain the same as in instance $\mathcal{I}$.
- The bound B_k of instance $\mathcal{I}_k$ is reduced by $\sum_{J_j \in \mathcal{J} \setminus \mathcal{J}_k} e_j$, i.e., $B_k = B - \sum_{J_j \in \mathcal{J} \setminus \mathcal{J}_k} e_j$.

Let OPT_k be the optimal value for instance $\mathcal{I}_k$. If $B_k < 0$, set $OPT_k = +\infty$, reflecting the fact that there is no feasible schedule for instance $\mathcal{I}_k$. Otherwise, use the following ALGORITHM 1 to obtain a feasible schedule for $\mathcal{I}_k$.

ALGORITHM 1

Step 1. Find the maximum index τ_k satisfying $\sum_{j \geq \tau_k; J_j \in \mathcal{J}_k} e_j > B_k$, which implies that $\sum_{j > \tau_k; J_j \in \mathcal{J}_k} e_j \leq B_k$, and reject the jobs in $\mathcal{J}_k$ with index bigger than τ_k. Let $\mathcal{R}_k = \{J_j \mid j > \tau_k, j \in \mathcal{J}_k\}$ denote the set of all rejected jobs;

Step 2. As in LS algorithm, we repeatedly assign the next job in $\mathcal{J}_k \setminus \mathcal{R}_k$ to a machine with currently smallest load. Let F_k denote the feasible schedule produced and OUT_k the makespan of F_k.

Lemma 1. *The value of OUT_k produced by* ALGORITHM 1 *is no more than* $2OPT_k$.

Proof. Due to space constraints, the proof is omitted.

Now, ALGORITHM 2 could be presented as follows.

ALGORITHM 2

Step 1. For each $k = 1, 2, \cdots, n$, construct the restricted instance $\mathcal{I}_k = (\mathcal{M}, \mathcal{J}_k, p, e, B_k)$ of $\mathcal{I}$. Use ALGORITHM 1 as a subroutine to get a feasible solution F_k with objective value OUT_k for each $\mathcal{I}_k$.

Step 2. Choose the solution F with minimum objective value OUT among $\{F_1, \ldots, F_n\}$.

To facilitate better understanding of ALGORITHM 2, we demonstrate an example below.

Example 1. We are given an instance $\mathcal{I}$ for $P \mid \sum_{J_j \in \mathcal{R}} e_j \leq B \mid C_{\max}$ with two machines, four jobs and $B = 3$. The sizes and penalties of jobs are given in the following table, where ϵ denotes a small positive number.

Jobs	J_1	J_2	J_3	J_4
p_j	3	3	3	2ϵ
e_j	3	3	3	ϵ

In this example, we construct four auxiliary instances. The first auxiliary instance is $\mathcal{I}_1 = (\mathcal{M}, \mathcal{J}_1; p, e, B_1)$, where $\mathcal{J}_1 = \{J_1, J_2, J_3, J_4\}$ and $B_1 = 3$. Obviously, $\tau_1 = 3$ and $\mathcal{R}_1 = \{J_j \mid j > 3, j \in \mathcal{J}_1\} = \{J_4\}$, which implies that we have to assign the jobs in $\mathcal{J}_1 \setminus \mathcal{R}_1 = \{J_1, J_2, J_3\}$ to two machines. Hence, $OUT_1 = 6$. Similarly, we have $OUT_2 = OUT_3 = 6$. The last auxiliary instance is

$\mathcal{I}_4 = (\mathcal{M}, \mathcal{J}_4, p, e, B_4)$, where $\mathcal{J}_4 = \{J_4\}$ and $B_4 = -6$. It implies that $OPT_4 = +\infty$. Hence, by ALGORITHM 2, the best solution is $OUT_1 = 6$, i.e. $OUT = 6$.

Theorem 1. *The* ALGORITHM 2 *is a strongly polynomial time 2-approximation algorithm for the problem* $P|\sum_{J_j \in \mathcal{R}} e_j \leq B|C_{\max}$, *and the ratio is tight.*

Proof. By Lemma 1 and the definition of OUT, we have $OUT = \min\{OUT_k | k = 1, 2, \ldots, n\} \leq 2\min\{OPT_k | k = 1, 2, \ldots, n\}$. Clearly, $OPT = \min\{OPT_k | k = 1, 2, \ldots, n\}$. Thus, $OUT \leq 2OPT$. This shows that the ALGORITHM 2 is indeed a 2-approximation algorithm for the problem $P|\sum_{J_j \in \mathcal{R}} e_j \leq B|C_{\max}$. Clearly, the running time of the ALGORITHM 2 is $O(mn^2)$, where m is the number of machines and n is the number of jobs.

In Example 1, we have shown that $OUT = 6$, and it is easy to see that an optimal schedule, with J_1 being rejected, gives that $OPT = 3 + 2\epsilon$. Since $\lim_{\epsilon \to 0} \frac{6}{3+2\epsilon} = 2$, the ratio is tight, completing the proof.

3   Approximation Schemes

In this section, based on the ALGORITHM 2 in the last section, we are devoted to presenting a PTAS whose running time is $O(nm^{O(\frac{1}{\epsilon^2})})$ for the problem $P|\sum_{J_j \in \mathcal{R}} e_j \leq B|C_{\max}$. Our construction of the PTAS here essentially exploits two facts. Firstly, any instance of the problem $P|\sum_{J_j \in \mathcal{R}} e_j \leq B|C_{\max}$ can be transformed into a so-called corresponding instance, which can be solved to optimality in polynomial time. Secondly, the optimal values of the corresponding instance and the original one enjoy a good property that they are almost equivalent in the sense that the difference of the optimal values of the corresponding instance and the original one can be bounded by the parameters we choose, and hence any algorithm that can solve the corresponding instance to the optimality can be used to construct a desired solution for the original instance. When the number of machines is fixed, we can solve the corresponding instance even more efficient, and therefore a FPTAS for the problem $Pm|\sum_{J_j \in \mathcal{R}} e_j \leq B|C_{\max}$ can be obtained.

3.1   The Corresponding Instance

In this subsection, we focus on the construction of the auxiliary instance, the *corresponding instance*, from any instance $\mathcal{I} = (\mathcal{J}, \mathcal{M}, p, e, B)$ for the problem $P|\sum_{J_j \in \mathcal{R}} e_j \leq B|C_{\max}$. Our construction is motivated by the work of Alon et al. [1], yet the difficult part in the construction is to assign appropriate penalties for jobs in the new instance so that the new instance can be solved in polynomial time.

Recall that based on what we have developed in Section 2, the ALGORITHM 2 produces a feasible schedule F with the makespan OUT for instance $\mathcal{I}$. For convenience, let L denote OUT. It is clear from the construction of the algorithm that $OPT \leq L \leq 2OPT$, or equivalently, $L/2 \leq OPT \leq L$, where, as before, OPT denotes the optimal value for instance $\mathcal{I}$. Since L is an upper bound for sizes of all accepted jobs in any optimal schedule for instance $\mathcal{I}$, it follows that we can reject jobs with sizes greater than L and reduce the upper bound B accordingly without loss of generality. Hence, from now on we may assume that $p_j \leq L$, $j = 1, 2, \ldots, n$, for instance $\mathcal{I} = (\mathcal{J}, \mathcal{M}, p, e, B)$.

Let δ be a small positive number such that $1/\delta$ is an integer. For convenience, the jobs in instance $\mathcal{I}$ with sizes $> \delta L$ will be called *large* jobs, and jobs with sizes $\leq \delta L$ will be called *small* jobs. Let $l = \frac{1-\delta}{\delta^2}$. It should be clear that l is an integer. We partition the jobs in $\mathcal{J}$ into l *large* job subsets $\mathcal{L}_k = \{J_j | (\delta + (k-1)\delta^2)L < p_j \leq (\delta + k\delta^2)L\}$, $k = 1, 2, \ldots, l$, and one *small* job subset $\mathcal{S} = \{J_j | p_j \leq \delta L\}$. Sort the jobs in $\mathcal{L}_k = \{J_1^k, J_2^k, \ldots, J_{|\mathcal{L}_k|}^k\}$ such that $e(J_1^k) \geq e(J_2^k) \geq \cdots \geq e(J_{|\mathcal{L}_k|}^k)$ for each $k \in \{1, 2, \ldots, l\}$. Sort the jobs in $\mathcal{S} = \{J_1^{\mathcal{S}}, J_2^{\mathcal{S}}, \ldots, J_{|\mathcal{S}|}^{\mathcal{S}}\}$ such that $e(J_1^{\mathcal{S}})/p(J_1^{\mathcal{S}}) \geq e(J_2^{\mathcal{S}})/p(J_2^{\mathcal{S}}) \geq \cdots \geq e(J_{|\mathcal{S}|}^{\mathcal{S}})/p(J_{|\mathcal{S}|}^{\mathcal{S}})$. Let $ss = p(\mathcal{S}) = \sum_{J_j \in \mathcal{S}} p_j$ denote the total size of the jobs in $\mathcal{S}$. The corresponding instance $\hat{\mathcal{I}}_\delta = (\hat{\mathcal{J}}, \mathcal{M}, \hat{p}, \hat{e}, B)$ with respect to δ is defined as follows:

- For every large job J_j in $\mathcal{I}$ with $J_j \in \mathcal{L}_k$, instance $\hat{\mathcal{I}}_\delta$ contains a corresponding job $\hat{J}_j \in \hat{\mathcal{L}}_k$ whose processing time $\hat{p}_j$ equals $(\delta + k\delta^2)L$ for $k = 1, 2, \ldots, l$, and whose penalty $\hat{e}_j$ remains the same, i.e., $\hat{e}_j = e_j$.
- For the small job set $\mathcal{S}$, instance $\hat{\mathcal{I}}_\delta$ contains a corresponding small job set $\hat{\mathcal{S}}$, where the small job set $\hat{\mathcal{S}}$ consists of $\lceil \frac{ss}{\delta L} \rceil$ new jobs, each of size δL, i.e., $\hat{\mathcal{S}} = \{\hat{J}_1^{\mathcal{S}}, \hat{J}_2^{\mathcal{S}}, \ldots, \hat{J}_{\lceil \frac{ss}{\delta L} \rceil}^{\mathcal{S}}\}$ with $\hat{p}(\hat{J}_j^{\mathcal{S}}) = \delta L$, for $j = 1, \ldots, \lceil \frac{ss}{\delta L} \rceil$. By the construction, it is not difficult to see that $|\hat{\mathcal{S}}| \leq |\mathcal{S}| + 1$. Now, we have to assign the penalty for each job in $\hat{\mathcal{S}}$. The penalty of job $\hat{J}_1^{\mathcal{S}}$ satisfies

$$\hat{e}(\hat{J}_1^{\mathcal{S}}) = \sum_{k=1}^{t_1 - 1} e(J_k^{\mathcal{S}}) + \frac{\delta L - \sum_{k=1}^{t_1 - 1} p(J_k^{\mathcal{S}})}{p(J_{t_1}^{\mathcal{S}})} e(J_{t_1}^{\mathcal{S}}),$$

where t_1 is the minimum integer satisfying $\sum_{j=1}^{t_1} p(J_j^{\mathcal{S}}) \geq \delta L$. The penalty of job $\hat{J}_j^{\mathcal{S}}$ satisfies

$$\hat{e}(\hat{J}_j^{\mathcal{S}}) = \sum_{k=1}^{t_j - 1} e(J_k^{\mathcal{S}}) + \frac{j\delta L - \sum_{k=1}^{t_j - 1} p(J_k^{\mathcal{S}})}{p(J_{t_j}^{\mathcal{S}})} e(J_{t_j}^{\mathcal{S}}) - \sum_{q=1}^{j-1} \hat{e}(\hat{J}_q^{\mathcal{S}}),$$

where t_j is the minimum integer such that $\sum_{j=1}^{t_j} p(J_j^{\mathcal{S}}) \geq j\delta L$, for $j = 2, 3, \ldots, \lceil \frac{ss}{\delta L} \rceil - 1$. The penalty of job $\hat{J}_{\lceil \frac{ss}{\delta L} \rceil}^{\mathcal{S}}$ satisfies

$$\hat{e}(\hat{J}_{\lceil \frac{ss}{\delta L} \rceil}^{\mathcal{S}}) = \sum_{k=1}^{|\mathcal{S}|} e(J_k^{\mathcal{S}}) - \sum_{q=1}^{\lceil \frac{ss}{\delta L} \rceil - 1} \hat{e}(\hat{J}_q^{\mathcal{S}}).$$

Clearly, $\hat{e}(\hat{\mathcal{S}}) = e(\mathcal{S})$ and $\hat{e}(\hat{J}_1^{\mathcal{S}}) \geq \hat{e}(\hat{J}_2^{\mathcal{S}}) \geq \cdots \geq \hat{e}(\hat{J}_{\lceil \frac{ss}{\delta L} \rceil}^{\mathcal{S}})$, as $e(J_1^{\mathcal{S}})/p(J_1^{\mathcal{S}}) \geq e(J_2^{\mathcal{S}})/p(J_2^{\mathcal{S}}) \geq \cdots \geq e(J_{|\mathcal{S}|}^{\mathcal{S}})/p(J_{|\mathcal{S}|}^{\mathcal{S}})$.
- The machine set is $\mathcal{M}$.
- The upper bound is B.

Consider an optimal schedule $(\hat{S}_1^*, \hat{S}_2^*, \ldots, \hat{S}_m^*; R^*)$ for $\hat{\mathcal{I}}_\delta$ and let v_k denote the number of accepted jobs with size $(\delta + k\delta^2)L$ in the schedule, for $k = 0, 1, \cdots, l$. It is easy to see that there is a unique $(l+1)$-dimensional vector $\mathbf{v} = (v_0, v_1, \cdots, v_l)$ associated with each optimal schedule $(\hat{S}_1^*, \hat{S}_2^*, \ldots, \hat{S}_m^*; R^*)$. Let T denote $\hat{p}(\cup_{i=1}^m \hat{S}_i^*)$, the total size of the accepted jobs in the schedule. Clearly, $\sum_{k=0}^l v_k(\delta + k\delta^2)L = T$.

On the other hand, for a given $(l+1)$-dimensional vector $\mathbf{v} = (v_0, v_1, \cdots, v_l)$, there may be several optimal schedules for $\hat{\mathcal{I}}_\delta$ such that the vector $\mathbf{v} = (v_0, v_1, \cdots, v_l)$ is the unique vector associated with them. However, due to the "nice" structure of instance $\hat{\mathcal{I}}$ in which each job has size $(\delta + k\delta^2)L$ for some $k \in \{0, 1 \ldots, l\}$, if we get two ways to partition $\hat{\mathcal{J}}$ into two sets $\hat{A}$ and $\hat{R}$ such that $\hat{A}$ containing v_0 jobs in $\hat{\mathcal{S}}$ as well as v_k jobs in $\hat{\mathcal{L}}_k$ for all k and $\hat{e}(\hat{R}) \leq B$, then either way is fine as each way leads to an optimal schedule with the same makespan. So using this kind of vectors to represent optimal schedules that have the same makespan should be appropriate. Moreover, it is obvious that scheduling the job set $\hat{\mathcal{J}}$ on m machines for the problem $P|\sum_{J_j \in \mathcal{R}} e_j \leq B|C_{\max}$ such that the vector $\mathbf{v} = (v_0, v_1, \cdots, v_l)$ is the unique vector associated to an optimal schedule of $\hat{\mathcal{I}}$ for problem $P|\sum_{J_j \in \mathcal{R}} e_j \leq B|C_{\max}$ is equivalent to scheduling the job set $\hat{A}$ on m machines to the optimality for the problem $P||C_{\max}$.

Now, let us focus on the vector and make some observations. We call a vector $\mathbf{v} = (v_0, v_1, \cdots, v_l)$ *feasible* if $\hat{\mathcal{J}}$ admits a partition $(\hat{A}, \hat{R})$ such that the followings hold:

(i) $\hat{A} \cup \hat{R} = \hat{\mathcal{J}}$ and $\hat{A} \cap \hat{R} = \emptyset$;
(ii) $\hat{A}$ contains v_0 jobs in $\hat{\mathcal{S}}$ and v_k jobs in $\hat{\mathcal{L}}_k$, for $k = 1, 2, \ldots, l$; and
(iii) $\hat{e}(\hat{R}) \leq B$.

For a given vector $\mathbf{v} = (v_0, v_1, \cdots, v_l)$, consider jobs $\hat{J}_1^{\mathcal{S}}, \hat{J}_2^{\mathcal{S}}, \cdots, \hat{J}_{v_0}^{\mathcal{S}} \in \hat{\mathcal{S}}$ and $\hat{J}_1^k, \hat{J}_2^k, \cdots, \hat{J}_{v_k}^k \in \hat{\mathcal{L}}_k$, for $k = 1, 2, \cdots, l$. We claim that

(1) the vector $\mathbf{v} = (v_0, v_1, \cdots, v_l)$ is feasible, if and only if, $v_0 \leq |\hat{\mathcal{S}}|$, $v_k \leq |\hat{\mathcal{L}}_k|$ for all k, and $\sum_{k=1}^l \sum_{j=1}^{v_k} \hat{e}(\hat{J}_j^k) + \sum_{j=1}^{v_0} \hat{e}(\hat{J}_j^{\mathcal{S}}) \geq \hat{e}(\hat{\mathcal{J}}) - B$.

When it comes to T, the total size of the accepted jobs in an optimal schedule $(\hat{S}_1^*, \hat{S}_2^*, \ldots, \hat{S}_m^*; R^*)$, we claim that

(2) $\delta L \leq T \leq m(1 + 2\delta)L$ and $T = (\delta + k'\delta^2)L$ for some $k' \in \{0, 1, \ldots, \frac{m(1+2\delta)-\delta}{\delta^2}\}$.

Now, we can state our main result in this subsection as follows.

Theorem 2. *There exists an optimal algorithm with a running time of $O(nm^{O(\frac{1}{\delta^2})})$ for $\hat{\mathcal{I}}_\delta$ of $P|\sum_{J_j \in \mathcal{R}} e_j \leq B|C_{\max}$.*

Proof. We will show the existence of such an algorithm by a method of exhaustion in the sense that we will correctly solve any given instance $\hat{\mathcal{I}}_\delta$ of $P|\sum_{J_j \in \mathcal{R}} e_j \leq B|C_{\max}$ in polynomial time $O(nm^{O(\frac{1}{\delta^2})})$ by trying all the possibilities. Our proof here relies heavily on the work of Alon et al. [1].

For any small positive number δ with $\frac{1}{\delta}$ being integer, we construct the corresponding instance $\hat{\mathcal{I}}_\delta = (\hat{\mathcal{J}}, \mathcal{M}, \hat{p}, \hat{e}, B)$ with respect to δ as we described earlier.

Let T be the total size of the accepted jobs in the optimal solution. For any $T \in \{(\delta + k\delta^2)L | k = 0, 1, \cdots, \frac{m(1+2\delta)-\delta}{\delta^2}\}$, let $\mathbf{V}_T = \{\mathbf{v} = (v_0, v_1, \cdots, v_l) | \sum_{k=0}^{l} v_k(\delta + k\delta^2)L = T$ and $\mathbf{v}$ is feasible$\}$ be the set of all possible feasible vectors whose total size of the accepted jobs is exactly T. Clearly, $|\mathbf{V}_T| = O((\frac{T}{\delta^2 L} + 1 + l)^l) = O((\frac{m(1+2\delta)}{\delta^2} + 1 + \frac{1}{\delta^2})^{\frac{1}{\delta^2}}) = O(m^{\frac{1}{\delta^2}})$, as δ is a constant.

For any $\mathbf{v} = (v_0, v_1, \cdots, v_l) \in \mathbf{V}_T$, we construct an instance $\hat{\mathcal{I}}_{\mathbf{v}}$ for the classical parallel machine scheduling problem $P||C_{\max}$, where $\hat{\mathcal{I}}_{\mathbf{v}}$ contains v_0 small jobs in $\hat{\mathcal{S}}$, and v_k large jobs in $\hat{\mathcal{L}}_k$ for $k = 1, 2, \cdots, l$, say, $\hat{\mathcal{I}}_{\mathbf{v}}$ contains jobs $\hat{J}_1^{\mathcal{S}}, \hat{J}_2^{\mathcal{S}}, \cdots, \hat{J}_{v_0}^{\mathcal{S}} \in \hat{\mathcal{S}}$ and $\hat{J}_1^k, \hat{J}_2^k, \cdots, \hat{J}_{v_k}^k \in \hat{\mathcal{L}}_k$, for $k = 1, 2, \cdots, l$. Denote the set of all jobs in $\hat{\mathcal{I}}_{\mathbf{v}}$ by $\hat{A}_{\mathbf{v}}$. Now, a theorem of Alon et al. (See Theorem 2.5 in [1] as well as the paper itself for a detailed account) enables us to find an optimal schedule of makespan $C_{\mathbf{v}}$ for instance $\hat{\mathcal{I}}_{\mathbf{v}}$ of problem $P||C_{\max}$ in time $O(n)$, and definitely the optimal schedule obtained from the method of Alon et al. together with $\hat{\mathcal{J}} \setminus \hat{A}_{\mathbf{v}}$ gives us a feasible schedule of the same makespan $C_{\mathbf{v}}$ for $\hat{\mathcal{I}}$ of the problem $P|\sum_{J_j \in \mathcal{R}} e_j \leq B|C_{\max}$. For a fixed T, try all possible $\mathbf{v}$ for T and we will get an optimal schedule corresponding to T whose makespan is $C_T = \min_{\mathbf{v} \in \mathbf{V}_T} C_{\mathbf{v}}$; Try all possible values of T and eventually we will get an optimal schedule for $\hat{\mathcal{I}}$ whose makespan is $\widehat{OPT} = \min\{C_T | T = (\delta + k\delta^2)L, k = 0, 1, \cdots, \frac{m(1+2\delta)-\delta}{\delta^2}\}$.

It is a routine matter to check that the method described above can be done within $O(\frac{m(1+2\delta)-\delta}{\delta^2} \cdot m^{\frac{1}{\delta^2}} \cdot n) = O(nm^{O(\frac{1}{\delta^2})})$ time, which is polynomial in the size of input data.

3.2   Corresponding Instance versus Original Instance

In this subsection, we study the relationships between the optimal values of the corresponding instance and the original instance. It can be shown that the difference of the optimal values of the corresponding instance and the original one can be bounded by the parameters we choose. Due to space constraints, the proofs of the following lemmas are omitted in the extended abstract.

Lemma 2. *Let OPT and $\widehat{OPT}$ denote the optimal values for original instance $\mathcal{I}$ and the corresponding instance $\hat{\mathcal{I}}_\delta$ with respect to δ, respectively. Then the following holds: $\widehat{OPT} \leq OPT + 2\delta L \leq (1 + 2\delta)L$.*

Lemma 3. *Let OPT and $\widehat{OPT}$ denote the optimal values for original instance $\mathcal{I}$ and the corresponding instance $\hat{\mathcal{I}}_\delta$ with respect to δ, respectively. Then the following holds: $OPT \leq \widehat{OPT} + \delta L$.*

3.3   Approximation Schemes

In this subsection, we present an approximation scheme for the problem $P|\sum_{J_j \in \mathcal{R}} e_j \leq B|C_{\max}$. By estimating the running time of our approximation scheme, we draw two conclusions as follows:

1. Our approximation scheme is a PTAS for the problem $P|\sum_{J_j \in \mathcal{R}} e_j \leq B|C_{\max}$;
2. Our approximation scheme is a FPTAS for the problem $P_m|\sum_{J_j \in \mathcal{R}} e_j \leq B|C_{\max}$, where m is a fixed number.

For a given instance $\mathcal{I}$ of $P|\sum_{J_j \in \mathcal{R}} e_j \leq B|C_{\max}$ and a given positive constant $\epsilon \in (0, 1)$, our approximation scheme performs the following.

Step 1. Set $\delta \leq \epsilon/6$ such that $1/\delta$ is a positive integer. Construct the corresponding instance $\hat{\mathcal{I}}_\delta$ with respect to δ from the instance $\mathcal{I}$.

Step 2. Solve instance $\hat{\mathcal{I}}_\delta$ to optimality; call the resulting schedule $(\hat{S}_1^*, \hat{S}_2^*, \ldots, \hat{S}_m^*; R^*)$;

Step 3. Transform schedule $(\hat{S}_1^*, \hat{S}_2^*, \ldots, \hat{S}_m^*; R^*)$ into a feasible schedule $(S_1, S_2, \ldots, S_m; R)$ for the original instance $\mathcal{I}$.(The constructive method will be presented in the full version)

Theorem 3. *There exists a PTAS with $O(nm^{O(\frac{1}{\epsilon^2})} + mn^2)$ running time for the problem $P|\sum_{J_j \in \mathcal{R}} e_j \leq B|C_{\max}$.*

Proof. For any given constant $\epsilon \in (0, 1)$, we have $\delta \leq \epsilon/6$ such that $1/\delta$ is a positive integer. The above mentioned scheme firstly constructs the corresponding instance $\hat{\mathcal{I}}_\delta$ with respect to δ, then secondly solves $\hat{\mathcal{I}}_\delta$ to the optimality, and finally obtains a feasible schedule $(S_1, S_2, \ldots, S_m; R)$ for instance $\mathcal{I}$. Based on Lemma 2 and Lemma 3, we have that the objective value of $(S_1, S_2, \ldots, S_m; R)$ is at most $OPT + 3\delta L$, which, by the inequality $L \leq 2OPT$, is no more than $(1 + 6\delta)OPT = (1 + \epsilon)OPT$.

When it comes to running time, by Theorem 1 and Lemma 3, it is not difficult to see that we need $O(mn^2)$ time for computing L, and no more than $O(mn^2)$ time for constructing the corresponding instance $\hat{\mathcal{I}}_\delta$ as well as for transforming an optimal schedule $(\hat{S}_1^*, \hat{S}_2^*, \ldots, \hat{S}_m^*; R^*)$ for $\hat{\mathcal{I}}$ into a feasible schedule $(S_1, S_2, \ldots, S_m; R)$ for $\mathcal{I}$. Moreover, by Theorem 2 and the fact that $\delta = \frac{\epsilon}{6}$, the running time for solving the instance $\hat{\mathcal{I}}_\delta$ to optimality is $O(nm^{O(\frac{1}{\delta^2})})$, which is also $O(nm^{O(\frac{1}{\epsilon^2})})$. Thus, the running time of the scheme is equal to $O(nm^{O(\frac{1}{\epsilon^2})} + mn^2)$, completing the proof.

A closer look at the proof above gives us that the running time of the scheme is actually equal to $O(mn^2)$ plus the running time of any algorithm for solving instance $\hat{\mathcal{I}}_\delta$. So we may restate Theorem 3 into a more general form as the following.

Theorem 4. *There exists an approximation scheme for instance $\mathcal{I}$, provided that the instance $\hat{\mathcal{I}}_\delta$ with respect to certain δ can be solved to optimality. Moreover, the running time of the scheme is equal to the running time of any algorithm for solving instance $\hat{\mathcal{I}}$ plus $O(mn^2)$ time.*

In view of Theorem 4, the improvement for the running time of our approximation scheme could be made either if a better algorithm than what we described in the proof of Theorem 2 for solving instance $\hat{\mathcal{I}}_\delta$ can be provided, or a somewhat restricted version of instance $\hat{\mathcal{I}}_\delta$ is here for us to handle. A good example is that when the number of machines is fixed, the number of jobs in instance $\hat{\mathcal{I}}_\delta$ is bounded. Based on dynamic programming, we are able to design an optimal algorithm for solving instance $\hat{\mathcal{I}}_\delta$ from a instance $\mathcal{I}$ of $Pm|\sum_{J_j \in \mathcal{R}} e_j \leq B|C_{\max}$ even more efficient.

Theorem 5. *Let $\mathcal{I}$ be an instance of $P_m|\sum_{J_j \in \mathcal{R}} e_j \le B|C_{\max}$ and $\hat{\mathcal{I}}_\delta$ the corresponding instance from $\mathcal{I}$ with respect to some small constant $\delta \in (0, 1)$. Then there exists an optimal algorithm whose running time is $O(\frac{1}{\delta^{2m+3}})$ for $\hat{\mathcal{I}}_\delta$.* ∎

Proof. Due to space constraints, the proof is omitted.

Based on Theorem 4 and Theorem 5, we have the following theorem.

Theorem 6. *There exists a FPTAS with $O(1/\epsilon^{2m+3} + mn^2)$ running time for the problem $P_m|\sum_{J_j \in \mathcal{R}} e_j \le B|C_{\max}$.*

Proof. It is immediate by noting that $O(\frac{1}{\delta^{2m+3}}) = O(\frac{1}{\epsilon^{2m+3}})$, as m is a constant and $\delta = \epsilon/6 < 1/6$.

4   Conclusion and Future Work

In this paper, the problem $P|\sum_{J_j \in \mathcal{R}} e_j \le B|C_{\max}$ is under consideration. We design a strongly polynomial 2-approximation algorithm for it. We focus on presenting a PTAS for it and as a consequence, a FPTAS for its restricted version $P_m|\sum_{J_j \in \mathcal{R}} e_j \le B|C_{\max}$. The running time of the FPTAS we presented here improves previous best result from $O(n^{m+2}/\epsilon^m)$ to $O(1/\epsilon^{2m+3} + mn^2)$ in this direction. It is worth pointing out again that the problem $P|\sum_{J_j \in \mathcal{R}} e_j \le B|C_{\max}$ either can be viewed as generalization of the classical problem $P||C_{\max}$, or can be viewed as a special case of the bicriteria problem on unrelated parallel machines introduced in [2]. It would be interesting to generalize our method to the bicriteria problem on unrelated parallel machines.

Acknowledgement. The work is supported in part by the National Natural Science Foundation of China [Nos. 11301466, 61170222, 11101193], the Tianyuan Fund for Mathematics of the National Natural Science Foundation of China [No. 11126315], the Project of First 100 High-level Overseas Talents of Yunnan Province, and the Natural Science Foundation of Yunnan Province of China [No. 2011FZ065].

References

1. Alon, N., Azar, Y., Woeginger, G.J., Yadid, T.: Approximation schemes for scheduling on parallel machines. Journal of Scheduling 1, 55–66 (1998)
2. Angel, E., Bampis, E., Kononov, A.: A FPTAS for approximating the unrelated parallel machines scheduling problem with costs. In: Meyer auf der Heide, F. (ed.) ESA 2001. LNCS, vol. 2161, pp. 194–205. Springer, Heidelberg (2001)
3. Bartal, Y., Leonardi, S., Spaccamela, A.M., Sgall, J., Stougie, L.: Multiprocessor scheduling with rejection. SIAM Journal on Discrete Mathematics 13, 64–78 (2000)
4. Cao, Z., Yang, X.: A PTAS for parallel batch scheduling with rejection and dynamic job arrivals. Theoretical Computer Science 410, 2732–2745 (2009)
5. Cao, Z., Zhang, Y.: Scheduling with rejection and non-identical job arrivals. Journal of Systems Science and Complexity 20, 529–535 (2007)

6. Cheng, Y., Sun, S.: Scheduling linear deteriorating jobs with rejection on a single machine. European Journal of Operational Research 194, 18–27 (2009)
7. Engels, D.W., Karger, D.R., Kolliopoulos, S.G., Sengupta, S., Uma, R.N., Wein, J.: Techniques for scheduling with rejection. Journal of Algorithms 49, 175–191 (2003)
8. Graham, R.L.: Bounds for certain multiprocessing anomalies. Bell System Technical Journal 45, 1563–1581 (1966)
9. Hochbaum, D.S., Shmoys, D.B.: Using dual approximation algorithms for scheduling problems: theoretical and practical results. Journal of Association for Computing Machinery 34, 144–162 (1987)
10. Hoogeveen, H., Skutella, M., Woeginger, G.J.: Preemptive scheduling with rejection. Mathematics Programming 94, 361–374 (2003)
11. Horowitz, E., Sahni, S.: Exact and approximate algorithms for scheduling nonidentical processors. Journal of the ACM 23, 317–327 (1976)
12. Jansen, K., Porkolab, L.: Improved approximation schemes for scheduling unrelated parallel machines. In: Proceedings of STOS 1999, 408- 417 (1999)
13. Lawler, E.L., Lenstra, J.K., Rinnooy Kan, A.H.G., Shmoys, D.B.: Sequencing and scheduling: Algorithms and complexity. Handbooks in Operations Research and Management Science 4, 445–452 (1993)
14. Lin, J.H., Vitter, J.S.: ϵ-Approximation algorithms with minimum packing constraint violation. In: Proceedings of STOS 1992, pp. 771–782 (1992)
15. Lu, L., Zhang, L., Yuan, J.: The unbounded parallel batch machine scheduling with release dates and rejection to minimize makespan. Theoretical Computer Science 396, 283–289 (2008)
16. Lu, L., Cheng, T.C.E., Yuan, J., Zhang, L.: Bounded single-machine parallel-batch scheduling with release dates and rejection. Computers & Operation Research 36, 2748–2751 (2009)
17. Lu, S., Feng, H., Li, X.: Minimizing the makespan on a single parallel batching machine. Theoretical Computer Science 411, 1140–1145 (2010)
18. Seiden, S.: Preemptive multiprocessor scheduling with rejection. Theoretical Computer Science 262, 437–458 (2001)
19. Shabtay, D., Gaspar, N., Yedidsion, L.: A bicriteria approach to scheduling a single machine with job rejection and positional penalties. Journal of Combinatorial Optimization 23, 395–424 (2012)
20. Shabtay, D., Gaspar, N., Kaspi, M.: A survey on offline scheduling with rejection. Journal of Scheduling 16, 3–28 (2013)
21. Shmoys, D.B., Tardos, E.: An approximation algorithm for the generalized assignment problem. Mathematical Programming 62, 461–474 (1993)
22. Zhang, L., Lu, L., Yuan, J.: Single machine scheduling with release dates and rejection. European Journal of Operational Research 198, 975–978 (2009)
23. Zhang, L., Lu, L., Yuan, J.: Single-machine scheduling under the job rejection constraint. Theoretical Computer Science 411, 1877–1882 (2010)
24. Zhang, Y., Ren, J., Wang, C.: Scheduling with rejection to minimize the makespan. In: Du, D.-Z., Hu, X., Pardalos, P.M. (eds.) COCOA 2009. LNCS, vol. 5573, pp. 411–420. Springer, Heidelberg (2009)

Approximation Algorithms on Consistent Dynamic Map Labeling

Chung-Shou Liao[1,*], Chih-Wei Liang[1,*], and Sheung-Hung Poon[2,**,***]

[1] Department of Industrial Engineering and Engineering Management,
National Tsing Hua University, Hsinchu 30013, Taiwan, R.O.C.
[2] Department of Computer Science &
Institute of Information Systems and Applications,
National Tsing Hua University, Hsinchu 30013, Taiwan, R.O.C.

Abstract. We consider the dynamic map labeling problem: given a set of rectangular labels on the map, the goal is to appropriately select visible ranges for all the labels such that no two consistent labels overlap at every scale and the sum of total visible ranges is maximized. We propose approximation algorithms for several variants of this problem. For the *simple ARO problem*, we provide a $3c \log n$-approximation algorithm for the unit-width rectangular labels if there is a c-approximation algorithm for unit-width label placement problem in the plane; and a randomized polynomial-time $O(\log n \log \log n)$-approximation algorithm for arbitrary rectangular labels. For the *general ARO problem*, we prove that it is NP-complete even for congruent square labels with equal selectable scale range. Moreover, we contribute 12-approximation algorithms for both arbitrary square labels and unit-width rectangular labels, and a 6-approximation algorithm for congruent square labels.

1 Introduction

Online maps have been widely used in recent years, especially on portable devices. Such geographical visualization systems provide user-interactive operations such as continuous zooming. Thus, the interfaces provide to a new model in map labeling problems. Been *et al.* [3] initiated an interesting consistent dynamic map labeling problem whose objective is to maximize the sum of total visible ranges, each of which corresponds to the consistent interval of scales at which the label is visible; in other words, the aim is to maximize the number of consistent labels at every scale. In contrast with the static map labeling problem, the dynamic map labeling problem can be considered a traditional map labeling by incorporating *scale* as an additional dimension. During zooming in and out operations on the map, the labeling is regarded as a function of the zoom scale and the map area.

* Supported in part by grants NSC 100-2221-E-007-108-MY3, NSC 102-2221-E-007-075-MY3, MOE 102N2073E1, and MOE 103N2073E1, Taiwan, R.O.C.
** Supported in part by grant NSC 100-2628-E-007-020-MY3 in Taiwan, R.O.C.
*** Corresponding author.

J. Chen, J.E. Hopcroft, and J. Wang (Eds.): FAW 2014, LNCS 8497, pp. 170–181, 2014.

Several desiderata [2] are provided by Been *et al.* to define this problem. We adopt all desiderata to our problem. Labels are selected to display at each scale and labels should be visible continuously without intersection. Moreover, labels could change their sizes as a function during monotonic zooming at some specific scale. For a specific label, the union of all its scaled labels in a specific scale range (s_E, S_E) is called an *extrusion E*. In order to maintain the consistence of notations, we also follow the definition by Been *et al.*'s work [2,3], and define *active* (visible) range (a_E, A_E) to be a continuous interval lying between the minimum scale s_E and the maximum scale S_E where a label could be exactly displayed. Our goal is to maximize the number of consistent labels at every scale, and thus we maximize the sum of total *active* ranges to achieve this goal. The detailed problem definition is described in the following.

Problem Definition. Given a set of n extrusions $\mathcal{E}$, and each extrusion $E \in \mathcal{E}$ with an open interval $(s_E, S_E) \subseteq (0, S_{\max})$, which we call *selectable range*, among the scale s. Note that $S_{\max}$ is an universal maximum scale for all extrusions. The goal is to compute a suitable active range $(a_E, A_E) \subseteq (s_E, S_E)$, for each E (see Figure 1). Actually, when an extrusion E intersects a horizontal plane at s, it forms a cross-section. We say that this cross-section is a label L. Here, we consider invariant point placements with axis-aligned rectangular labels, in which labels always map to the same location, so labels do not slide and rotate.

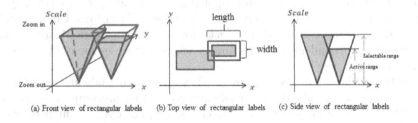

(a) Front view of rectangular labels   (b) Top view of rectangular labels   (c) Side view of rectangular labels

Fig. 1. Two unit-width rectangular labels with selectable ranges and active ranges

According to [3], we consider two models in this problem—*general* and *simple*. The general active range optimization (ARO) problem is to choose the active ranges (a_E, A_E) so as to maximize the sum of total active ranges. For the simple ARO problem, it is a variant in which the active range are restricted so that a label is never deselected when zooming in. That is, the active range of a selected extrusion $E \in \mathcal{E}$ is $(0, A_E) \subseteq (0, S_{\max})$.

Moreover, we consider two types of dilation cases in this paper—*proportional dilation* and *constant dilation*. We say that labels have *proportional dilation* if their sizes could change with scale proportionally. In contrast, if the sizes of labels are fixed at every scale, we say that labels have *constant dilation*. For the simple ARO problem with proportional dilation, because we consider rectangular labels, the shapes of extrusions are in fact rectangular pyramids. Let $\pi(s)$ be the hyperplane at scale s. Also let the width and length of the rectangular label

$E \cap \pi(s)$ of an (pyramid) extrusion E at scale s be functions $w_E(s) = \frac{s}{S_{max}} w_E$ and $l_E(s) = \frac{s}{S_{max}} l_E$, respectively, where w_E and l_E are the width and length of E, respectively, at scale S_{max}. Then, for the general ARO problem with constant dilation, the shapes of extrusions are rectangular prisms. Let width and length be $w_E(s) = w_E$ and $l_E(s) = l_E$, respectively, where $s \in (s_E, S_E)$, because the sizes of all labels are fixed at every scale. In addition, we say that E and $E' \in \mathcal{E}$ intersect at scale s, if and only if $s \subset (s_E, S_E) \cap (s_{E'}, S_{E'})$, $|x_E - x_{E'}| \le \frac{1}{2}(l_E(s) + l_{E'}(s))$ and $|y_E - y_{E'}| \le \frac{1}{2}(w_E(s) + w_{E'}(s))$ are satisfied, where (x_E, y_E) is the central point of a pyramid extrusion E.

Accordingly, our goal is to compute a set of pairwise disjoint truncated extrusions $\mathcal{T} = \{T_E : (a_E, A_E) \mid E \in \mathcal{E}\}$, where T_E is the truncated extrusion of E, so as to maximize the sum of total active range height $\mathcal{H}(\mathcal{T}) = \sum_{E \in \mathcal{E}} |A_E - a_E|$.

Previous Work. Map labeling is an important application [9] and a popular research topic during the past three decades [16]. The labeling problems which were proposed before dynamic labeling problems are mostly static labeling problems [3]. There are various settings for static labeling problems [10] and they have been shown to be NP-hard [11]. One of major topics and its typical goal is to select and place labels without intersection and its objective is to maximize the total number of labels. Agarwal *et al.* presented a PTAS for the unit-width rectangular label placement problem and a $\log n$-approximation algorithm for the arbitrary rectangle case [1]; Berman *et al.* [5] improved the latter result and obtained a $\lceil \log_k n \rceil$-factor algorithm for any integer constant $k \ge 2$. Then, Chan [7,8] improved the running time of these algorithms. Chalermsook and Chuzhoy [6] showed an $O(\log^{d-2} n \log \log n)$-approximation algorithm for the maximum independent set of rectangles where rectangles are d-dimensional hyper-rectangles.

In addition, there have been a few studies on dynamic labeling. Poon and Shin [15] developed an algorithm for labeling points that precomputes a hierarchical data structure for a number of specific scales. For dynamic map labeling problems, Been *et al.* [2] proposed several consistency desiderata and presented several algorithms for one-dimensional (1D) and two-dimensional (2D) labeling problems [3]. Note that labels in 1D problems are open intervals; labels in 2D problems are open rectangles. They showed NP-completeness of the general 1D ARO problem with "constant" dilation with square extrusions of distinct sizes, and the simple 2D ARO problem with "proportional" dilation with congruent square cone extrusions. They focused on dynamic label selection, i.e., assuming a 1-position model for label placement. Moreover, Gemsa *et al.* [12] provided a FPTAS for general sliding models of the 1D dynamic map labeling problem. Since dynamic map labeling is still a new research topic, there are still many unsolved problems. Yap [17] summarized some open problems.

Our Contribution. In this paper, we consider simple ARO with proportional dilation and general ARO with constant dilation. We design a list of approximation algorithms as shown in Table 1. Moreover, we also prove that the general ARO problem with constant dilation for congruent square prisms is NP-complete.

Table 1. Summary of our approximation results

Problem	Extrusion Shape	Approximation Ratio	Time Complexity
Simple ARO	unit-width rectangular pyramids	$6 \log n$	$O(n \log^2 n)$
		$3\frac{k+1}{k} \log n$	$O(n \log^2 n + n \triangle^{k-1} \log n)$
	rectangular pyramids	$O(\log n \log \log n)$	Polynomial
General ARO	unit-width rectangular prisms	12	$O(n \log^3 n)$
	arbitrary square prisms	12	$O(n \log^3 n)$
	congruent square prisms	6	$O(n \log^3 n)$

2   Approximation for the Simple ARO Problem

In this section, we investigate the simple ARO problem with proportional dilation and present two approximation algorithms for a given set $\mathcal{E}$ of axis-aligned rectangular pyramids, where the intersection of a pyramid E with the horizontal plane at scale s is a rectangular label whose width and length are $\frac{s}{S_{max}} w_E$ and $\frac{s}{S_{max}} l_E$, respectively. First, we explore the simple ARO problem for an input set of unit-width rectangular pyramids, in which the rectangular label of a pyramid at scale S_{max} is associated with a given uniform width and an arbitrary length. In particular, we propose a $3c \log n$-approximation algorithm for this problem, where c is an approximation factor for the unit-width rectangular label placement problem in the plane. The best known-to-date approximation ratio for this two-dimensional label placement problem is $c = \frac{k+1}{k}$, derived by Agarwal *et al.* [1] and Chan [8], for any integer $k \geq 1$. Subsequently, we extend the technique to the arbitrary rectangular pyramid case and obtain an expected $O(\log n \log \log n)$-approximation algorithm.

2.1   Approximation for Unit-Width Rectangular Pyramids

Given a set $\mathcal{E}$ of n unit-width rectangular pyramids for the simple ARO problem with proportional dilation, where the uniform label width of each pyramid at scale s is $\frac{s}{S_{max}} w$, the objective is to select a set of truncated pyramids such that they are pairwise disjoint and the total sum of their active range height is maximized. Note that the maximum unit-width label placement problem at scale s can be approximated well by using the famous shifting technique [13]. However, the major challenge is that a feasible label placement at each scale cannot be merged into a feasible solution for the ARO problem; that is, an integrated solution may cause inconsistent active range for a pyramid, even if an optimal label placement can be derived at each scale s.

The rationale behind the proposed approach is described as follows. We divide the scale into $(\log n + 1)$ heights for $(\log n + 1)$ restricted simple ARO problems such that in each of the problems, every rectangular pyramid has an upper bound on the selectable range that cannot exceed $s_j = S_{max}/2^{\log n-j+1}$, $1 \leq j \leq (\log n + 1)$, where $s_0 = 0$ and the $(\log n + 1)$'th scale $s_{\log n+1}$ is in fact the universal maximum scale S_{max}. Then, for each restricted simple ARO problem, we devise a good approximation solution $\mathcal{S}$ for the unit-width rectangular label placement problem in the hyperplane at scale s_j, and select the whole rectangular pyramids whose labels at scale s_j are selected in $\mathcal{S}$. That is, we take the complete selectable ranges $(0, s_j)$ of those pyramids in $\mathcal{S}$ as their active ranges. Finally, we choose the largest approximation solution among all the $(\log n + 1)$ restricted ARO problems and analyze its ratio.

First, we recall the unit-width rectangular label placement problem in the plane. Agarwal *et al.* [1] presented a $\frac{k+1}{k}$-approximation algorithm based on the shifting technique [13]. For ease of exposition, we refer to Agarwal *et al.*'s method and use a simple 2-approximation algorithm to derive a label placement solution M_j in the hyperplane at scale s_j. We draw a set of horizontal lines from top to bottom of y-axis using an incremental approach, i.e., from $y_{\max} = \max_{E \in \mathcal{E}}\{y_E\}$ to $y_{\min} = \min_{E \in \mathcal{E}}\{y_E\}$ of y-axis. The separation between two horizontal lines is larger than the uniform width, i.e., $\frac{s_j}{s_{\max}}w$, and the lines that do not intersect any labels, if any, are skipped. For each line ℓ_k^j at scale s_j, where $1 \le k \le m_j \le n$, let a subset of labels intersected by line ℓ_k^j be denoted by R_k^j. The lines are drawn such that the next two properties hold: every line ℓ_k^j intersects at least one unit-width rectangular label, i.e., $R_k^j \neq \emptyset$, and each label is intersected by exactly one line. Hence, we have $\sum_{k=1}^{m_j} |R_k^j| = n$, where $|R_k^j|$ is the cardinality of R_k^j, i.e., the number of labels that are intersected by ℓ_k^j. Moreover, for every line ℓ_k^j, $R_k^j \cap R_i^j = \emptyset$, when $k + 1 < i \le m_j$ and $1 \le i < k - 1$. Subsequently, we apply a greedy algorithm to compute a *one-dimensional* maximum independent set, denoted by M_k^j, for each subset R_k^j of labels that are intersected by line ℓ_j^k. The greedy strategy, which takes $O(|R_k^j| \log |R_k^j|)$ time, proceeds as follows: Sort the right boundaries of all labels in R_k^j by their x-axis and scan the labels from left to right. Select the label whose right boundary is the smallest, say L, into the independent set, M_k^j, and remove labels that overlap L. Repeat the argument until each label is scanned. The correctness of this simple greedy algorithm is straightforward. Then, consider two sets $M_{odd} = \{M_1^j, M_3^j, ..., M_{2\lceil m_j/2 \rceil - 1}^j\}$ and $M_{even} = \{M_2^j, M_4^j, ..., M_{2\lceil m_j/2 \rceil}^j\}$; clearly, both of them are independent sets, i.e., feasible label placement at scale s_j. Let the larger one of M_{odd} and M_{even} be M_j, which implies $|M_j| \ge \frac{1}{2}(|M_{odd}| + |M_{even}|)$. Thus, a 2-approximation algorithm follows for the unit-width rectangular label placement problem at scale s_j.

In the restricted ARO problem for an input set of unit-width rectangular pyramids whose selectable range cannot exceed scale s_j, we select the whole pyramid E whose label at scale s_j is in M_j; that is, we set the active range of E as $A_E = s_j$, for every $E \in M_j$, and the solution at scale s_j, denoted by $\mathcal{S}_j = \{T_E : (0, A_E) \mid E \in M_j\}$, has the sum of active ranges $\mathcal{H}(\mathcal{S}_j) = \sum_{E \in M_j} A_E$. As mentioned earlier, we select the maximum among all the $(\log n + 1)$ approximation solutions at scale s_j, $1 \le j \le \log n + 1$, denoted by $\mathcal{S}$. That is $\mathcal{H}(\mathcal{S}) = \max_j \{\mathcal{H}(\mathcal{S}_j)\}$. Note that the running time of the overall algorithm is $O(n \log^2 n)$, because there are $(\log n + 1)$ restricted ARO problems.

In the following theorem, we analyze the approximation ratio of the proposed algorithm.

Theorem 1. *Given a set of n unit-width rectangular pyramids in the simple ARO problem, there exists a $6 \log n$-approximation algorithm, which takes $O(n \log^2 n)$ time, for this problem.*

Proof. Given a set $\mathcal{E}$ of n unit-width rectangular pyramids, let $\mathcal{S}^* = \{T_E^* : (0, A_E^*) \mid E \in \mathcal{E}, A_E^* > 0\}$ be the optimum solution for the problem and $|\mathcal{S}^*|$ be its cardinality. The sum of active range height of $\mathcal{S}^*$ is $\mathcal{H}(\mathcal{S}^*) = \sum_{E \in \mathcal{S}^*} A_E^*$. We consider the intersection of those pyramids in $\mathcal{S}^*$ with the hyperplane $\pi(s_j)$ at scale $s_j = S_{\max}/2^{\log n - j + 1}$, denoted by $\mathcal{S}_j^*$, $1 \le j \le \log n + 1$. We define two subsets $P_{1,j}^k$ and $P_{2,j}^k$ recursively as follows.

$$P_{1,j}^k = \begin{cases} \{T_E^* : (0, A_E^*) \mid A_E^* \ge s_j\}, & \text{if } k = 1; \\ \{T_E^* : (0, A_E^*) \mid A_E^* > s_j \text{ and } T_E^* \in \mathcal{S}^* \setminus \{\bigcup_{k'=1}^{k-1} P_{2,j+1}^{k'}\}\}, & \text{if } k > 1; \end{cases}$$

$$P_{2,j}^k = \begin{cases} \{T_E^* : (0, A_E^*) \mid A_E^* < s_j\}, & \text{if } k = 1; \\ \{T_E^* : (0, A_E^*) \mid A_E^* \le s_j \text{ and } T_E^* \in \mathcal{S}^* \setminus \{\bigcup_{k'=1}^{k-1} P_{2,j+1}^{k'}\}\}, & \text{if } k > 1; \end{cases}$$

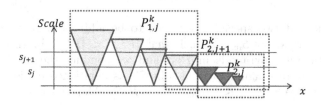

Fig. 2. Side view of an illustration of $P_{1,j}^k$, $P_{2,j}^k$ and $P_{2,j+1}^k$

Figure 2 illustrates the definition of $P_{1,j}^k$ and $P_{2,j}^k$. Initially, let $k = 1$, and we locate the scale s_j such that

$$|P_{1,j}^1| \ge |\mathcal{S}^*|/2, \quad |P_{2,j}^1| \le |\mathcal{S}^*|/2, \text{ and}$$
$$|P_{1,j+1}^1| < |\mathcal{S}^*|/2, \quad |P_{2,j+1}^1| > |\mathcal{S}^*|/2.$$

As mentioned above, because M_j is a 2-approximation solution at scale s_j, we have $2|M_j| \ge |\mathcal{S}_j^*|$, which implies

$$2|\mathcal{S}_j| \ge |\mathcal{S}^* \setminus P_{2,j}^1| = |P_{1,j}^1| \ge |P_{2,j}^1|, \text{ and}$$

$$2\mathcal{H}(\mathcal{S}_j) = 2|\mathcal{S}_j| \times s_j \ge |P_{1,j}^1| \times s_j \ge |P_{2,j}^1| \times s_j \ge \mathcal{H}(P_{2,j}^1). \tag{1}$$

Because $P_{2,j+1}^1 \setminus P_{2,j}^1 \subseteq P_{1,j}^1$ and $2s_j = s_{j+1}$, we have

$$2\mathcal{H}(\mathcal{S}_j) \ge |P_{1,j}^1| \times s_j \ge |P_{2,j+1}^1 \setminus P_{2,j}^1| \times \frac{s_{j+1}}{2} \ge \frac{1}{2}\mathcal{H}(P_{2,j+1}^1 \setminus P_{2,j}^1). \tag{2}$$

Based on the equations (1) and (2), clearly, $6\mathcal{H}(\mathcal{S}_j) \geq \mathcal{H}(P^1_{2,j+1})$. Then, we remove the rectangular pyramids in $P^1_{2,j+1}$ from $\mathcal{S}^*$. Since $|P^1_{2,j+1}| > |\mathcal{S}^*|/2$, more than half of pyramids in $\mathcal{S}^*$ are removed at this step. Now we set the index j for this first round to be j_1. Next, for the remaining active ranges in $\mathcal{S}^*$, we proceed to locate another scale s_j in the same fashion. Then we consider $P^2_{1,j}$ and $P^2_{2,j}$ for the new scale s_j. Similarly we remove $P^2_{2,j+1}$ from $\mathcal{S}^*$, and set this index j to be j_2. We repeat the above step until we locate the scale $s_j = S_{\max}$ satisfying the property or all the rectangular pyramids in $\mathcal{S}^*$ are removed. Suppose that the number of rounds considered in the above process in k, and we set last scale considered to be s_{j_k}. Note that for each of the above steps, more than half of remaining rectangular pyramids in $\mathcal{S}^*$ are removed. Thus, the above step repeats at most $\log n$ times, i.e., $k \leq \log n$. According to the above reasoning,

$$6\sum_{\ell=1}^{k} \mathcal{H}(\mathcal{S}_{j_\ell}) \geq \sum_{\ell=1}^{k} \mathcal{H}(P^\ell_{2,j_\ell+1}) \geq \mathcal{H}(\mathcal{S}^*),$$

$$\Rightarrow \mathcal{H}(\mathcal{S}) = \max_{j=1}^{\log+1}\{\mathcal{H}(\mathcal{S}_j)\} \geq \max_{\ell=1}^{k}\{\mathcal{H}(\mathcal{S}_{j_\ell})\} \geq \frac{\sum_{\ell=1}^{k}\mathcal{H}(\mathcal{S}_{j_\ell})}{\log n} \geq \frac{\mathcal{H}(\mathcal{S}^*)}{6\log n}.$$

We remark that two special cases need to be addressed. Assume the first scale we reach is $s_1 = S_{\max}/n$ such that $|P^1_{1,1}| < |P^1_{2,1}|$. In this case, we skip the rectangular pyramids in $P^1_{2,1}$, and we conduct the similar analysis on the remaining rectangular pyramids in $\mathcal{S}^*$, as mentioned earlier. Moreover, $\mathcal{H}(P^1_{2,1}) \leq n \times \frac{S_{\max}}{n} = S_{\max} \leq \mathcal{H}(\mathcal{S}_{\log n+1}) \leq \mathcal{H}(\mathcal{S})$. Thus, we have

$$(6(\log n - 1) + 1)\mathcal{H}(\mathcal{S}) \geq \mathcal{H}(\mathcal{S}^*)$$

$$\Rightarrow \mathcal{H}(\mathcal{S}) \geq \frac{\mathcal{H}(\mathcal{S}^*)}{6\log n - 5} \geq \frac{\mathcal{H}(\mathcal{S}^*)}{6\log n}.$$

In addition, consider the other case that the first scale we reach is $s_1 = S_{\max}$ such that $|P^1_{1,\log n+1}| \geq |P^1_{2,\log n+1}|$. Then clearly,

$$2|\mathcal{S}_{\log n+1}| \geq |P^1_{1,\log n+1}| \geq |P^1_{2,\log n+1}|$$

$$\Rightarrow 2\mathcal{H}(\mathcal{S}_{\log n+1}) \geq \mathcal{H}(P^1_{1,\log n+1}) \geq \mathcal{H}(P^1_{2,\log n+1})$$

$$\Rightarrow 4\mathcal{H}(\mathcal{S}) \geq 4\mathcal{H}(\mathcal{S}_{\log n+1}) \geq \mathcal{H}(P^1_{1,\log n+1}) + \mathcal{H}(P^1_{2,\log n+1}) = \mathcal{H}(\mathcal{S}^*),$$

which implies a better approximation ratio for $\mathcal{S}$. The proof is complete.   □

In fact, the analysis of the $O(\log n)$-approximation algorithm is tight, whose proof detail is omitted.

Moreover, if there is a c-approximation algorithm for the unit-width label placement problem in the plane, the approximation factor can be improved to $3c\log n$, according to the equations (1) and (2). The currently best ratio, obtained by Agarwal et al. [1] and Chan [8], is $\frac{k+1}{k}$, for any integer $k \geq 1$. Therefore, the unit-width case of the simple ARO problem can be approximated with a $\frac{3\log n(k+1)}{k}$-approximation factor, for any integer $k \geq 1$, though, in $O(n \log^2 n + n\Delta^{k-1} \log n)$ time [8], where Δ is the maximum intersection number of rectangular pyramids. We remark that the tight example, as described above, can be applied to $3c \log n$-factor approximation algorithm as well.

Corollary 1. *The simple ARO problem with unit-width rectangular pyramids can be approximated with $3c \log n$-factor, where c is an approximation ratio for the unit-width label placement problem in the plane.*

2.2 Approximation for Arbitrary Rectangular Pyramids

Given a set $\mathcal{E}$ of n arbitrary rectangular pyramids for the simple ARO problem with proportional dilation, the objective aims for a set of truncated pyramids such that they are pairwise disjoint and the total sum of their active range height is maximized. The currently best randomized algorithm, which produces an $O(\log \log n)$-approximation solution in polynomial time with high probability, has been proposed by Chalermsook and Chuzhoy [6]. Here, we extend their proposed technique to solve the simple ARO problem for a given set of arbitrary rectangular pyramids. Similarly, we divide the scale into $(\log n + 1)$ heights and consider these restricted ARO problem, in each of which there is an upper bound s_j on the selectable range of every pyramid, where $s_j = S_{\max}/2^{\log n - j + 1}$, $1 \le j \le (\log n + 1)$. Then, we use the currently best approximation algorithm to obtain a feasible label placement $\mathcal{S}_j$ at scale s_j for each restricted ARO problem, and select the whole pyramids whose labels at s_j are contained in $\mathcal{S}_j$. The selected truncated pyramids are pairwise disjoint and thus, a feasible solution for the restricted ARO problem. Finally, we choose the largest solution among all the $(\log n + 1)$ restricted ARO problems and the ratio can be proved in a similar manner according to the equations (1) and (2). The next theorem follows immediately.

Theorem 2. *Given a set of n axis-aligned arbitrary rectangular pyramids for the simple ARO problem, there exists a randomized polynomial time $O(\log n \log \log n)$-approximation algorithm.*

3   Complexity and Approximation for General ARO Problems

In this section, we first prove that the general ARO problem with constant dilation for a given set of axis-aligned congruent square prisms of equal selectable scale range (or height, for short) is NP-complete. We then proceed to present a greedy algorithm that yields constant-factor approximation for the general ARO problem with constant dilation for an input set of general axis-aligned square prisms of equal height, or for an input set of axis-aligned unit-width rectangular prisms of equal height.

3.1   Complexity of General ARO for Congruent Square Prisms

Been *et al.* [2] showed the NP-completeness of the simple 2D ARO problem with "proportional" dilation for congruent square cone extrusions. Here, we show that even the general ARO problem with "constant dilation" for congruent square prisms is also NP-complete, whose proof uses a reduction from the known NP-complete problem, the planar 3SAT problem [14].

Theorem 3. *The general ARO problem with constant dilation is NP-complete. That is, given a set $\mathcal{E}$ of axis-aligned congruent square prisms of equal selectable range of height and a real number $K > 0$, it is NP-complete to decide whether there is a set of pairwise disjoint truncated prisms $\mathcal{T}$ from the prisms in $\mathcal{E}$ such that $\mathcal{H}(\mathcal{T}) \geq K$. Moreover, the problem remains NP-complete even when restricted to instances where not all the spans on the scale dimension of the input prisms are the same.*

Proof.   Clearly, the problem is in NP. To show its hardness, we reduce the planar 3SAT problem [14] to our problem. The input instance for the *planar 3SAT problem* is a set $\{x_1, x_2, \ldots, x_n\}$ of n variables, and a Boolean expression $\Phi = c_1 \wedge c_2 \wedge \ldots \wedge c_m$ of m clauses, where each clause consists of exactly three literals, such that the variable clause graph of the input instance is planar. The *planar 3SAT problem* asks for whether there exists a truth assignment to the variables so that the Boolean expression Φ is satisfied. In the following, we will describe our polynomial-time reduction. In our construction, each prism in $\mathcal{E}$ is a prism with unit-square base and with height h, and not all prisms have the same span along the scale dimension.

Variable gadgets. The gadget of a variable x consists of a horizontal chain G_x of $4m$ pairs of congruent square prisms, where every four consecutive pairs of square prisms are dedicated for connecting to one literal

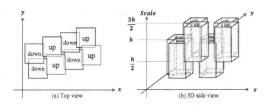

Fig. 3. Top view and side view of the variable gadget

of a clause in Φ. Every pair of square prisms intersects and locates either at the span $[0, h]$, called at *down location*, or at the span $[h/2, 3h/2]$, called at *up location*, on the scale dimension (see Figure 4).

Along the chain of the variable gadget, the prism pairs jump up and down alternately. We observe that every pair of overlapping square prisms along the chain of a variable gadget can contribute at most h in total to the final solution. We let variable x be true corresponding to that the upper prism of first prism pair along G_x is selected; the lower prism of second prism pair along G_x is selected; the upper prism of third prism pair along G_x is selected; etc. On the other hand, variable x being false corresponds to the selection of the remaining prisms in contrast to the selected prisms for variable x being true. It is not hard to see that either of these two solutions for prisms in G_x are the best possible. Whatever variable x is true or false, G_x contributes $4m$ full prisms, and thus total height $4mh$ units, to the final solution $\mathcal{H}(\mathcal{T})$.

Literal gadgets. A literal gadget connects a variable gadget to a clause gadgets. Again the gadget G_λ of a literal λ consists of a chain of square prism pairs such that all its prims locate at up or down positions. However, the literal gadget consists of a vertical part and a horizontal part, if λ corresponds to the left or right literal of the corresponding clause gadget. See Figure 4 for an example.

Suppose that λ is positive, say being x, and its dedicated chain of four consecutive square pairs from the corresponding variable gadgets is $G_{x,\lambda}$; then we connect G_λ to the two middle pairs of square prisms of $G_{x,\lambda}$ (see Figure 4). Otherwise, suppose that λ is negative, say being $\overline{x}$; then we connect G_λ to the two rightmost pairs of square prisms of $G_{x,\lambda}$. The literal gadget G_λ propagates with square prisms at up or down locations in the fashion as shown in Figure 4. When a literal gadget needs to turn left or right, we have to modify the location of one square prism at the turning corner from up to down or vice versa so that the propagation of up and down prisms can proceed. If a literal λ is true, the prism that connects the clause gadget is not selected into the set $\mathcal{T}$; otherwise, if λ is false, then the prism which connects the clause gadget is selected into the set $\mathcal{T}$. For a literal λ, let n_λ be the number of square prism pairs in G_λ. Then, λ contributes n_λ full prisms, and thus $n_\lambda h$ units of height, to the final solution $\mathcal{H}(\mathcal{T})$, no matter whether literal λ is true or false.

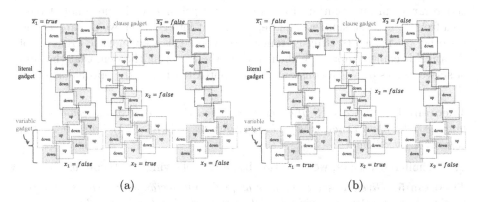

(a)   (b)

Fig. 4. (a)The gadget G_c for the clause $c = (\overline{x_1} \vee x_2 \vee \overline{x_3})$ when c is true; (b)The gadget G_c for the clause $c = (\overline{x_1} \vee x_2 \vee \overline{x_3})$ when c is false

Clause gadgets. One prism of the ending square prism pair of a literal gadget connects to a clause gadget. A gadget G_c for clause c consists of three mutually intersecting square prisms all at up locations (see Figure 4(a)). Thus the gadget

G_c can contribute at most one full square prism, and thus h units of height, to the final solution $\mathcal{H}(\mathcal{T})$.

Equivalence proof. The variable, literal, and clause gadgets form the set $\mathcal{E}$ of all input prisms representing Φ. It remains for the proof to set the threshold K such that Φ is satisfiable iff $\mathcal{H}(\mathcal{T}) \geq K$. All variable gadgets contribute $4mn$ full prisms, and thus $4mnh$ units of height, to the final solution $\mathcal{H}(\mathcal{T})$. On the other hand, all literal gadgets contribute $\sum_{\lambda \in \mathrm{lit}(\Phi)} n_\lambda$ full prisms, and thus $h(\sum_{\lambda \in \mathrm{lit}(\Phi)} n_\lambda)$ units of height, to the final solution $\mathcal{H}(\mathcal{T})$, where $\mathrm{lit}(\Phi)$ is the set of literals in clauses of formula Φ.

Since at least one literal of a clause c is true if and only if a clause gadget G_c can contribute one full square prism, and thus h units of height, to the final solution $\mathcal{H}(\mathcal{T})$. If all literals of clause c are false, then G_c contributes zero prism, and thus zero units, to $\mathcal{H}(\mathcal{T})$ (see Figure 4(b)). Hence we conclude that Φ is satisfiable, *i.e.*, all clauses are satisfied, if and only if $\mathcal{H}(\mathcal{T}) \geq K$, where

$$K = h(4mn + \sum_{\lambda \in \mathrm{lit}(\Phi)} n_\lambda + m).$$

This completes the NP-hardness proof. □

3.2 Approximation Algorithms

Given a set $\mathcal{E}$ of general axis-aligned square prisms of equal height, we propose a 12-approximation algorithm for such a general ARO problem with constant dilation. Our algorithm runs in a greedy fashion as follows. We greedily select a subset of prisms $\mathcal{S}$ from $\mathcal{E}$, and we take their complete selectable ranges as their active ranges in our solution. First we select a prism E with the smallest base area from $\mathcal{E}$, i.e., the smallest square, and put it into $\mathcal{S}$. Then we discard E and the other prisms intersecting E from $\mathcal{E}$. We repeat this step on the current set of $\mathcal{E}$ until $\mathcal{E}$ becomes empty. We then show in the following theorem that the sum of active ranges of $\mathcal{S}$ is a constant-factor approximation to the optimal solution $\mathcal{S}^*$.

Theorem 4. *Given a set $\mathcal{E}$ of general axis-aligned square prisms of equal height, there is a 12-approximation algorithm which takes $O(n \log^3 n)$ time for such a general ARO problem.*

Using similar strategies, we can also obtain the following two theorems.

Theorem 5. *Given a set $\mathcal{E}$ of axis-aligned unit rectangular prisms of equal height, there is a 12-approximation algorithm which takes $O(n \log^3 n)$ time for such a general ARO problem.*

Theorem 6. *Given a set $\mathcal{E}$ of axis-aligned congruent square prisms of equal selectable of height, there is a 6-approximation algorithm which takes $O(n \log^3 n)$ time for such a general ARO problem.*

Furthermore, we can show that the approximation factors in the above three theorems are, in fact, tight in the worst case, whose details are omitted.

References

1. Agarwal, P.K., van Kreveld, M., Suri, S.: Label placement by maximum independent set in rectangles. Computational Geometry: Theory and Application 11, 209–218 (1998)
2. Been, K., Daiches, E., Yap, C.: Dynamic map labeling. IEEE Transactions on Visualization and Computer Graphics 12(5), 773–780 (2006)
3. Been, K., Nöllenburg, M., Poon, S.-H., Wolff, A.: Optimizing active ranges for consistent dynamic map labeling. Computational Geometry: Theory and Application 43(3), 312–328 (2010)
4. de Berg, M., Cheong, O., van Kreveld, M., Overmars, M.: Computational Geometry: Algorithms and Applications, 3rd edn. Springer, Berlin (2008)
5. Berman, P., DasGupta, B., Muthukrishnan, S., Ramaswami, S.: Efficient approximation algorithms for tiling and packing problems with rectangles. Journal of Algorithms 41, 443–470 (2001)
6. Chalermsook, P., Chuzhoy, J.: Maximum independent set of rectangles. In: Proc. 20th Annual ACM-SIAM Symposium on Discrete Algorithms (SODA 2009), pp. 892–901 (2009)
7. Chan, T.M.: Polynomial-time approximation schemes for packing and piercing fat objects. Journal of Algorithms 46, 178–189 (2003)
8. Chan, T.M.: A note on maximum independent sets in rectangle intersection graphs. Information Processing Letters 89(1), 19–23 (2004)
9. Chazelle, B., et al.: The computational geometry impact task force report. In: Advances in Discrete and Computational Geometry, vol. 223, pp.407–463. American Mathematical Society, Providence (1999)
10. Doddi, S., Marathe, M.V., Mirzaian, A., Moret, B.M.E., Zhu, B.: Map labeling and generalizations. In: Proc.8th Annual ACM-SIAM Symposium on Discrete Algorithms (SODA 1997), pp. 148–157 (1997)
11. Formann, M., Wagner, F.: A packing problem with applications to lettering of maps. In: Proc. 7th Annual Symposium on Computational Geometry (SoCG 1991), pp. 281–288 (1991)
12. Gemsa, A., Nöllenburg, M., Rutter, I.: Sliding labels for dynamic point labeling. In: Proc. 23th Canadian Conference on Computational Geometry (CCCG 2011), pp. 205–210 (2011)
13. Hochbaum, D.S., Maas, W.: Approximation schemes for covering and packing problems in image processing and VLSI. Journal of the ACM (JACM) 32(1), 130–136 (1985)
14. Knuth, D.E., Raghunathan, A.: The problem of compatible representatives. SIAM J. Discrete Math. 5(3), 422–427 (1992)
15. Poon, S.-H., Shin, C.-S.: Adaptive zooming in point set labeling. In: Liśkiewicz, M., Reischuk, R. (eds.) FCT 2005. LNCS, vol. 3623, pp. 233–244. Springer, Heidelberg (2005)
16. Wolff, A., Strijk, T.: The map-labeling bibliography (2009),
    http://i11www.iti.uni-karlsruhe.de/map-labeling/bibliography/
17. Yap, C.K.: Open problem in dynamic map labeling. In: Proc. International Workshop on Combinatoral Algorithms, IWOCA 2009 (2009)

The PoA of Scheduling Game
with Machine Activation Costs[*]

Ling Lin[1,2,3], Yujie Yan[1], Xing He[1], and Zhiyi Tan[1,**]

[1] Department of Mathematics,
Zhejiang University, Hangzhou 310027, P.R. China
tanzy@zju.edu.cn
[2] School of Computer & Computing Science,
Zhejiang University City College, Hangzhou 310015, P.R. China
[3] Department of Fundamental Education, Ningbo Institute of Technology,
Zhejiang University, Ningbo 315100, P.R. China

Abstract. In this paper, we study the scheduling game with machine activation costs. A set of jobs is to be processed on identical parallel machines. The number of machines available is unlimited, and an activation cost is needed whenever a machine is activated in order to process jobs. Each job chooses a machine on which it wants to be processed. The cost of a job is the sum of the load of the machine it chooses and its shared activated cost. The social cost is the total cost of all jobs. Representing PoA as a function of the number of jobs, we get the tight bound of PoA. Representing PoA as a function of the smallest processing time of jobs, improved lower and upper bound are also given.

1 Introduction

In this paper, we study the scheduling game with machine activation costs. There is a set $\mathcal{J} = \{J_1, J_2, \cdots, J_n\}$ of jobs to be processed on identical parallel machines. The processing time of J_j is p_j, $j = 1, \cdots, n$. The number of machines available is unlimited, and an activation cost B is needed whenever a machine is activated in order to process jobs. Each job chooses a machine on which it wants to be processed. The choices of all jobs determine a schedule. The *load* of a machine M_i in a schedule is the sum of the processing time of all jobs selecting M_i. The activation cost of an activated machine is shared by the jobs selecting M_i, and the amount of each job shares is proportional to its processing time. The cost of a job in the schedule is the sum of the load of the machine it chooses and its shared activated cost. A schedule is a *Nash Equilibrium* (NE) if no job can reduce its cost by neither moving to a different machine, nor activating a new machine. The game model was first proposed by [7], and it was proved that the NE always exists for any job set $\mathcal{J}$.

---

[*] Supported by the National Natural Science Foundation of China (10971191, 11271324), Zhejiang Provincial Natural Science Foundation of China (LR12A01001) and Fundamental Research Funds for the Central Universities.

[**] Corresponding author.

J. Chen, J.E. Hopcroft, and J. Wang (Eds.): FAW 2014, LNCS 8497, pp. 182–193, 2014.

Though the behavior of each job is influenced by individual costs, the performance of the whole system is measured by certain social cost. It is well known that in most situation NE are not optimal from this perspective due to lack of coordination. The inefficiency of NE can be measured by the *Price of Anarchy* (PoA for short) [9]. The PoA of an instance is defined as the ratio between the maximal social cost of a NE and the optimal social cost. The PoA of the game is the supremum value of the PoA of all instances.

The most favorite utilitarian social cost is the total cost of all jobs. Unfortunately, it is easily to show that the PoA of above game is infinity [2], which makes no sense. However, since PoA of the game is a kind of worst-case measure, it does not imply that the NE behaviors poorly for each job set. A common method to reveal the complete characteristic of NE in such situation is as follows: select a parameter and represent the PoA as a function of it. In [2], Chen and Gurel regard the PoA as a function of $\rho = \frac{B}{\min_{1 \le j \le n} p_j}$, and prove that the PoA is at least $\frac{1}{4}(\sqrt{\rho} + 2)$ and at most $\frac{1}{2}(\rho + 1)$. However, the bounds are not tight.

Scheduling games with machine activation costs with different social costs were also studied in the literature. For the egalitarian social cost of minimizing the maximum cost among all jobs. Feldman and Tamir [7] proved that the PoA is $\frac{\tau+1}{2\sqrt{\tau}}$ when $\tau > 1$ and 1 when $0 < \tau \le 1$, where $\tau = \frac{B}{\max_{1 \le j \le n} p_j}$. Fruitful results on scheduling games without machine activation costs can be found in [9],[6], [1], [3], [8], [4], [5].

In this paper, we revisit the scheduling game with machine activation cost with social cost of minimizing the total cost of jobs. Representing the PoA as a function of n, the number of jobs. We show that the PoA is $\frac{n+1}{3}$, and the bound is tight. We also improve the lower and upper bounds on the PoA with respect to ρ. The PoA is at most $\max\{1, \frac{\rho+1}{3}\}$, and at least $\frac{\rho+1}{2\sqrt{\rho}}$.

The paper is organized as follows. In Section 2, we give some preliminary results. In Sections 3 and 4, we present the PoA as a function of the number of jobs and the smallest processing time of the jobs, respectively.

2   Preliminaries

Let $\mathcal{J} = \{J_1, J_2, \cdots, J_n\}$ be a job set. W.l.o.g., we assume $n \ge 2$, $p_1 \ge p_2 \ge \cdots \ge p_n$. By scaling the processing times we can assume that $B = 1$. Denote $P = \sum_{j=1}^{n} p_j$. Write $\rho = \frac{1}{p_n}$ and $\tau = \frac{1}{p_1}$ for simplicity. Given a schedule σ^A, the number of machines activated in σ^A is denoted m^A. Denote by $\mathcal{J}_i^A$ the set of jobs processing on M_i, $i = 1, \cdots, m^A$. The number of jobs and the total processing time of jobs of $\mathcal{J}_i^A$ are denoted n_i^A and L_i^A, respectively. Let $n_{min}^A = \min_{1 \le i \le m^A} n_i^A$ and $n_{max}^A = \max_{1 \le i \le m^A} n_i^A$. For any $J_j \in \mathcal{J}$, the cost of J_j in σ^A is denoted C_j^A, and the total cost of jobs of $\mathcal{J}$ in σ^A is denoted $C^A(\mathcal{J})$. Let σ^* be the optimal schedule with minimal social cost, and σ^{NE} be the worst NE, i.e., a NE with maximal social cost. W.l.o.g., we assume $J_1 \in \mathcal{J}_1^{NE}$ and $J_1 \in \mathcal{J}_1^*$. A job is called *separate* in σ^A if it is processed separately on the machine it selects.

The following three lemmas are given in [2], and are relevant to our study.

Lemma 1. *[2] (i) Any job of processing time no less than 1 must be a separate job in σ^{NE}.*

(ii) If each job selecting machine M_i has processing time no more than 1, then $L_i^{NE} \leq 1$.

Lemma 2. *[2] (i) Any job of processing time no less than 1 must be a separate job in σ^*.*

(ii) If $n_i^* \geq 2$, then $L_i^* \leq 1$.

Lemma 3. *[2] (i) $C^{NE} \leq m + P \leq n + P$,*

(ii) For any job set $\mathcal{J}$, $\frac{C^{NE}(\mathcal{J})}{C^*(\mathcal{J})} \leq \frac{n+P}{2P\sqrt{\tau}}$.

Lemma 4. *If there exist M_i in σ^{NE} and M_k in σ^*, such that $\mathcal{J}_i^{NE} = \mathcal{J}_k^*$, then $\frac{C^{NE}(\mathcal{J})}{C^*(\mathcal{J})} \leq \frac{C^{NE}(\mathcal{J}\backslash\mathcal{J}_i^{NE})}{C^*(\mathcal{J}\backslash\mathcal{J}_i^{NE})}$.*

Proof. Denote by n_0 and P_0 the number and the total processing time of jobs of $\mathcal{J}_i^{NE}$, respectively. Let σ' be the schedule resulting from σ^{NE} by deleting $\mathcal{J}_i^{NE}$ and M_i. Since the set of jobs selecting any machine other than M_i is not change, σ' is also a NE, and the cost of any job of $\mathcal{J}\backslash\mathcal{J}_i^{NE}$ in σ' remains the same as that in σ^{NE}. Hence,

$$C^{NE}(\mathcal{J}) = \sum_{J_j \in \mathcal{J}_i^{NE}} C_j + \sum_{J_j \in \mathcal{J}\backslash\mathcal{J}_i^{NE}} C_j = 1 + n_0 P_0 + \sum_{J_j \in \mathcal{J}\backslash\mathcal{J}_i^{NE}} C_j'$$
$$\leq 1 + n_0 P_0 + C^{NE}(\mathcal{J}\backslash\mathcal{J}_i^{NE}),$$

where C_j' is the cost of J_j in σ'. On the other hand, construct a schedule $\sigma^{*'}$ from σ^* by deleting $\mathcal{J}_k^*$ and M_k. Clearly, $\sigma^{*'}$ is a feasible schedule of $\mathcal{J}\backslash\mathcal{J}_k^*$, and the cost of any job of $\mathcal{J}\backslash\mathcal{J}_k^*$ in $\sigma^{*'}$ remains the same as that in σ^*. Hence

$$C^*(\mathcal{J}) = \sum_{J_j \in \mathcal{J}_k^*} C_j^* + \sum_{J_j \in \mathcal{J}\backslash\mathcal{J}_k^*} C_j^* = 1 + n_0 P_0 + \sum_{J_j \in \mathcal{J}\backslash\mathcal{J}_k^*} C_j^{*'}$$
$$\geq 1 + n_0 P_0 + C^*(\mathcal{J}\backslash\mathcal{J}_k^*),$$

where $C_j^{*'}$ is the cost of J_j in $\sigma^{*'}$. Recall that $\mathcal{J}_i^{NE} = \mathcal{J}_k^*$. Thus,

$$\frac{C^{NE}(\mathcal{J})}{C^*(\mathcal{J})} \leq \frac{1 + n_0 P_0 + C^{NE}(\mathcal{J}\backslash\mathcal{J}_i^{NE})}{1 + n_0 P_0 + C^*(\mathcal{J}\backslash\mathcal{J}_i^{NE})} \leq \frac{C^{NE}(\mathcal{J}\backslash\mathcal{J}_i^{NE})}{C^*(\mathcal{J}\backslash\mathcal{J}_i^{NE})}.$$

$\square$

The main results of this paper are the following two theorems.

Theorem 1. *For any job set $\mathcal{J}$, $\frac{C^{NE}(\mathcal{J})}{C^*(\mathcal{J})} \leq \frac{n+1}{3}$, and the bound is tight.*

Theorem 2. *For any job set $\mathcal{J}$, $\frac{C^{NE}(\mathcal{J})}{C^*(\mathcal{J})} \leq \max\{1, \frac{P+1}{3}\}$.*

Both theorems will be proved by contradiction. Suppose that there exist counterexamples. Let $\mathcal{J}$ be a minimal counterexample with the smallest number of jobs. By Lemmas 1(i), 2(i) and 4, all jobs of $\mathcal{J}$ must have processing time less than 1. Otherwise, delete any job of processing time no less than 1 results a counterexample with smaller number of jobs. Consequently, $L_i^{NE} \leq 1, i = 1, \cdots, m$ by Lemma 1(ii). On the other hand, no matter whether or not there exists a machine in σ^* which processes exactly one job, we have $L_i^* \leq 1, i = 1, \cdots, m^*$ by Lemma 2(ii). Thus

$$P \leq m^*. \tag{1}$$

Recall that $J_1 \in \mathcal{J}_1^{NE}$ and $J_1 \in \mathcal{J}_1^*$, the following lemma compares the social cost of the NE and optimal schedule of $\mathcal{J}$ and $\mathcal{J} \backslash \mathcal{J}_1^{NE}$.

Lemma 5. (i) $C^{NE}(\mathcal{J}) \leq 1 + n_1^{NE} L_1^{NE} + C^{NE}(\mathcal{J} \backslash \mathcal{J}_1^{NE})$.
(ii) $C^*(\mathcal{J}) \geq 2L_1^{NE} + C^*(\mathcal{J} \backslash \mathcal{J}_1^{NE})$.
(iii) If $\mathcal{J}_1^* = \{J_1\}$, then $C^*(\mathcal{J}) \geq 1 - p_1 + 2L_1^{NE} + C^*(\mathcal{J} \backslash \mathcal{J}_1^{NE})$.

Proof. (i) is obvious. Construct a feasible schedule $\sigma^{*'}$ of $\mathcal{J} \backslash \mathcal{J}_1^{NE}$ from σ^* by deleting all jobs of $\mathcal{J}_1^{NE}$, and all machines which only processing jobs of $\mathcal{J}_1^{NE}$. Obviously, $C_j^{*'} \leq C_j^*$ for any $J_j \in \mathcal{J} \backslash \mathcal{J}_1^{NE}$, where $C_j^{*'}$ is the cost of J_j in $\sigma^{*'}$. Thus

$$C^*(\mathcal{J}) = \sum_{J_j \in \mathcal{J} \backslash \mathcal{J}_1^{NE}} C_j^* + \sum_{J_j \in \mathcal{J}_1^{NE}} C_j^* \geq \sum_{J_j \in \mathcal{J} \backslash \mathcal{J}_1^{NE}} C_j^{*'} + \sum_{J_j \in \mathcal{J}_1^{NE}} C_j^*$$

$$\geq C^*(\mathcal{J} \backslash \mathcal{J}_1^{NE}) + \sum_{J_j \in \mathcal{J}_1^{NE}} C_j^*.$$

For any $J_j \in \mathcal{J}_1^{NE}$, if J_j is a separate job of σ^*, then $C_j^* = 1 + p_j \geq 2p_j$. Otherwise, J_j is processed on the machine together with at least one other job, we also have $C_j^* \geq 2p_j$. Thus (ii) and (iii) are proved.   □

3   PoA with Respect to the Number of Jobs

In this section, we will show the tight PoA as a function of n. We first give some more lemmas revealing properties of NE schedule.

Lemma 6. If there exist machine M_i and M_k such that $L_i^{NE} < L_k^{NE} + p_j$, where J_j is the job of $\mathcal{J}_i^{NE}$ with the largest processing time, then $n_i^{NE} L_k^{NE} + L_i^{NE} \geq 1$.

Proof. Note that $p_j \geq \frac{L_i^{NE}}{n_i^{NE}}$. If J_j moves to M_k, its new cost would be $C_j'' = L_k^{NE} + p_j + \frac{p_j}{L_k^{NE} + p_j}$. If $n_i^{NE} L_k^{NE} + L_i^{NE} < 1$, then

$$C_j'' - C_j^{NE} = \left(L_k^{NE} + p_j + \frac{p_j}{L_k^{NE} + p_j} \right) - \left(L_i^{NE} + \frac{p_j}{L_i^{NE}} \right)$$

$$= (L_k^{NE} + p_j - L_i^{NE}) \left(1 - \frac{p_j}{(L_k^{NE} + p_j) L_i^{NE}} \right)$$

$$\leq (L_k^{NE} + p_j - L_i^{NE}) \left(1 - \frac{\frac{L_i^{NE}}{n_i^{NE}}}{(L_k^{NE} + \frac{L_i^{NE}}{n_i^{NE}})L_i^{NE}}\right)$$

$$= (L_k^{NE} + p_j - L_i^{NE}) \left(1 - \frac{1}{n_i^{NE}L_k^{NE} + L_i^{NE}}\right) < 0.$$

This contradicts that σ^{NE} is a NE. Therefore, $n_i^{NE}L_k^{NE} + L_i^{NE} \geq 1$. □

Lemma 7. *If J_j is a separate job in σ^{NE} selecting M_i, then*
(i) $L_k^{NE} \geq 1 - p_j$ *for any $1 \leq k \leq m^{NE}$ and $k \neq i$.*
(ii) *For any job J_l with $p_l > p_j$, J_l is also a separate job in σ^{NE}.*

Proof. (i) Note that $L_i^{NE} = p_j < L_k^{NE} + p_j$. By Lemma 6, $L_k^{NE} \geq \frac{1-L_i^{NE}}{n_i^{NE}} = 1 - p_j$.

(ii) Assume that J_l selects M_k together with at least one another job, say J_t. By (i), $L_k^{NE} \geq 1 - p_j$. Hence,

$$L_i^{NE}L_k^{NE} + (L_k^{NE} - 1)p_t \geq L_i^{NE}L_k^{NE} - p_jp_t = p_j(L_k^{NE} - p_t) \geq p_jp_l > 0. \quad (2)$$

If J_t moves to M_i, its new cost would be $C_t'' = L_i^{NE} + p_t + \frac{p_t}{L_i^{NE}+p_t}$. Thus by (2) and $L_i^{NE} + p_t = p_j + p_t < p_l + p_t \leq L_k^{NE}$, we have

$$C_t'' - C_t^{NE} = \left(L_i^{NE} + p_t + \frac{p_t}{L_i^{NE}+p_t}\right) - \left(L_k^{NE} + \frac{p_t}{L_k^{NE}}\right)$$

$$= (L_i^{NE} + p_t - L_k^{NE}) \left(1 - \frac{p_t}{(L_i^{NE}+p_t)L_k^{NE}}\right)$$

$$= (L_i^{NE} + p_t - L_k^{NE}) \frac{L_i^{NE}L_k^{NE} + (L_k^{NE} - 1)p_t}{(L_i^{NE}+p_t)L_k^{NE}} < 0.$$

It contradicts that σ^{NE} is a NE. Hence, J_l is also a separate job. □

Lemma 8. *If there exist M_i and M_k, such that $L_i^{NE} + L_k^{NE} < 1$, then $n_i^{NE} \geq 2$ and $n_k^{NE} \geq 2$.*

Proof. Assume $n_i^{NE} = 1$ and let the unique job selecting M_i be J_j. By Lemma 7(i), $L_k^{NE} \geq 1 - p_j = 1 - L_i^{NE}$, a contradiction. Hence, $n_i^{NE} \geq 2$. Similarly, we also have $n_k^{NE} \geq 2$. □

Lemma 9. (i) *If $P < 1$, then $n_i^{NE} \geq 2$ for any $1 \leq i \leq m^{NE}$ and $m^{NE} \leq \frac{n}{2}$.*
(ii) *If $P < \frac{1}{n}$, then $m^{NE} = 1$.*

Proof. (i) We only need to consider the case of $m^{NE} \geq 2$ due to $n \geq 2$. Since $P < 1$, the sum of the loads of any two machines are less than 1. By Lemma 8, $n_i^{NE} \geq 2$ for any $1 \leq i \leq m^{NE}$. It follows $n \geq 2m^{NE}$.

(ii) Assume that $m^{NE} \geq 2$. Let M_i and M_k be any two machines and $L_i^{NE} \leq L_k^{NE}$. By Lemma 6,

$$1 \leq n_i^{NE}L_k^{NE} + L_i^{NE} < n_i^{NE}P + P = (n_i^{NE} + 1)P < nP,$$

a contradiction. □

Lemma 10. *If $P \leq 1$, then $\frac{C^{NE}(\mathcal{J})}{C^*(\mathcal{J})} \leq \frac{n+1}{3}$.*

Proof. We distinguish several cases according to the value of m^*. If $m^* \geq 3$, then $C^*(\mathcal{J}) > m^* \geq 3$. Applying Lemma 3(i), $\frac{C^{NE}(\mathcal{J})}{C^*(\mathcal{J})} < \frac{n+P}{3} \leq \frac{n+1}{3}$.

Now we turn to the case of $m^* = 2$. Clearly, $C^{NE}(\mathcal{J}) \leq m^{NE} + n_{max}^{NE} P$ and $C^*(\mathcal{J}) \geq 2 + n_{min}^* P$. Define

$$\Delta_1 = (n+1)(2 + n_{min}^* P) - 3(m^{NE} + n_{max}^{NE} P).$$

To prove that $\frac{C^{NE}(\mathcal{J})}{C^*(\mathcal{J})} \leq \frac{n+1}{3}$, it is sufficient to show $\Delta_1 \geq 0$. If $m^{NE} = 1$, then

$$\Delta_1 \geq (n+1)(2+P) - 3(1+nP) = (2n-1)(1-P) \geq 0. \tag{3}$$

Otherwise, $m^{NE} \geq 2$ and $n \geq n_{max}^{NE} + 2(m^{NE} - 1)$ by Lemma 9(i). Hence,

$$\begin{aligned}
\Delta_1 &\geq (n_{max}^{NE} + 2m^{NE} - 1)(2 + n_{min}^* P) - 3(m^{NE} + n_{max}^{NE} P) \\
&= n_{max}^{NE}(2 + (n_{min}^* - 3)P) + (m^{NE} - 2) + (2m^{NE} - 1)n_{min}^* P \\
&= n_{max}^{NE}(3 - n_{min}^*)(1 - P) + n_{max}^{NE}(n_{min}^* - 1) \\
&\quad + (m^{NE} - 2) + (2m^{NE} - 1)n_{min}^* P.
\end{aligned}$$

The last two equalities of above formula indicate that $\Delta_1 \geq 0$ no matter whether $n_{min}^* \geq 3$ or not. The proof of this case is thus completed.

For the remaining case of $m^* = 1$, $C^*(\mathcal{J}) = 1 + nP$. If $P < \frac{1}{n}$, $m^{NE} = 1$ by Lemma 9(ii). Thus $C^{NE}(\mathcal{J}) = C^*(\mathcal{J})$. Otherwise, $nP \geq 1$, and we have $n \geq 2m^{NE}$ by Lemma 9(i). Therefore,

$$\frac{C^{NE}(\mathcal{J})}{C^*(\mathcal{J})} \leq \frac{m^{NE} + nP}{1 + nP} \leq \frac{m^{NE} + 1}{2} \leq \frac{n+1}{3}.$$

$\square$

By Lemma 10, in order to prove the Theorem 1, only the situation of $P > 1$ would be considered. If $n = n_1^{NE} + 1$, then $n_1^{NE} = 1$ by Lemma 7(ii). Thus $n = 2$ and $P = p_1 + p_2 > 1$. By Lemmas 1(ii) and 2(ii), $m^{NE} = m^* = 2$. Thus $C^{NE}(\mathcal{J}) = C^*(\mathcal{J})$. Hence, we have

$$n \geq n_1^{NE} + 2. \tag{4}$$

Before proving Theorem 1, we give a technique lemma about the optimal schedule of the partial job set $\mathcal{J} \backslash \mathcal{J}_1^{NE}$.

Lemma 11. *If $P > 1$ and $n \geq n_1^{NE} + 2$, then $C^*(\mathcal{J} \backslash \mathcal{J}_1^{NE}) \geq 3 - 2L_1^{NE} \geq 1$.*

Proof. Let the number of machines activated in the optimal schedule of $\mathcal{J} \backslash \mathcal{J}_1^{NE}$ be m'. If $m' \geq 3$, then $C^*(\mathcal{J} \backslash \mathcal{J}_1^{NE}) > m' > 3 - 2L_1^{NE}$. If $m' = 2$, then $C^*(\mathcal{J} \backslash \mathcal{J}_1^{NE}) \geq 2 + (P - L_1^{NE}) \geq 3 - 2L_1^{NE}$. If $m' = 1$, then

$$C^*(\mathcal{J} \backslash \mathcal{J}_1^{NE}) \geq 1 + (n - n_1^{NE})(P - L_1^{NE}) \geq 1 + 2(P - L_1^{NE}) > 3 - 2L_1^{NE}.$$

Proof of Theorem 1. Since $\mathcal{J}$ is the minimal counterexample, we have $\frac{C^{NE}(\mathcal{J})}{C^*(\mathcal{J})} > \frac{n+1}{3}$ and $\frac{C^{NE}(\mathcal{J}\backslash\mathcal{J}_1^{NE})}{C^*(\mathcal{J}\backslash\mathcal{J}_1^{NE})} \leq \frac{n-n_1^{NE}+1}{3}$. Define

$$\Delta_2 = n_1^{NE}C^*(\mathcal{J}\backslash\mathcal{J}_1^{NE}) + (2n+2-3n_1^{NE})L_1^{NE} - 3. \tag{5}$$

Then by Lemma 5(i) and (ii),

$$\Delta_2 = (n+1)(2L_1^{NE} + C^*(\mathcal{J}\backslash\mathcal{J}_1^{NE}))$$
$$-3\left(1 + n_1^{NE}L_1^{NE} + \frac{n-n_1^{NE}+1}{3}C^*(\mathcal{J}\backslash\mathcal{J}_1^{NE})\right)$$
$$\leq (n+1)(2L_1^{NE} + C^*(\mathcal{J}\backslash\mathcal{J}_1^{NE})) - 3(1 + n_1^{NE}L_1^{NE} + C^{NE}(\mathcal{J}\backslash\mathcal{J}_1^{NE}))$$
$$\leq (n+1)C^*(\mathcal{J}) - 3C^{NE}(\mathcal{J}) < 0.$$

However, we will show below that $\Delta_2 \geq 0$, which leads contradiction.

Substituting (4) into (5), we have

$$\Delta_2 = n_1^{NE}C^*(\mathcal{J}\backslash\mathcal{J}_1^{NE}) + (2n_1^{NE}+4+2-3n_1^{NE})L_1^{NE} - 3$$
$$= n_1^{NE}(C^*(\mathcal{J}\backslash\mathcal{J}_1^{NE}) - L_1^{NE}) + 6L_1^{NE} - 3.$$

If $n_1^{NE} \geq 3$, then

$$\Delta_2 \geq 3(C^*(\mathcal{J}\backslash\mathcal{J}_1^{NE}) - L_1^{NE}) + 6L_1^{NE} - 3$$
$$= 3C^*(\mathcal{J}\backslash\mathcal{J}_1^{NE}) + 3L_1^{NE} - 3 \geq 0.$$

Otherwise, by Lemma 11,

$$\Delta_2 \geq n_1^{NE}(3 - 3L_1^{NE}) + 6L_1^{NE} - 3$$
$$= 3n_1^{NE} + (6 - 3n_1^{NE})L_1^{NE} - 3 \geq 3n_1^{NE} - 3 \geq 0.$$

Both are contradictions.

The following instance shows that the bound is tight. Consider a job set $\mathcal{J}$ consisting of n jobs. The processing times of the jobs are $p_i = \varepsilon$, $i = 1, \cdots, n-1$ and $p_n = 1 - (n-1)\varepsilon$, where $0 < \varepsilon < \frac{1}{n-1}$. All jobs select the same machine forms a NE σ. In fact, the cost of any job of processing time ε in σ is $1 + \varepsilon$, which equals to the cost that it activates a new machine. The cost of the job of processing time $1 - (n-1)\varepsilon$ in σ is $2 - (n-1)\varepsilon$, which also equals to the cost that it activates a new machine. Hence, $C^{NE}(\mathcal{J}) \geq 1 + n$. On the other hand, consider a schedule that J_n is processed on one machine, and the other jobs are processed on the other machine. Clearly,

$$C^*(\mathcal{J}) \leq 2 + 1 - (n-1)\varepsilon + (n-1)^2\varepsilon = 3 + (n-1)^2\varepsilon - (n-1)\varepsilon.$$

Consequently, we have

$$\frac{C^{NE}(\mathcal{J})}{C^*(\mathcal{J})} \geq \frac{1+n}{3 + (n-1)^2\varepsilon - (n-1)\varepsilon} \to \frac{n+1}{3}(\varepsilon \to 0).$$

$\square$

4   PoA with Respect to the Smallest Processing Time

In this section, we will show the PoA as a function of ρ. We begin with some lemmas revealing properties of the optimal schedule.

Lemma 12. *For any $1 \leq i, k \leq m^*$, if there exists a job $J_j \in \mathcal{J}_k^*$ such that $L_k^* - p_j > L_i^*$, then $n_i^* \geq n_k^*$.*

Proof. Assume for the sake of contradiction that $n_i^* < n_k^*$. Thus $n_i^* \leq n_k^* - 1$. Note that the total cost of jobs of $\mathcal{J}_i^*$ and $\mathcal{J}_k^*$ in σ^* are $1 + n_i^* L_i^*$ and $1 + n_k^* L_k^*$, respectively. Construct a schedule $\sigma^{*'}$ from σ^* by moving J_j from M_k to M_i, while the assignment of the other jobs remains unchanged. The new total cost of jobs assigned to M_i and M_k in $\sigma^{*'}$ are $1 + (n_i^* + 1)(L_i^* + p_j)$ and $1 + (n_k^* - 1)(L_k^* - p_j)$, respectively. The cost of any job of $\mathcal{J} \backslash (\mathcal{J}_i^* \cup \mathcal{J}_k^*)$ in $\sigma^{*'}$ is the same as that in σ^*. Since

$$
\begin{aligned}
C^{*'}(\mathcal{J}) - C^*(\mathcal{J}) &= (1 + (n_i^* + 1)(L_i^* + p_j)) + (1 + (n_k^* - 1)(L_k^* - p_j)) \\
&\quad - ((1 + n_i^* L_i^*) + (1 + n_k^* L_k^*)) \\
&= n_i^* p_j + L_i^* + p_j - (n_k^* - 1)p_j - L_k^* \\
&= (n_i^* - n_k^* + 1)p_j + (L_i^* + p_j - L_k^*) < 0,
\end{aligned}
$$

a contradiction.   □

Lemma 13. *If J_j is a separate job in σ^*, then for any job J_l with $p_l > p_j$, J_l is also a separate job in σ^*.*

Proof. Let the machine which processes J_j be M_i. Assume that J_l is processed on M_k together with at least one another job, say J_t. Then $L_k^* - p_t \geq p_l > p_j = L_i^*$. By Lemma 12, $n_k^* \leq n_i^* = 1$, a contradiction. Hence J_l is also a separate job.   □

Recall that $\frac{1}{\rho} = p_n \leq p_j \leq p_1 = \frac{1}{\tau}$ for any j. Thus

$$
P \geq np_n = \frac{n}{\rho}. \tag{6}
$$

If $\rho < 2$, then the sum of the processing times of any two jobs of $\mathcal{J}$ is greater than 1. Thus all jobs are separate jobs both in σ^{NE} and σ^* by Lemmas 1(ii) and 2(ii). Thus $C^{NE}(\mathcal{J}) = C^*(\mathcal{J})$. Hence, we have $\rho \geq 2$ and thus $\frac{\rho+1}{3} \geq 1$. We will analysis and exclude the situation which can not appear in the minimal counterexample in the following three lemmas.

Lemma 14. *If any one of the following conditions: (i) $\tau \geq \frac{9}{4}$; (ii) $P \leq 1$; (iii) $n_{min}^* \geq 2$ holds, then $\frac{C^{NE}(\mathcal{J})}{C^*(\mathcal{J})} \leq \frac{\rho+1}{3}$.*

Proof. (i)By Lemma 3(ii), $\tau \geq \frac{9}{4}$ and (6), $\frac{C^{NE}(\mathcal{J})}{C^*(\mathcal{J})} \leq \frac{n+P}{2P\sqrt{\tau}} \leq \frac{n+P}{3P} = \frac{\frac{n}{P}+1}{3} \leq \frac{\rho+1}{3}$.

(ii)By Theorem 2, (6) and $P \leq 1$, $\frac{C^{NE}(\mathcal{J})}{C^*(\mathcal{J})} \leq \frac{n+1}{3} \leq \frac{\rho P+1}{3} \leq \frac{\rho+1}{3}$.

(iii)Clearly, $C^*(\mathcal{J}) \geq m^* + n^*_{min}P$. By Lemma 3(i), (1) and $n^*_{min} \geq 2$,

$$\frac{C^{NE}(\mathcal{J})}{C^*(\mathcal{J})} \leq \frac{n+P}{m^* + n^*_{min}P} = \frac{\frac{n}{P}+1}{\frac{m^*}{P} + n^*_{min}} \leq \frac{\rho+1}{1 + n^*_{min}} \leq \frac{\rho+1}{3}.$$

□

By Lemma 14, we assume that $\tau < \frac{9}{4}$, $P > 1$ and $n^*_{min} = 1$ in the following. Since J_1 is the job with the largest processing time, and there is at least one machine in σ^* which processes exactly one job. We have

$$\mathcal{J}_1^* = \{J_1\} \tag{7}$$

by Lemma 13. Similarly, if $n^{NE}_{min} = 1$, then $\mathcal{J}_1^{NE} = \{J_1\}$ according to Lemma 7(ii). Thus $\mathcal{J}_1^{NE} = \mathcal{J}_1^*$ and $\mathcal{J}\backslash\mathcal{J}_1^{NE}$ is also a counterexample by Lemma 4, which contradicts the definition of $\mathcal{J}$. Hence, we have $n_1^{NE} \geq n^{NE}_{min} \geq 2$, and

$$1 \geq L_1^{NE} \geq p_1 + (n_1^{NE} - 1)p_n = p_1 + (n_1^{NE} - 1)\frac{1}{\rho} = \frac{1}{\tau} + \frac{n_1^{NE} - 1}{\rho}. \tag{8}$$

Lemma 15. *If any one of the following two conditions: (i) $2 \leq \rho < 3$; (ii) $3 \leq \rho < 4$ and $n_1^{NE} = 3$ holds, then $n_i^* \leq 2$ for any $1 \leq i \leq m^*$.*

Proof. If $2 \leq \rho < 3$, then $3p_n = \frac{3}{\rho} > 1 \geq L_i^*$. The result clearly follows. If $3 \leq \rho < 4$ and $n_1^{NE} = 3$, then $L_1^* = p_1 \leq 1 - \frac{2}{\rho} < \frac{2}{\rho} = 2p_n$ by (8). Assume that there exists a machine M_k, $k \geq 2$ such that $n_k^* \geq 3$, and job $J_j \in \mathcal{J}_k^*$. Then $L_k^* - p_j \geq 2p_n \geq L_i^*$, and $n_1^* \geq n_k^* = 3$ by Lemma 12, contradicts (7). □

Lemma 16. *If any one of the following two conditions: (i) $2 \leq \rho < 3$; (ii) $3 \leq \rho < 4$ and $n_1^{NE} = 3$ holds, then $\frac{C^{NE}(\mathcal{J})}{C^*(\mathcal{J})} \leq \frac{\rho+1}{3}$.*

Proof. When $2 \leq \rho < 4$, then $4p_n = \frac{4}{\rho} > 1$. Hence $2 \leq n^{NE}_{min} \leq n_i^{NE} \leq 3$ for any $1 < i \leq m^{NE}$. Consider a subset of machines $\mathcal{M}_3$ which consists of machines that exactly select three jobs select in σ^{NE}. Denote by m_3 and P_3 the number of machines of $\mathcal{M}_3$ and the total processing time of jobs selecting one machine of $\mathcal{M}_3$, respectively. Then

$$n = 3m_3 + 2(m^{NE} - m_3) \geq 3m_3 \tag{9}$$

and

$$C^{NE}(\mathcal{J}) = m_3 + 3P_3 + (m^{NE} - m_3) + 2(P - P_3)$$
$$= m_3 + 3P_3 + \frac{n - 3m_3}{2} + 2(P - P_3) = \frac{n}{2} + 2P + P_3 - \frac{m_3}{2}. \tag{10}$$

On the other hand, $1 \leq n_i^* \leq 2$ for any $1 \leq i \leq m^*$ by Lemma 15. Consider the subset of machines $\mathcal{M}_1^*$ which consists of machines processing exactly one job in σ^*. Denote by m_1^* and P_1^* the number of machines of $\mathcal{M}_1^*$ and the total processing time of jobs processed on one machine of $\mathcal{M}_1^*$, respectively. Then $n = m_1^* + 2(m^* - m_1^*)$ and

$$C^*(\mathcal{J}) = m_1^* + P_1^* + (m^* - m_1^*) + 2(P - P_1^*)$$

$$= m_1^* + P_1^* + \frac{n - m_1^*}{2} + 2(P - P_1^*) = \frac{n}{2} + 2P + \frac{m_1^*}{2} - P_1^*. \quad (11)$$

Define

$$\Delta_3 = (\rho + 1)C^*(\mathcal{J}) - 3C^{NE}(\mathcal{J}).$$

In order to prove $\frac{C^{NE}(\mathcal{J})}{C^*(\mathcal{J})} \leq \frac{\rho+1}{3}$, it is sufficient to prove that $\Delta_3 \geq 0$. By (10), (11),

$$\Delta_3 = (\rho + 1)\left(\frac{n}{2} + 2P + \frac{m_1^*}{2} - P_1^*\right) - 3\left(\frac{n}{2} + 2P + P_3 - \frac{m_3}{2}\right)$$

$$= (\rho - 2)\left(\frac{n}{2} + 2P\right) + (\rho + 1)\left(\frac{m_1^*}{2} - P_1^*\right) - 3\left(P_3 - \frac{m_3}{2}\right). \quad (12)$$

(i) If $2 \leq \rho < 3$, then $3p_n = \frac{3}{\rho} > 1$. Thus $\mathcal{M}_3 = \emptyset$, $P_3 = 0$ and $n_1^{NE} = 2$. By (8), $p_1 \leq 1 - \frac{1}{\rho}$. Thus,

$$P_1^* \leq p_1 m_1^* \leq \frac{\rho - 1}{\rho} m_1^*, \quad (13)$$

and

$$P \geq P_1^* + (n - m_1^*)p_n = P_1^* + \frac{n - m_1^*}{\rho} \quad (14)$$

Substituting (13), (14) to (12), we have

$$\Delta_3 = (\rho - 2)\left(\frac{n}{2} + 2P\right) + (\rho + 1)\left(\frac{m_1^*}{2} - P_1^*\right)$$

$$\geq (\rho - 2)\left(\frac{n}{2} + 2P_1^* + \frac{2(n - m_1^*)}{\rho}\right) + (\rho + 1)\left(\frac{m_1^*}{2} - P_1^*\right)$$

$$= (\rho - 2)\left(\frac{n}{2} + \frac{2(n - m_1^*)}{\rho}\right) + (\rho - 5)P_1^* + \frac{\rho + 1}{2}m_1^*$$

$$\geq (\rho - 2)\left(\frac{n}{2} + \frac{2(n - m_1^*)}{\rho}\right) + (\rho - 5)P_1^* + \frac{\rho + 1}{2}\frac{\rho}{\rho - 1}P_1^*$$

$$= (\rho - 2)\left(\frac{n}{2} + \frac{2(n - m_1^*)}{\rho}\right) + \frac{(\rho - 2)(3\rho - 5)}{2(\rho - 1)}P_1^* \geq 0.$$

(ii) If $3 \leq \rho < 4$ and $n_1^{NE} = 3$, then $p_1 \leq 1 - \frac{2}{\rho} < \frac{1}{2}$ by (8). Thus $P_1^* \leq \frac{m_1^*}{2}$. Moreover, $P_3 \leq m_3$ since $L_i^{NE} \leq 1$. Together with (9), we have

$$\Delta_3 \geq (\rho - 2)\left(\frac{n}{2} + 2P\right) - 3\left(P_3 - \frac{m_3}{2}\right) \geq (\rho - 2)\left(\frac{n}{2} + 2P\right) - \frac{3m_3}{2}$$

$$\geq \frac{n}{2} - \frac{3m_3}{2} + 2P \geq 0.$$

$\square$

Proof of Theorem 2. Since $\mathcal{J}$ is the minimal counterexample, $\frac{C^{NE}(\mathcal{J})}{C^*(\mathcal{J})} > \frac{\rho+1}{3}$.
Moreover, the smallest processing time of jobs of $\mathcal{J}\backslash\mathcal{J}_1^{NE}$ is no smaller than p_n.
Hence, $\frac{C^{NE}(\mathcal{J}\backslash\mathcal{J}_1^{NE})}{C^*(\mathcal{J}\backslash\mathcal{J}_1^{NE})} \le \frac{\rho+1}{3}$. Define

$$\Delta_4 = (\rho+1)(L_1^{NE} - p_1) + (\rho+1-3n_1^{NE})L_1^{NE} + \rho - 2.$$

Then by Lemma 5(i), (iii),

$$\begin{aligned}
\Delta_4 &= (\rho+1)(1 + 2L_1^{NE} - p_1 + C^*(\mathcal{J}\backslash\mathcal{J}_1^{NE})) \\
&\quad -3\left(1 + n_1^{NE}L_1^{NE} + \frac{\rho+1}{3}C^*(\mathcal{J}\backslash\mathcal{J}_1^{NE})\right) \\
&\le (\rho+1)(1 + 2L_1^{NE} - p_1 + C^*(\mathcal{J}\backslash\mathcal{J}_1^{NE})) \\
&\quad -3(1 + n_1^{NE}L_1^{NE} + C^{NE}(\mathcal{J}\backslash\mathcal{J}_1^{NE})) \\
&\le (\rho+1)C^*(\mathcal{J}) - 3C^{NE}(\mathcal{J}) < 0.
\end{aligned}$$

However, we will show below that $\Delta_4 \ge 0$, which leads contradiction.
If $\rho+1 \ge 3n_1^{NE}$, then $\Delta_4 \ge 0$. Otherwise, $\rho < 3n_1^{NE} - 1$. We distinguish
several cases according to the value of n_1^{NE}. If $n_1^{NE} = 2$, then $\rho < 5$. By (8) and
$\rho \ge 3$,

$$\Delta_4 \ge (\rho+1)\frac{1}{\rho} + \rho - 5 + \rho - 2 = \frac{2\rho^2 - 6\rho + 1}{\rho} \ge 0.$$

If $n_1^{NE} = 3$, then $\rho < 8$. We only need to consider the case of $\rho \ge 4$ by Lemma
16. By (8),

$$\Delta_4 \ge (\rho+1)\frac{2}{\rho} + \rho - 8 + \rho - 2 = \frac{2\rho^2 - 8\rho + 2}{\rho} \ge 0.$$

If $n_1^{NE} \ge 4$, we have $\rho \ge \frac{n_1^{NE}-1}{1-\frac{1}{7}} \ge \frac{9}{5}(n_1^{NE} - 1) \ge \frac{27}{5}$ by (8). Hence,

$$\begin{aligned}
\Delta_4 &\ge (\rho+1)\frac{n_1^{NE} - 1}{\rho} + (\rho+1-3n_1^{NE}) + \rho - 2 \\
&= \frac{1-2\rho}{\rho}n_1^{NE} + 2\rho - 1 - \frac{\rho+1}{\rho} \\
&\ge \frac{1-2\rho}{\rho}\left(\frac{5\rho}{9}+1\right) + 2\rho - 1 - \frac{\rho+1}{\rho} = \frac{8\rho - 31}{9} \ge 0.
\end{aligned}$$

□

The bound given in Theorem 2 may not be tight. Currently, we only have the
following lower bound on the PoA.

Theorem 3. *For any square number ρ, there exists a job set $\mathcal{J}$, such that*
$\frac{C^{NE}(\mathcal{J})}{C^*(\mathcal{J})} \ge \frac{\rho+1}{2\sqrt{\rho}}.$

Proof. Consider a job set $\mathcal{J}$ consists of ρ jobs. All jobs have processing times $\frac{1}{\rho}$. All jobs select the same machine forms a NE σ. In fact, the cost of any job is $1 + \frac{1}{\rho}$, which equals to the cost that it activates a new machine. Hence, $C^{NE}(\mathcal{J}) \geq \rho + 1$. On the other hand, consider a schedule that every $\sqrt{\rho}$ jobs are processed on one machine. Clearly, $C^*(\mathcal{J}) \leq \sqrt{\rho} + \rho \frac{\sqrt{\rho}}{\rho} = 2\sqrt{\rho}$. Consequently, we have $\frac{C^{NE}(\mathcal{J})}{C^*(\mathcal{J})} \geq \frac{\rho+1}{2\sqrt{\rho}}$. □

References

1. Berenbrink, P., Goldberg, L., Goldberg, P., Martin, R.: Utilitarian resource assignment. Journal of Discrete Algorithm 4, 567–587 (2006)
2. Chen, B., Gürel, S.: Resource allocation games of utilitarian social objectives. Journal of Scheduling 15, 157–164 (2012)
3. Czumaj, A., Vöcking, B.: Tight bounds for worst-case equilibria. ACM Transactions on Algorithms 3(4) (2007)
4. Epstein, L.: Equilibria for two parallel links: the strong price of anarchy versus the price of anarchy. Acta Informatica 47, 375–389 (2010)
5. Epstein, L., van Stee, R.: The price of anarchy on uniformly related machines revisited. Information and Computation 212, 37–54 (2012)
6. Feldmann, R., Gairing, M., Lücking, T., Monien, B., Rode, M.: Nashification and the coordination ratio for a selfish routing game. In: Baeten, J.C.M., Lenstra, J.K., Parrow, J., Woeginger, G.J. (eds.) ICALP 2003. LNCS, vol. 2719, pp. 514–526. Springer, Heidelberg (2003)
7. Feldman, M., Tamir, T.: Conflicting congestion effects in resource allocation games. Operations Research 60, 529–540 (2012)
8. Fotakis, D., Kontogiannis, S., Koutsoupias, E., Mavronicolas, M., Spirakis, P.: The structure and complexity of Nash equilibria for a selfish routing game. Theoretical Computer Science 410, 3305–3326 (2009)
9. Koutsoupias, E., Papadimitriou, C.H.: Worst-case equilibria. Computer Science Review 3, 65–69 (2009)

Approximation Algorithms
for Bandwidth Consecutive Multicolorings
(Extended Abstract)

Yuji Obata and Takao Nishizeki

Kwansei Gakuin University, 2-1 Gakuen, Sanda 669-1337, Japan
{bnb86950,nishi}@kwansei.ac.jp

Abstract. Let G be a graph in which each vertex v has a positive integer weight $b(v)$ and each edge (v, w) has a nonnegative integer weight $b(v, w)$. A bandwidth consecutive multicoloring, simply called a b-coloring of G, assigns each vertex v a specified number $b(v)$ of consecutive positive integers as colors of v so that, for each edge (v, w), all integers assigned to vertex v differ from all integers assigned to vertex w by more than $b(v, w)$. The maximum integer assigned to vertices is called the span of the coloring. The b-coloring problem asks to find a b-coloring of a given graph G with the minimum span. In the paper, we present four efficient approximation algorithms for the problem, which have theoretical performance guarantees for the computation time, the span of a found b-coloring and the approximation ratio. We also obtain several upper bounds on the minimum span, expressed in terms of the maximum b-degrees, one of which is an extension of Brooks' theorem on an ordinary coloring.

1 Introduction

An ordinary coloring of a graph G assigns each vertex a color so that, for each edge (v, w), the color assigned to v differs from the color assigned to w [13]. The problem of finding a coloring of a graph G with the minimum number $\chi(G)$ of colors often appears in the scheduling, task-allocation, etc.[4,11,13]; $\chi(G)$ is called the *chromatic number* of G. However, the problem is NP-hard, and it is difficult to find a good approximate solution. More precisely, for all $\epsilon > 0$, approximating the chromatic number $\chi(G)$ within $n^{1-\epsilon}$ is NP-hard [14], where n is the number of vertices in G.

In this paper we deal with a generalized coloring, called a "bandwidth consecutive multicoloring"[10]. Each vertex v of a graph G has a positive integer weight $b(v)$, while each edge (v, w) of G has a non-negative integer weight $b(v, w)$. A *bandwidth consecutive multicoloring* F of G is an assignment of positive integers to vertices such that

(a) each vertex v of G is assigned a set $F(v)$ of $b(v)$ consecutive positive integers; and

J. Chen, J.E. Hopcroft, and J. Wang (Eds.): FAW 2014, LNCS 8497, pp. 194–204, 2014.

(b) for each edge (v, w) of G, all integers assigned to vertex v differ from all integers assigned to vertex w by more than $b(v, w)$, that is, $b(v, w) < |i - j|$ for any integers $i \in F(v)$ and $j \in F(w)$. (Note that our edge weight $b(v, w)$ is one less than the conventional edge weight.)

A bandwidth consecutive multicoloring F is simply called a *b-coloring* for a weight function b. The maximum integer assigned to vertices is called the *span* of a *b*-coloring F, and is denoted by span(F). The *b-chromatic number* $\chi_b(G)$ of a graph G is the minimum span over all *b*-colorings F of G. A *b*-coloring F is called *optimal* if span(F) = $\chi_b(G)$. The *b-coloring problem* asks to find an optimal *b*-coloring of a given graph G. The ordinary vertex-coloring is merely a *b*-coloring when $b(v) = 1$ for every vertex v and $b(v, w) = 0$ for every edge (v, w). The "bandwidth coloring" or "channel assignment" is a *b*-coloring when $b(v) = 1$ for every vertex v [8,9,10]. In Fig. 1(a), a vertex is drawn as a circle, in which the weight is written, while an edge is drawn as a straight line segment, to which the weight is attached. The *b*-chromatic number $\chi_b(G)$ of the graph G in Fig. 1(a) is 10, and an optimal *b*-coloring F of G with span(F) = 10 is drawn in Fig. 1(a), where a set $F(v)$ is attached to each vertex v.

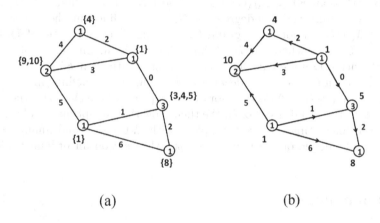

(a)   (b)

Fig. 1. (a)A graph G and its optimal *b*-coloring F, and (b)the optimal *b*-coloring f corresponding to F and an acyclic orientation of G

A *b*-coloring problem often arises in the assignment of radio channels of cellular communication systems [8,9,10] and in the non-preemptive task scheduling [11]. The $b(v)$ consecutive integers assigned to a vertex v correspond to the contiguous bandwidth of a channel v or a consecutive time period of a task v. The weight $b(v, w)$ assigned to edge (v, w) represents the requirement that the frequency band or time period of v must differ from that of w by more than $b(v, w)$. The span of an optimal *b*-coloring corresponds to the minimum total bandwidth or the minimum makespan.

Since the *b*-coloring problem is strongly NP-hard, there is no FPTAS (fully polynomial-time approximation scheme) for general graphs unless P = NP. On the other hand, the problem is NP-hard even for graphs with bounded tree-width, and there is an FPTAS for graphs with bounded tree-width [10]. However, the computation time is very large; the FPTAS takes time $O(n^4/\varepsilon^3)$ even for series-parallel graphs, where n is the number of vertices in a graph and ϵ is the approximation error rate. For a bandwidth coloring or multicoloring, several heuristics using tabu search and genetic methodologies have been proposed and experimentally compared on their performances [2,6,7]. Thus, it is desired to obtain an efficient approximation algorithm with theoretical performance guarantees for the b-coloring problem, which runs in linear time or $O(m \log n)$ time, where m is the number of edges in a graph.

In this paper we first present four efficient approximation algorithms for the b-coloring problem, which have theoretical performance guarantees for the computation time, the span of a found b-coloring and the approximation ratio. The first algorithm **Color-1** finds a b-coloring F with span$(F) \leq (c-1)\chi_b(G)$ in linear time when a given graph G is ordinarily vertex-colored with $c(\geq 2)$ colors. Hence, the approximation ratio of **Color-1** is at most $c-1$, and is at most three particularly for planar graphs. The second algorithm **Color-2** is a variant of **Color-1**. The third algorithm **Delta** finds a b-coloring F of a given graph G with span$(F) \leq \Delta_{1b}(G) + 1$ in time $O(m \log n)$, where $\Delta_{1b}(G)$ is the "maximum uni-directional b-degree" of G, newly defined in the paper. Thus $\chi_b(G) \leq \Delta_{1b}(G) + 1$ for every graph G. The approximation ratio of **Delta** is at most $\Delta(G) + 1$, where $\Delta(G)$ is the ordinary maximum degree of G. The fourth algorithm **Degenerate** finds a b-coloring F with span$(F) \leq k+1$ in time $O(m \log \Delta(G))$ if G is a "(k, b)-degenerated graph," newly defined in the paper. It implies that $\chi_b(G) \leq \Delta_{2b}(G)+1$ for every graph G, where $\Delta_{2b}(G)$ is the "maximum bi-directional b-degree" of G. We then show that an optimal b-coloring can be found in linear time for every graph G with $\Delta(G) \leq 2$, and finally present a b-coloring analogue of the famous Brooks' theorem on an ordinary coloring [4,13].

2   Preliminaries

In this section, we define several terms and present a known result.

Let $G = (V, E)$ be a simple undirected graph with vertex set V and edge set E. Let $n=|V|$ and $m=|E|$ throughout the paper. For two integers α and β, we denote by $[\alpha, \beta]$ the set of all integers z with $\alpha \leq z \leq \beta$. Let $\mathbb{N}$ be the set of all positive integers, that are regarded as colors.

A b-coloring $F : V \to 2^{\mathbb{N}}$ of a graph $G = (V, E)$ can be simply represented by a function $f \colon V \to \mathbb{N}$ such that $f(v) = \max F(v)$ for every vertex $v \in V$ [10]. Such a function f is also called a b-*coloring* of G. Obviously, span$(F) = \max_{v \in V} f(v)$, and hence we often denote span(F) by span(f). The optimal b-coloring F in Fig. 1(a) corresponds to the optimal b-coloring f in Fig. 1(b), where $f(v)$ is attached to each vertex $v \in V$.

Let f be a b-coloring of a graph G. For every vertex v of G, let

$$B(v) = [1, b(v) - 1].$$

Then $f(v) \notin B(v)$. In this sense, we call $B(v)$ the *forbidden base band of a vertex* v. We denote by $\delta(v)$ the *width* of $B(v)$, that is,

$$\delta(v) = |B(v)| = b(v) - 1.$$

For every edge (v, w) of G, let

$$\alpha(v, w) = f(w) - (b(w) - 1) - b(v, w)$$

and

$$\beta(v, w) = f(w) + b(v, w) + b(v) - 1,$$

and let

$$B(v, w) = [\alpha(v, w), \beta(v, w)].$$

Then $f(v) \notin B(v, w)$, and we call $B(v, w)$ the *forbidden band of a vertex* v *for a neighbor* w. We denote by $\delta(v, w)$ the *width* of $B(v, w)$, that is,

$$\delta(v, w) = |B(v, w)| = b(v) - 1 + 2b(v, w) + b(w).$$

The band $B(v, w)$ is not necessarily the same as $B(w, v)$, but $\delta(v, w) = \delta(w, v)$. Let

$$B_{\mathrm{h}}(v, w) = [f(w), \beta(v, w)],$$

and we call $B_{\mathrm{h}}(v, w)$ the *forbidden higher band of a vertex* v *for a neighbor* w. Clearly $B_{\mathrm{h}}(v, w) \subseteq B(v, w)$. We denote by $\delta_{\mathrm{h}}(v, w)$ the *width* of $B_{\mathrm{h}}(v, w)$, that is,

$$\delta_{\mathrm{h}}(v, w) = |B_{\mathrm{h}}(v, w)| = b(v, w) + b(v).$$

Only the $\delta_{\mathrm{h}}(v, w)$ colors in $B_{\mathrm{h}}(v, w)$ are forbidden for v if $f(v) > f(w)$. Note that $\delta(v) = 0$ for every vertex v and $\delta(v, w) = \delta_{\mathrm{h}}(v, w) = 1$ for every edge (v, w) if every vertex weight is 1 and every edge weight is 0.

Orient all edges of an undirected graph G so that the resulting directed graph $\overrightarrow{G}$ is acyclic. Such a directed graph $\overrightarrow{G}$ is called an *acyclic orientation* of G. Figure 1(b) depicts an acyclic orientation of the graph G in Fig. 1(a). The *length* $l(P, \overrightarrow{G})$ of a directed path P in $\overrightarrow{G}$ is defined to be the sum of weights of all vertices and edges in P. For a vertex v in $\overrightarrow{G}$, we denote by $l(v, \overrightarrow{G})$ the length of the longest directed path in $\overrightarrow{G}$ ending at v. We denote by $l_{\max}(\overrightarrow{G})$ the length of the longest directed path in $\overrightarrow{G}$, and hence

$$l_{\max}(\overrightarrow{G}) = \max_{v \in V} l(v, \overrightarrow{G}).$$

The following theorem [10] is known as an extension of the Gallai-Roy theorem on an ordinary coloring (see for example [13]).

Theorem 1. *The following (a) and (b) hold for a graph $G = (V, E)$ with a weight function b:*

(a) *If $\overrightarrow{G}$ is an acyclic orientation of G and $f(v) = l(v, \overrightarrow{G})$ for every vertex $v \in V$, then f is a b-coloring of G; and*

(b) *The b-chromatic number $\chi_b(G)$ of G satisfies*

$$\chi_b(G) = \min_{\overrightarrow{G}} l_{\max}(\overrightarrow{G}),$$

where the minimum is taken over all acyclic orientations $\overrightarrow{G}$ of G.

3   Algorithms Color-1 and Color-2

In this section we present two approximation algorithms **Color-1** and **Color-2** for the b-coloring problem.

One may assume without loss of generality that a given graph $G = (V, E)$ has no isolated vertex, and hence every vertex has degree one or more. Assume that G is ordinarily vertex-colored with c colors, color 1, color 2, ..., color c. Of course, $2 \le \chi(G) \le c$, where $\chi(G)$ is the ordinary chromatic number of G. Then the vertex set V of G is partitioned to c subsets $V_1, V_2, ..., V_c$, called the color classes, so that every vertex v in V_i, $1 \le i \le c$, is colored with color i and hence V_i is an independent set of G.

Orient every edge (v, w) of G from v to w if the color of v is smaller than the color of w. The resulting orientation $\overrightarrow{G}$ of G is obviously acyclic. Let $f(v) = l(v, \overrightarrow{G})$ for every vertex v in G, then Theorem 1(a) implies that f is a b-coloring of G. The first algorithm **Color-1** simply outputs such a b-coloring f of G.

One can obtain $\overrightarrow{G}$ in linear time from a given graph G colored with c colors. Furthermore, one can compute $l(v, \overrightarrow{G})$ for all vertices v in linear time since $\overrightarrow{G}$ is acyclic [1]. Thus **Color-1** takes linear time. We can show that $\mathrm{span}(f) \le (c - 1)\chi_b(G)$, as follows. Let P be the longest directed path in the acyclic orientation $\overrightarrow{G}$ above, then $\mathrm{span}(f) = l(P, \overrightarrow{G})$. Let $P = v_1, v_2, ..., v_p$ for some vertices $v_1, v_2, ..., v_p$, then $p \ge 2$ since G has no isolated vertex. Let $v_1 \in V_{i_1}, v_2 \in V_{i_2}, ..., v_p \in V_{i_p}$ for some indices $i_1, i_2, ..., i_p$, then the orientation implies that $1 \le i_1 < i_2 < ... < i_p \le c$, and hence $2 \le p \le c$. The definition of the length of a directed path implies

$$l(P, \overrightarrow{G}) = \sum_{j=1}^{p} b(v_j) + \sum_{j=1}^{p-1} b(v_j, v_{j+1}).$$

Clearly

$$b(v_j) + b(v_j, v_{j+1}) + b(v_{j+1}) \le \chi_b(G)$$

for every j, $1 \le j \le p$, and $1 \le b(v_j)$ for every j, $1 \le j \le p$. Therefore

$$\mathrm{span}(f) = l(P, \overrightarrow{G}) \le (p - 1)\chi_b(G) - (p - 2) \le (c - 1)\chi_b(G).$$

We thus have the following theorem.

Theorem 2. *For a graph G colored with $c(\geq 2)$ colors, Algorithm* **Color-1** *finds a b-coloring f with* $\mathrm{span}(f) \leq (c-1)\chi_b(G)$ *in linear time, and the approximation ratio of* **Color-1** *is at most $c-1$.*

Every bipartite graph can be colored by two colors, every series-parallel graph by three colors [12], and every planar graph by four colors [13]. Therefore, we have the following corollary.

Corollary 1
(a) *If $G = (V, E)$ is a bipartite graph without isolated vertices, then* **Color-1** *finds an optimal b-coloring and*

$$\chi_b(G) = \max\{b(v) + b(v, w) + b(w) \mid (v, w) \in E\};$$

(b) *The approximation ratio of* **Color-1** *is at most two for series-parallel graphs; and*
(c) *The approximation ratio of* **Color-1** *is at most three for planar graphs.*

The second algorithm **Color-2** is a simple variant of **Color-1**. Algorithm **Color-2** considers many acyclic orientations of a given graph G although **Color-1** considers only a single acyclic orientation $\overrightarrow{G}$. Let $q = c!$, and let $\pi_1, \pi_2, ..., \pi_q$ be all the permutations of set $[1, c]$ of colors $1, 2, ..., c$. For each permutation π_j, $1 \leq j \leq q$, let $\overrightarrow{G_j}$ be an acyclic orientation of G constructed as follows: replace colors $1, 2, ..., c$ of the given coloring of G by colors $\pi_j(1), \pi_j(2), ..., \pi_j(c)$, respectively; $\overrightarrow{G_j}$ is an acyclic orientation in which an edge (v, w) of G is oriented from v to w if the new color of v is smaller than that of w in the resulting coloring of G. Algorithm **Color-2** finds a number q of b-colorings f_j, $1 \leq j \leq q$, such that $f_j(v) = l(v, \overrightarrow{G_j})$ for every vertex v of G, and outputs one of them with the smallest span as a b-coloring f of G.

Of course, the span of a b-coloring f obtained by **Color-2** is no more than that by **Color-1**, and the approximation ratio of **Color-2** is at most $c-1$. There is a graph colored with c colors for which the approximation ratio of **Color-1** and **Color-2** approaches $c-1$. (The details are omitted in this extended abstract.)

4   Algorithm Delta

The *degree* $d(v, G)$ of a vertex v in a graph G is the number of neighbors of v in G. We denote by $\Delta(G)$ the *maximum degree* of G, and hence $\Delta(G) = \max_{v \in V} d(v, G)$. Then clearly the chromatic number $\chi(G)$ of G satisfies $\chi(G) \leq \Delta(G) + 1$ [4,13]. In this section we generalize this result to a b-coloring.

In an ordinary coloring of a graph G, the colors of all neighbors of a vertex v are "forbidden" for v, and the number of these colors is no more than the degree $d(v, G)$ of v. For a b-coloring of G, we define the *uni-directional b-degree* $d_{1b}(v, G)$ of a vertex v in a graph G as follows:

$$d_{1b}(v, G) = \delta(v) + \sum_w \delta_h(v, w),$$

where $\delta(v)(= b(v) - 1)$ is the width of the forbidden base band $B(v)$ of v, the summation above is taken over all neighbors w of v in G, and $\delta_h(v,w)(= b(v,w) + b(v))$ is the width of the forbidden higher band $B_h(v,w)$ of v for a neighbor w. If $f(v) = \text{span}(f)$ for a b-coloring f of G, then $f(v) > f(w)$ for every neighbor w of v and hence the number of forbidden colors of v is no more than $d_{1b}(v, G)$. The *maximum uni-directional b-degree* $\Delta_{1b}(G)$ of G is defined as follows:

$$\Delta_{1b}(G) = \max_{v \in V} d_{1b}(v, G).$$

For example, $\Delta_{1b}(G) = 19$ for the graph G in Fig. 1(a). If every vertex weight is 1 and every edge weight is 0, then a b-coloring is merely an ordinary coloring, $d_{1b}(v, G) = d(v, G)$ for every vertex v, and hence $\Delta_{1b}(G) = \Delta(G)$.

The third algorithm **Delta** finds a b-coloring f with $\text{span}(f) \leq \Delta_{1b}(G) + 1$ for a graph G. Every b-coloring f of G satisfies $f(w_1) \leq f(w_2) \leq ... \leq f(w_n)$ for some vertex ordering $w_1, w_2, ..., w_n$ of G. Clearly, for every vertex w_i, $1 \leq i \leq n$,

$$f(w_i) \geq \max\{b(w_i), \max_{w_j}\{f(w_j) + b(w_j, w_i) + b(w_i)\}\},$$

where the second maximum is taken over all lower indexed neighbors w_j of w_i with $j < i$. Algorithm **Delta** indeed finds such a vertex ordering $w_1, w_2, ..., w_n$ and computes $f(w_1), f(w_2), ..., f(w_n)$ in this order so that

$$f(w_i) = \max\{b(w_i), \max_{w_j}\{f(w_j) + b(w_j, w_i) + b(w_i)\}\}$$

for every vertex w_i, $1 \leq i \leq n$. The details of **Delta** are as follows, where set W consists of all vertices $w \in V$ such that the colors $f(w)$ have been decided.

Algorithm. Delta(G, f)

$W := \emptyset$;
for every vertex $v \in V$ **do**
    $f(v) := b(v)$; (initialization of $f(v)$)
end for
while $W \neq V$ **do**
    {
    let w be a vertex $v \in V/W$ with minimum $f(v)$; (choice of w)
    $W := W \cup \{w\}$; (insert the chosen vertex w to set W)
    for each vertex $v \in V/W$ adjacent to w **do**
        $f(v) := \max\{f(v), f(w) + b(w,v) + b(v)\}$; (update of $f(v)$)
    end for
    }
end while

We can show that **Delta** finds a b-coloring f of G, that $\text{span}(f) \leq \Delta_{1b}(G) + 1$, and that $\Delta_{1b}(G) + 1 \leq (\Delta(G) + 1)\chi_b(G)$. (The details are omitted in this extended abstract.) Using a binary heap [1] as a data structure to represent the set of vertices $v \in V/W$ having $f(v)$ as a key-attribute, one can implement **Delta** to run in time $O(m \log n)$, similarly as Dijkstra's shortest path algorithm. We thus have the following theorem.

Theorem 3. *Algorithm* **Delta** *finds a b-coloring f of a graph G with* span(f) $\leq \Delta_{1b}(G)+1$ *in time* $O(m \log n)$, *and the approximation ratio is at most* $\Delta(G)+1$.

One can immediately obtain the following corollary.

Corollary 2
(a) $\chi_b(G) \leq \Delta_{1b}(G) + 1$ *for every graph G; and*
(b) *The approximation ratio of* **Delta** *is at most* 4 *for graphs G with* $\Delta(G)$ ≤ 3.

5   Algorithm Degenerate

If a graph G has a vertex of degree at most $k \in \mathbb{N}$, then delete it from G. If the resulting graph has a vertex of degree at most k, then delete it from the graph. If all vertices of G can be deleted in this way, then clearly $\chi(G) \leq k+1$ and G is called k-*degenerated* [4]. In this section, we generalize this result to a b-coloring. We define the *bi-directional b-degree* $d_{2b}(v, G)$ of a vertex v in a graph $G = (V, E)$ as follows:

$$d_{2b}(v, G) = \delta(v) + \sum_w \delta(v, w),$$

where the summation is taken over all neighbors w of v in G and $\delta(v, w)(= b(v) - 1 + 2b(v, w) + b(w))$ is the width of the forbidden band $B(v, w)$ of v for a neighbor w. Clearly, the number of forbidden colors of v is no more than $d_{2b}(v, G)$. The *maximum bi-directional b-degree* $\Delta_{2b}(G)$ of G is defined as follows:

$$\Delta_{2b}(G) = \max_{v \in V} d_{2b}(v, G).$$

Clearly $d_{1b}(v, G) \leq d_{2b}(v, G)$ for every vertex v, and hence $\Delta_{1b}(G) \leq \Delta_{2b}(G)$. If every vertex weight is 1 and every edge weight is 0, then $d_{2b}(v, G) = d(v, G)$ for every vertex v and hence $\Delta_{2b}(G) = \Delta(G)$.

We define a graph G to be (k, b)-*degenerated* for an integer $k \in \mathbb{N}$ if G has a vertex ordering $v_1, v_2, ..., v_n$ such that $d_{2b}(v_i, G_i) \leq k$ for every index i, $1 \leq i \leq n$, where G_i is a subgraph of G induced by the first i vertices $v_1, v_2, ..., v_i$. The graph in Fig. 1(a) is $(15, b)$-degenerated.

One can prove by induction on i that if G is (k, b)-degenerated then G_i has a b-coloring f with span(f) $\leq k + 1$ and hence $\chi_b(G) \leq k + 1$.

The fourth algorithm **Degenerate** successively finds b-colorings of $G_1, G_2, ..., G_n (= G)$ in this order. Indeed it employs a simple greedy technique; when extending a b-coloring of G_{i-1} to that of G_i, $2 \leq i \leq n$, **Degenerate** always chooses, as $f(v_i)$, the *smallest* positive integer j that is not countained in B_i^+, where B_i^+ is the set of all positive integers contained in $B(v_i) \cup (\cup_w B(v_i, w))$ and the second union is taken over all neighbors w of v_i in G_i.

We then show that **Degenerate** can be implemented so that it runs in polynomial time. The set B_i^+ is a union of one or more pairwise disjoint sets $[l_1, u_1]$, $[l_2, u_2], ..., [l_p, u_p]$ such that

$$1 \leq l_1 \leq u_1$$
$$u_1 + 2 \leq l_2 \leq u_2$$
$$u_2 + 2 \leq l_3 \leq u_3$$
$$\cdots\cdots$$
$$u_{p-1} + 2 \leq l_p \leq u_p.$$

Then $u_1 + 1$ is the smallest positive integer $j \notin B_i^+$. Sorting the set $\{a(v_i, w) | (v_i, w) \in G_i\}$ of $d(v_i, G_i)$ integers, one can easily find u_1 and hence $j = u_1 + 1$ in time $O(d(v_i, G_i) \log d(v_i, G_i))$ from integer pairs $(1, b(v_i) - 1)$ and $(a(v_i, w), \beta(v_i, w))$ for all neighbors w of v_i in G_i. Since $d(v_i, G_i) \leq d(v_i, G) \leq \Delta(G)$ and $\sum_{v_i \in V} d(v_i, G) = 2m$, we have

$$\sum_{i=1}^{n} d(v_i, G_i) \log d(v_i, G_i) \leq 2m \log \Delta(G).$$

Thus **Degenerate** takes time $O(m \log \Delta(G))$.

One can easily find the vertex ordering $v_1, v_2, ..., v_n$ above, as follows. Let $G_n = G$, and let v_n be a vertex of the smallest bi-directional b-degree $d_{2b}(v_n, G_n)$ in G_n. Let G_{n-1} be the graph obtained from G_n by deleting v_n, and let v_{n-1} be a vertex of the smallest bi-directional b-degree $d_{2b}(v_{n-1}, G_{n-1})$ in G_{n-1}. Repeating the operation above, one can obtain a vertex ordering $v_1, v_2, ..., v_n$. Let

$$k_b(G) = \max_{1 \leq i \leq n} d_{2b}(v_i, G_i).$$

Then one can prove that G is (k, b)-degenerated for an integer $k \in \mathbb{N}$ if and only if $k_b(G) \leq k$; the proof is similar as the case of an ordinary coloring [3]. Thus $k_b(G)$ should be called the b-degeneracy of G. Using a Fibonacci heap to represent the set of vertices v in G_i having $d_{2b}(v, G_i)$ as a key-attribute, one can find the vertex ordering $v_1, v_2, ..., v_n$ and compute $k_b(G)$ in time $O(m + n \log n)$, similarly as the implementation of Dijkstra's shortest path algorithm using a Fibonacci heap [1,3].

The b-degeneracy $k_b(G)$ of G is not necessarily smaller than or equal to $\Delta_{1b}(G)$, but obviously $k_b(G) \leq \Delta_{2b}(G)$. Since $\delta(v) + 1 \leq \chi_b(G)$ for every vertex v and $\delta(v, w) < 2\chi_b(G)$ for every edge (v, w), we have $\Delta_{2b}(G) + 1 \leq (2\Delta(G) + 1)\chi_b(G)$. We thus have the following theorem.

Theorem 4. *Algorithm* **Degenerate** *finds a b-coloring f of G with span(f) $\leq k + 1$ for a (k, b)-degenerated graph G in time $O(m \log \Delta(G))$, and the approximation ratio of* **Degenerate** *is at most $2\Delta(G) + 1$ if $k = k_b(G)$.*

Since $k_b(G) \leq \Delta_{2b}(G)$, Theorem 4 immediately implies the following corollary.

Corollary 3 *For every graph G, $\chi_b(G) \leq \Delta_{2b}(G) + 1$.*

Since $\Delta_{1b}(G) \leq \Delta_{2b}(G)$, Corollary 3 is also an immediate consequence of Corollary 2(a). If every vertex weight is 1, every edge weight is 0 and G is either a complete graph or an odd cycle having an odd number of vertices, then $\chi_b(G) = \chi(G) = \Delta(G) + 1 = \Delta_{1b}(G) + 1 = \Delta_{2b}(G) + 1$ and hence G attains the upper bounds on $\chi_b(G)$ in Corollary 2(a) and Corollary 3.

We can show that an optimal b-coloring can be found in linear time for every graph G with $\Delta(G) \leq 2$. (The details are omitted in this extended abstract.)

We finally show, as an extension of Brooks' theorem, that $\chi_b(G) \leq \Delta_{3b}(G) + 1 \leq \Delta_{2b}(G)$ if G is a 2-connected graph with $\Delta(G) \geq 3$ and is not a complete graph, where $\Delta_{3b}(G) = \max_{v \in V} d_{3b}(v, G)$, $d_{3b}(v, G) = d_{2b}(v, G) - \min_w \delta(v, w)$ and w runs over all neighbors of v. (The details are omitted in this extended abstract.)

6   Conclusions

In this paper we presented four efficient approximation algorithms **Color-1**, **Color-2**, **Delta**, and **Degenerate** with theoretical performance guarantees. The first two algorithms **Color-1** and **Color-2** find a b-coloring f with span$(f) \leq (c-1)\chi_b(G)$ for a graph G colored with $c(\geq 2)$ colors, and hence their approximation ratio is at most $c - 1$ although there is a graph for which the ratio approaches $c - 1$. Algorithms **Color-1** and **Color-2** are useful when c is small, say $c \leq 4$. The third algorithm **Delta** finds a b-coloring f with span$(f) \leq \Delta_{1b}(G) + 1$, and the approximation ratio is at most $\Delta(G) + 1$. Algorithm **Delta** is useful especially when $\Delta_{1b}(G)$ is small. The fourth algorithm **Degenerate** finds a b-coloring f with span$(f) \leq k + 1$ for a (k, b)-degenerated graph. Although the approximation ratio of **Degenerate** with choosing the b-degeneracy $k_b(G)$ as k is at most $2\Delta(G) + 1$, **Degenerate** often finds a b-coloring of span smaller than one found by **Delta**. The greedy algorithm **Degenerate** is useful when $k_b(G)$ is small. Algorithm **Color-1** takes linear time, **Delta** takes time $O(m \log n)$, and **Degenerate** takes time $O(m \log \Delta(G))$. These computational complexities are much better than those of FPTASs in [10] for series-parallel graphs and graphs with bounded tree-width. Algorithms **Delta** and **Degenerate** imply upper bounds $\chi_b(G) \leq \Delta_{1b}(G) + 1$ and $\chi_b(G) \leq \Delta_{2b}(G) + 1$, respectively. We also showed that an optimal b-coloring can be found in linear time for every graph G with $\Delta(G) \leq 2$, and finally presented an analogue of Brooks' theorem for b-colorings: $\chi_b(G) \leq \Delta_{3b}(G) + 1 \leq \Delta_{2b}(G)$ if G is a 2-connected graph with $\Delta(G) \geq 3$ and is not a complete graph.

Acknowledgments. This work is partly supported by MEXT-supported Programs for the Strategic Research Foundation at Private Universities.

References

1. Corman, T.H., Leiserson, C.E., Rivest, R.L., Stein, C.: Introduction to Algorithms. MIT Press and McGraw Hill, Cambridge, MA (2001)
2. Fijuljamin, J.: Two genetic algorithms for the bandwidth multicoloring problem. Yugoslav Journal of Operation Research 22(2), 225–246 (2012)
3. Fredman, M.L., Tarjan, R.E.: Fibonacci heaps and their uses in improved network optimization. J. Assoc. Comput. Mach. 34, 596–615 (1987)
4. Jensen, T.R., Toft, B.: Graph Coloring Problems. John Wiley & Sons, New York (1995)
5. Lovász, L.: Three short proofs in graph theory. J. Combinatorial Theory (B) 19, 269–271 (1975)
6. Malaguti, E., Toth, P.: An evolutionary approach for bandwidth multicoloring problems. European Journal of Operation Research 189, 638–651 (2008)
7. Marti, R., Gortazar, F., Duarte, A.: Heuristics for the bandwidth colouring problem. Int. J. of Metaheuristics 1(1), 11–29 (2010)
8. McDiamid, C.: On the span in channel assignment problems: bounds, computing and counting. Discrete Math. 266, 387–397 (2003)
9. McDiamid, C., Reed, B.: Channel assignment on graphs of bounded treewidth. Discrete Math. 273, 183–192 (2003)
10. Nishikawa, K., Nishizeki, T., Zhou, X.: Algorithms for bandwidth consecutive multicolorings of graphs. In: Snoeyink, J., Lu, P., Su, K., Wang, L. (eds.) AAIM 2012 and FAW 2012. LNCS, vol. 7285, pp. 117–128. Springer, Heidelberg (2012); also Theoretical Computer Science (to appear)
11. Pinedo, M.L.: Scheduling: Theory, Algorithms and Systems. Springer Science, New York (2008)
12. Takamizawa, K., Nishizeki, T., Saito, N.: Linear-time computability of combinatorial problems on series-parallel graphs. J. Assoc. Comput. Mach. 29, 623–641 (1982)
13. West, D.B.: Introduction to Graph Theory. Prentice-Hall, Englewood Cliffs (1996)
14. Zuckerman, D.: Linear degree extractors and the inapproximability of max clique and chromatic number. Theoretical Computer Science 3, 103–128 (2007)

Improved Approximation Algorithm
for Maximum Agreement Forest of Two Trees[*]

Feng Shi, Jie You, and Qilong Feng

School of Information Science and Engineering, Central South University, China

Abstract. Given two rooted binary phylogenetic trees with identical leaf label-set, the MAXIMUM AGREEMENT FOREST (MAF) problem asks for a largest common subforest of these two trees. This problem is known to be NP-complete and MAX SNP-hard, and the previously best approximation algorithm for this problem has a ratio 3. In this paper, we present an improved 2.5-approximation algorithm for the MAF problem on two rooted binary phylogenetic trees.

1 Introduction

Phylogenetic (evolutionary) trees have been widely used in the study of evolutionary biology to represent the tree-like evolution of a collection of species. Due to hybridization events in evolution, given the same set of species, different gene data sets may result in the construction of different trees. In order to facilitate the comparison of these different phylogenetic trees, several distance metrics have been proposed [1–4]. Among them, the SPR (Subtree Prune and Regraft) distance [3, 4] has been most studied.

For the computation of the SPR distance between two rooted phylogenetic trees, a graph theoretical model, the *maximum agreement forest* (MAF), has been formulated [5]. Define the *order* of a forest to be the number of connected components in the forest.[1] Bordewich and Semple [6] proved that the rSPR distance between two rooted binary phylogenetic trees is equal to the order of the MAF minus 1. In terms of computational complexity, it is known that computing the order of an MAF is NP-hard and MAX SNP-hard for two rooted binary phylogenetic trees [5, 6].

Approximation algorithms have been studied extensively for the MAF problem on two rooted binary trees [5, 7–10]. To our knowledge, the best approximation algorithm for the MAF problem on two rooted binary trees is a linear-time 3-approximation algorithm [11, 12].

---

[*] This work is supported by the National Natural Science Foundation of China under Grants (61103033, 61173051, 61232001), and Hunan Provincial Innovation Foundation For Postgraduate (CX2013B073).

[1] The definitions for the study of maximum agreement forests have been kind of confusing. If *size* denotes the number of edges in a forest, then for a forest, the size is equal to the number of vertices minus the order. In particular, when the number of vertices is fixed, a forest of a large size means a small order of the forest.

J. Chen, J.E. Hopcroft, and J. Wang (Eds.): FAW 2014, LNCS 8497, pp. 205–215, 2014.

In this paper, we study the approximation algorithm for the MAF problem on two rooted binary trees, and present a 2.5-approximation algorithm for this problem. Our algorithm is an improvement over the previously best 3-approximation algorithm for the problem. Our method is based on careful analysis of the graph structures that takes advantage of special relations among leaves in the trees.

2 Definitions

2.1 Rooted X-Trees and X-Forests

A tree is a *single-vertex tree* if it consists of a single leaf vertex. A tree is a *single-edge tree* if it consists of a single edge. A tree is *binary* if either it is a single-vertex tree or each of its vertices has degree either 1 or 3. The degree-1 vertices are *leaves* and the degree-3 vertices are *non-leaves* of the tree. For a subset E' of edges in a graph G, we will denote by $G \setminus E$ the graph G with the edges in E removed.

Let X be a label-set. A *binary phylogenetic X-tree*, or simply an X-tree, is a binary tree whose leaves are labeled bijectively by the label-set X (all non-leaves are unlabeled). An X-tree is *rooted* if a particular leaf is designated as the root of the tree (thus, this vertex is both a root and a leaf), which specifies a unique ancestor-descendant relation in the tree. The root of a rooted X-tree will always be labeled by a special symbol ρ, which is always assumed to be in the label-set X.

A *subtree T'* of a rooted X-tree T is a connected subgraph of T that contains at least one leave. In order to preserve the ancestor-descendant relation in T, we should define the root of the subtree T'. If T' contains the root labeled ρ of T, then it is the root of T', otherwise, the node in T' that is in T the least common ancestor of the leaves in T' is defined to be the root of T'. A *subforest* of a rooted X-tree T is a subgraph of T. A rooted *X-forest* is a subforest of a rooted X-tree T that contains a collection of subtrees whose label-sets are disjoint such that union of the label-sets is equal to X. Define the *order* of a rooted X-forest F, denoted $\mathrm{Ord}(F)$, to be the number of connected components in F.

A subtree T' of a rooted X-tree may contains unlabeled vertices of degree less than 3. In this case, we apply the *forced contraction* on T', which replaces each degree-2 vertex v and its incident edges with a single edge connecting its two neighbors, and removes each unlabeled vertex that has degree smaller than 2. However, if the root r of a subtree T' of T is an unlabeled vertex of degree-2, then the operation will not be applied on r, in order to preserve the ancestor-descendant relation in T. Note that the forced contraction does not change the order of a rooted X-forest F, because each connected component of F contains at least one leaf. A rooted X-forest is *irreducible* if the forced contraction can not apply to F. We will assume that the forced contraction is applied whenever it is applicable. Thus, the X-forests in our discussion are always irreducible.

Two leaf-labeled forests F_1 and F_2 are isomorphic if there is a graph isomorphism between F_1 and F_2 in which each leaf of F_1 is mapped to a leaf of F_2 with the same label. We say that a leaf-labeled forest F' is a subforest of a rooted

X-forest F if up to the forced contraction, there is an isomorphism between F' and a subforest of F that preserves the ancestor-descendant relation.

2.2   Agreement Forest

Let F_1 and F_2 be two rooted X-forests. An X-forest F is an *agreement forest* for F_1 and F_2 if F is a common subforest of F_1 and F_2. A *maximum agreement forest* (abbr. MAF) for F_1 and F_2 is an agreement forest for F_1 and F_2 with the minimum order. The MAXIMUM AGREEMENT FOREST problem is formally given as follows.

> MAXIMUM AGREEMENT FOREST (MAF)
> *Input*:   Two rooted X-forests F_1 and F_2
> *Output*: a maximum agreement forest F^* for F_1 and F_2

3   Edge-Removal Meta-Step

For any edge subset E' of an X-forest F, we have $\mathrm{Ord}(F \setminus E') \leq \mathrm{Ord}(F) + |E'|$. An edge subset E' of an X-forest F is an *essential edge-subset* (abbr. ee-set) if $\mathrm{Ord}(F \setminus E') = \mathrm{Ord}(F) + |E'|$. It is easy to see that for any X-subforest F' of an X-forest F, there is an ee-set E' of $\mathrm{Ord}(F') - \mathrm{Ord}(F)$ edges in F such that $F' = F \setminus E'$ (up to forced contraction). Note that it is easy to see whether an edge subset of F is an ee-set for F.

The MAF problem on the instance (F_1, F_2) looks for an MAF for the instance. An *optimal-edge set* E for (F_1, F_2) is an ee-set of F_1 that $F_1 \setminus E$ is an MAF for (F_1, F_2). The *optimal order* for (F_1, F_2), denoted $O(F_1, F_2)$, is the order of an MAF for the instance.

Our approximation algorithm for MAF consists of a sequence of "meta-steps". An *edge-removal meta-step* (or simply meta-step) in an algorithm for MAF is a collection of consecutive computational steps in the algorithm that on an instance (F_1, F_2) of MAF removes an ee-set E_M of forest F_i, $1 \leq i \leq 2$ (and applies the forced contraction).

The performance of the algorithm for MAF heavily depends on the quality of the meta-steps we employ in the algorithm. For this, we introduce the following concept that measures the quality of a meta-step, where $t \geq 1$ is an arbitrary real number.

Definition 1. *Let (F_1, F_2) be an instance of MAF, and let M be a meta-step that removes an ee-set E_M of F_i, $1 \leq i \leq 2$, producing instance (F'_1, F'_2). Meta-step M keeps ratio t if $(O(F'_1, F'_2) - O(F_1, F_2)) \leq \frac{(t-1)}{t}|E_M|$.*

Note that by the definition if a meta-step does not change the optimal order for the instance, then it keeps ratio t for any $t \geq 1$. We define an edge-removal meta-step is *safe* if it does not change the optimal order for the instance, thus it keeps ratio t for any $t \geq 1$.

Due to the limit of space, we just briefly present the reduction rules and part of meta-steps in the algorithm in the following sections. The entire discussion for this algorithm will be given in a complete version.

4   Reduction Rules for MAF Problem

Because the bijection between the leaves of a rooted X-forest F and the elements in the label-set X, sometimes we will use, without confusion, a label in X to refer to the corresponding leaf in F, or vice versa. Two labels (and their corresponding leaves) in a rooted X-forest are *siblings* if they have a common parent.

Let (F_1, F_2) be an instance of MAF problem. In this section, we present several reduction rules for (F_1, F_2). Moreover, we assume that the reduction rules are applied in order, i.e., Reduction Rule j will not be applied unless all Reduction Rules i for $i < j$ become unapplicable.

Reduction Rule 1. If a label l is a single-vertex tree in one of F_1 and F_2, then remove the edge (if any) incident to l in the other forest.

For instance (F_1, F_2) on which Reduction Rule 1 is not applicable, we can assume F_2 has a sibling pair (a, b), otherwise, F_1 and F_2 are isomorphic. Let e_a and e_b be the edges incident to a and b in F_1, respectively.

Reduction Rule 2. If a and b are also siblings in F_1, then shrink (a, b) in both forests.

By *shrink* (a, b), we mean that we label the common parent of a and b by a new label $\underline{ab}$, and delete a and b in the forest. We also replace the label-set X by label-set $X' = (X \setminus \{a, b\}) \cup \{\underline{ab}\}$.

Reduction Rule 3. If a and b are in different connected components in F_1, then remove ee-set $\{e_a, e_b\}$ of F_1.

Let v_1 and v_2 be two vertices that in the same connected component in a rooted X-forest F, then let $P = \{v_1, c_1, c_2, \cdots, c_r, v_2\}$ be the path in F that connects v_1 and v_2. Denote by $E_F(v_1, v_2)$ the edge set that contains all pendant edges in F that not on the path P but incident to a internal vertex on P, except the one that incident to the least common ancestor of v_1 and v_2, $LCA_F(v_1, v_2)$. Let $e = [u, v]$ be an arbitrary edge on the path that connects v_1 and $LCA_F(v_1, v_2)$ in F. Then, $E_F(v_1, e)$ is equal to $E_F(v_1, v)$, where u is on the path that connects v_1 and v in F. Obviously, $E_F(v_1, e) \subseteq E_F(v_1, v_2)$.

Reduction Rule 4. If a and b are in the same connected component in F_1 that $|E_{F_1}(a, b)| = 1$, then remove ee-set $E_{F_1}(a, b)$.

Reduction Rule 5. If $LCA_{F_1}(a, b)$ is the root of the connected component which contains a and b in F_1, then remove ee-set $\{e_a, e_b\}$ of F_1.

For the reduction rules above, we have that Reduction Rules $1 - 2$ and 4 are safe, Reduction Rules 3 and 5 keep ratio 2. Due to limit space, we just give the analysis for the ratio of Reduction Rule 5 in the following.

Lemma 1. *Let (F_1, F_2) be an instance of MAF problem, and let e be an edge in F_1. Then, $O(F_1 \setminus \{e\}, F_2)$ is at most $O(F_1, F_2) + 1$.*

Lemma 2. *Reduction Rule 5 keeps ratio 2.*

Proof. Let F be an fixed MAF of order r for F_1 and F_2. There are three possible cases for labels a and b in F.

(1). Label a is a single-vertex tree in F. Then, F is also an MAF for $(F_1 \setminus \{e_a\}, F_2)$. By Lemma 1, $O(F_1 \setminus \{e_a, e_b\}, F_2)$ is at most $r + 1$.

(2). Label b is a single-vertex tree in F. The analysis for (2) is similar to that for (1). $O(F_1 \setminus \{e_a, e_b\}, F_2)$ is at most $r + 1$.

(3). Neither a nor b is a single-vertex tree in F. Then the two edges that incident to a and b in F_2 cannot be removed. Thus, a and b are siblings in F. Because $LCA_{F_1}(a, b)$ is the root of the connected component which contains a and b in F_1, so the common parent of a and b in F is the root of the connected component which contains a and b in F. Now the removal of edge e_a in F makes both a and b become single-vertex trees. Thus, $F \setminus \{e_a\}$ is also a subforest of $F_1 \setminus \{e_a, e_b\}$. $F \setminus \{e_a\}$ is an agreement forest for $(F_1 \setminus \{e_a, e_b\}, F_2)$, $O(F_1 \setminus \{e_a, e_b\}, F_2)$ is at most $r + 1$.

Therefore, $O(F_1 \setminus \{e_a, e_b\}, F_2)$ is always at most $r + 1$. By the definition of the ratio for meta-step, we have Reduction Rule 5 keeps ratio 2.   □

During our process, we will exhaustively apply Reduction Rules 1-5 on the instance (F_1, F_2) whenever the rules are applicable, and work on the reduced instance. An instance is *strongly reduced* if these reduction rules are not applicable on the instance. Therefore, through out the discussion, we will assume that the instance (F_1, F_2) is always strongly reduced.

For the strongly reduced instance (F_1, F_2), if F_2 has a sibling pair, then we need do further analysis for a better ratio for our approximation algorithm. At first, we give some related definitions. The *distance* of a sibling pair is the number of vertices on the path that connects the parent of the sibling pair and the root of the connected component which the sibling pair belongs to. A sibling pair (a, b) of a rooted X-forest is *bottom* if the distance of (a, b) is not less than that of the other sibling pairs which are in the same connected component with (a, b).

Let (a, b) be a bottom sibling pair of F_2. Let p_{ab} be the parent of (a, b) in F_2, and let C_{ab} be the connected component that contains (a, b) in F_2. The following discussion will be divided into 4 cases. Case 1: vertex p_{ab} is the root of C_{ab} in F_2; Case 2: label ρ is the parent of p_{ab} in F_2; Case 3: the node which has a common parent with p_{ab} in F_2 is a leaf labeled c; Case 4: the node which has a common parent with p_{ab} in F_2 is the parent of a bottom sibling pair (c, d). See Figure 1 for an illustration.

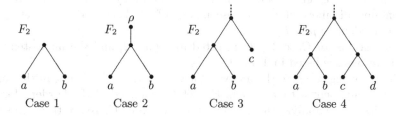

Fig. 1. Four cases

5   Analysis for Case 1 − 4

In this section, we briefly enumerate various cases we divided, and present part of meta-steps for these cases.

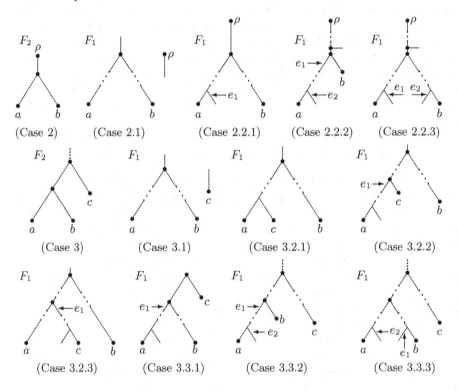

Fig. 2. Situations for Case 2 − 3

Let v_1 and v_2 be two arbitrary vertices that are in the same connected component in a rooted X-forest F. Denote by $L(v_1)$ the label set that contains all labels which are descendants of v_1 in F, and denote by $L(v_1, v_2)$ the label set that contains all labels which are descendants of $LCA_F(v_1, v_2)$ in F. Let e_ρ be the edge incident to ρ in F_1.

The situations for Case 2 − 3 are listed in Figure 2, and the meta-steps for these situations are listed in Table 1.

In the following, we give the analysis for Case 2.1: label ρ is not in the same connected component with a and b in F_1. Let F be a fixed MAF of order r for F_1 and F_2. Firstly, we analyze the conflict about a and b. There are three possible situations for a and b in F.

Situation (i): a is a single-vertex tree in F. Obviously, F is also an MAF for $(F_1 \setminus \{e_a\}, F_2 \setminus \{e_a\})$, $O(F_1 \setminus \{e_a\}, F_2 \setminus \{e_a\}) = r$. Since labels b and ρ are connected by an edge in $F_2 \setminus \{e_a\}$, but are in different connected components

Table 1. Meta-steps for Case $1-3$

Cases			Meta-step	Ratio
Case 1: vertex p_{ab} is be the root of C_{ab} in F_2.			Remove ee-set $\{e_a, e\}$ of F_1, where e is an arbitrary edge of $E_{F_1}(a,b)$.	2
Case 2: label ρ is the parent of p_{ab} in F_2.				
Case 2.1: ρ is not in the same connected component with a in F_1.			Remove ee-set $E_M = \{e_a, e_b, e_\rho, e_1, e_2\}$ of F_1, where e_1 and e_2 are two arbitrary edges of $E_{F_1}(a,b)$.	2.5
Case 2.2: ρ is in the same connected component with a in F_1.(W.l.o.g., assume that $E_{F_1}(a, LCA_{F_1}(a,b)) \neq \emptyset$.)	**Case 2.2.1:** $E_{F_1}(\rho, LCA_{F_1}(a,b)) = \emptyset$.		Remove ee-set $E_M = \{e_a, e_b, e_\rho, e_1\}$ of F_1.	2
	Case 2.2.2: $E_{F_1}(\rho, LCA_{F_1}(a,b)) \neq \emptyset$, $E_{F_1}(b, LCA_{F_1}(a,b)) = \emptyset$.		Remove ee-set $E_M = \{e_a, e_b, e_\rho, e_1, e_2\}$ of F_1, where e_1 and e_2 are the edges that marked in Figure 2 (Case 2.2.2).	2.5
	Case 2.2.3: $E_{F_1}(\rho, LCA_{F_1}(a,b)) \neq \emptyset$, $E_{F_1}(b, LCA_{F_1}(a,b)) \neq \emptyset$.		Remove ee-set $E_M = \{e_a, e_b, e_\rho, e_1, e_2\}$ of F_1, where e_1 and e_2 are the edges that marked in Figure 2 (Case 2.2.3).	2.5
Case 3: Vertex p_{ab} and label c have a common parent in F_2.				
Case 3.1: c is not in the same connected component with a in F_1.			Remove ee-set $E_M = \{e_a, e_b, e_c, e_{ab}\}$ of F_1, where e_{ab} is the edge between $LCA_{F_1}(a,b)$ and its parent in F_1.	2
Case 3.2: c belongs to $L(a,b)$ in F_1 (W.l.o.g., assume that $LCA_{F_1}(a,c)$ is a descendant of $LCA_{F_1}(a,b)$.)	**Case 3.2.1:** a and c are siblings in F_1.		Remove ee-set $E_M = \{e_b\}$ of F_1.	safe
	Case 3.2.2: a and c are not siblings, c's parent is on the path from a to b in F_1.		Remove ee-set $E_M = \{e_a, e_b, e_c, e_1\}$ of F_1, where e_1 is the edge that marked in Figure 2 (Case 3.2.2).	2
	Case 3.2.3: c's parent is not on the path from a to b in F_1.		Remove ee-set $E_M = \{e_a, e_b, e_c, e_1, e_2\}$ of F_1, where e_1 is the edge that marked in Figure 2 (Case 3.2.3) and e_2 is an arbitrary edge of $E_{F_1}(a,b) \setminus \{e_1\}$.	2.5
Case 3.3: c does not belong to $L(a,b)$ in F_1. (W.l.o.g., assume that $E_{F_1}(a, LCA_{F_1}(a,b)) \neq \emptyset$.)	**Case 3.3.1:** $E_{F_1}(LCA_{F_1}(a,b), c) = \emptyset$ and c's parent is the root of the component in F_1.		Remove ee-set $E_M = \{e_a, e_b, e_c, e_1\}$ of F_1, where e_1 is the edge that marked in Figure 2 (Case 3.3.1).	2
	Case 3.3.2: $E_{F_1}(LCA_{F_1}(a,b), c) \neq \emptyset$ or c's parent is not the root of the component in F_1, $LCA_{F_1}(a,b)$ is the parent of b in F_1.		Remove ee-set $E_M = \{e_a, e_b, e_c, e_1, e_2\}$ of F_1, where e_1 and e_2 are the edges that marked in Figure 2 (Case 3.3.2).	2.5
	Case 3.3.3: $E_{F_1}(LCA_{F_1}(a,b), c) \neq \emptyset$ or c's parent is not the root of the component in F_1, $LCA_{F_1}(a,b)$ is not the parent of b in F_1.		Remove ee-set $E_M = \{e_a, e_b, e_c, e_1, e_2\}$ of F_1, where e_1 and e_2 are the edges that marked in Figure 2 (Case 3.3.3).	2.5

in $F_1 \setminus \{e_a\}$, so both b and ρ would be single-vertex trees in F. Thus, in this situation, all a, b, and ρ are single-vertex trees in F. We have that F is also an MAF for $(F_1 \setminus \{e_a, e_b, e_\rho\}, F_2 \setminus \{e_a\})$. Obviously, $(F_1 \setminus \{e_a, e_b, e_\rho\}, F_2 \setminus \{e_a\})$ and $(F_1 \setminus \{e_a, e_b, e_\rho\}, F_2)$ have the same collection of solutions, thus, $O(F_1 \setminus \{e_a, e_b, e_\rho\}, F_2)$ is at most r.

Situation (ii): b is a single-vertex tree in F. The analysis for this situation is similar to that for Situation (i). $O(F_1 \setminus \{e_a, e_b, e_\rho\}, F_2) \leq r$.

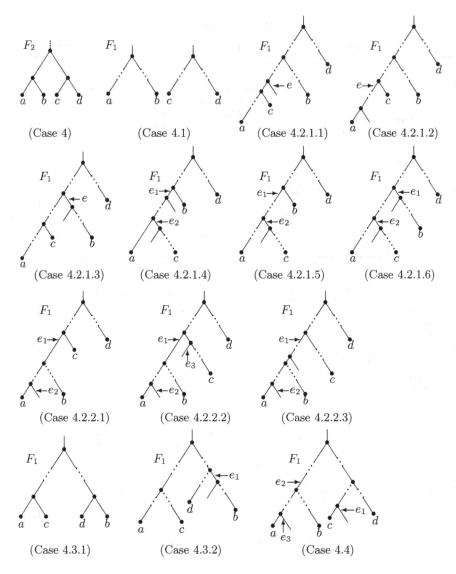

Fig. 3. Situations for Case 4

Situation (iii): a and b are siblings in F. In order to make a and b siblings in F_1, all edges in $E_{F_1}(a, b)$ should be removed. Let e_1 and e_2 be two arbitrary edges of $E_{F_1}(a, b)$, then $O(F_1 \setminus \{e_1, e_2\}, F_2)$ is at most r.

In the following, we enumerate all feasible approaches for resolving Case 2.1 by removing edges from F_1, which have been discussed above.

$$\begin{cases} remove\ e_a; & \begin{cases} remove\ e_b,\ e_\rho; & (a) \\ remove\ e_a,\ e_\rho; & (b) \end{cases} \\ remove\ e_b; \\ remove\ E_{F_1}(a, b); & (c) \end{cases}$$

Meta-step 2.1. Remove ee-set $E_M = \{e_a, e_b, e_\rho, e_1, e_2\}$ of F_1, where e_1 and e_2 are two arbitrary edges of $E_{F_1}(a, b)$.

Lemma 3. *Meta-step 2.1 keeps ratio 2.5.*

Proof. In Case 2.1, F can be constructed by one of approaches (a-c). Whatever F is, by Lemma 1, we always have that $O(F_1 \setminus E_M, F_2)$ is at most $r+3$. Therefore, Meta-step 2.1 keeps ratio 2.5. □

The analysis for the other cases are similar to that for Case 2.1. Due to limit space, we do not present the details here.

For Case 4: vertex p_{ab} and the parent of (c, d) have a common parent in F_2, c and d may not be in the same connected component with a and b in F_1 (recall that c and d are in the same connected component in F_1 and $|E_{F_1}(c, d)| \geq 2$), this situation will be analyzed in Case 4.1. If c and d are in the same connected component with a and b in F_1, 3 possible subcases are discussed. Case 4.2: the least common ancestor of one label pair is the ancestor of that of the other label pair in F_1 (i.e., $LCA_{F_1}(a, b)$ is the ancestor of $LCA_{F_1}(c, d)$, or vice versa); Case 4.3: $LCA_{F_1}(a, b)$ and $LCA_{F_1}(c, d)$ are the same vertex in F_1; Case 4.4: $LCA_{F_1}(a, b)$ and $LCA_{F_1}(c, d)$ have no ancestor-descendant relation in F_1. Due to limit space, we just enumerate the situations for Case 4 in Figure 3. For each meta-steps in Case 4, we also ensure that it keeps ratio not greater than 2.5.

6   Approximation Algorithms for MAF

Now we are ready to present our algorithm for the MAF problem, which is given in Figure 4.

Our algorithm consists of a sequence of meta-steps (including reduction rules). For each meta-step in the algorithm, we ensure that it keeps ratio bounded by 2.5. Therefore, we have the following theorem.

Apx-MAF
INPUT: two rooted X-forests F_1 and F_2
OUTPUT: an agreement forest for F_1 and F_2
1.   apply Reduction Rules 1-5 on (F_1, F_2) until they are not applicable;
2.   **repeat** until F_2 has no sibling pair
2.1.   Let (a, b) be a bottom sibling pair in F_2
2.2.   apply the corresponding meta-steps according to the cases;
2.3.   apply Reduction Rules 1-5 on (F_1, F_2) if possible;
3.   Let F' be the MAF for F_1 and F_2; return F'.

Fig. 4. An approximation algorithm for MAF

Theorem 1. *Algorithm Apx-MAF is a 2.5-approximation algorithm for the* MAF *problem that runs in time* $O(n \log n)$, *where* $n = |X|$.

Proof. Due to limit space, we briefly analyze the ratio and time complexity of the algorithm in the following.

Suppose the sequence of meta-steps in the algorithm is $S = \{M_1, M_2, \ldots, M_h\}$, where for each i, $1 \leq i \leq h$, meta-step M_i removes an ee-set $E_{M,i}$ from the instance $I_i = (F_{1,i}, F_{2,i})$ produces an instance $I_{i+1} = (F_{1,i+1}, F_{2,i+1})$. Suppose that t_i is the optimal order for the instances I_i and meta-step M_i keeps ratio r_i. Thus, forest $F_{1,h+1}$ is an MAF for the instance I_{h+1}, and $\text{Ord}(F_{1,h+1}) = t_{h+1}$.

Let S_1 be a set that contains all meta-steps of S that remove edges from F_1, and let S_2 be a set that contains all meta-steps of S that remove edges from F_2 (note that Reduction Rule 1 may remove edges in F_2).

For each meta-step $M_i \in S_1$, by the analysis given in the previous sections, there is $r_i \leq 2.5$. Thus, we have $(t_{i+1} - t_i) \leq \frac{r_i - 1}{r_i} |E_{M,i}| \leq \frac{3}{5} |E_{M,i}|$, where $|E_{M,i}| = \text{Ord}(F_{1,i+1}) - \text{Ord}(F_{1,i})$ (because every meta-step removes an ee-set). Therefore, for each meta-step $M_i \in S_1$, we have the inequality $(t_{i+1} - t_i) \leq \frac{3}{5}(\text{Ord}(F_{1,i+1}) - \text{Ord}(F_{1,i}))$.

For each meta-step $M_i \in S_2$, which is an application of Reduction Rule 1 on F_2, by the analysis given in Section 4, M_i is safe, so $t_{i+1} - t_i = 0$. Since M_i does not change F_1, $\text{Ord}(F_{1,i+1}) - \text{Ord}(F_{1,i}) = 0$. Therefore, for each meta-step $M_i \in S_2$, we also have $(t_{i+1} - t_i) \leq \frac{3}{5}(\text{Ord}(F_{1,i+1}) - \text{Ord}(F_{1,i}))$.

Then, we add up these inequalities for all meta-steps in S, and get $(t_{h+1} - t_1) \leq \frac{3}{5}(\text{Ord}(F_{1,h+1}) - \text{Ord}(F_{1,1}))$, where $t_{h+1} = \text{Ord}(F_{1,h+1})$. From this, we can easily get that $t_{h+1} \leq \frac{5}{2} t_1$. Therefore, the ratio of the approximation algorithm Apx-MAF is at most $\frac{5}{2}$.

Finally, we consider the time complexity of the algorithm. Suppose $n = |X|$, each X-forest has a size (i.e., the number of edges) $O(n)$. Therefore, the size of (F_1, F_2) is $O(n)$. Because each meta-step decreases the number of edges in the instance by at least one, so the total number of times these meta-steps can be applied is bounded by $O(n)$. Moreover, it is not diffcult to see that with careful implementation of the data structure representing X-forests, the running time of each meta-step can be bounded by $O(\log n)$. Therefore, the running time of the algorithm is $O(n \log n)$. □

7 Conclusion

In this paper, we presented a 2.5-approximation algorithm for the MAXIMUM AGREEMENT FOREST problem on two rooted binary phylogenetic trees. To our knowledge, this is the best approximation algorithm for the problem on two rooted binary phylogenetic trees.

References

1. Robinson, D., Foulds, L.: Comparison of phylogenetic trees. Mathematical Biosciences 53(1-2), 131–147 (1981)
2. Li, M., Tromp, J., Zhang, L.: On the nearest neighbour interchange distance between evolutionary trees. Journal on Theoretical Biology 182(4), 463–467 (1996)
3. Hodson, F., Kendall, D., Tauta, P. (eds.): The recovery of trees from measures of dissimilarity. Mathematics in the Archaeological and Historical Sciences, pp. 387–395. Edinburgh University Press, Edinburgh (1971)
4. Swofford, D., Olsen, G., Waddell, P., Hillis, D.: Phylogenetic inference. In: Molecular Systematics, 2nd edn., pp. 407–513. Sinauer, Associates (1996)
5. Hein, J., Jiang, T., Wang, L., Zhang, K.: On the complexity of comparing evolutionary trees. Discrete Applied Mathematics 71, 153–169 (1996)
6. Bordewich, M., Semple, C.: On the computational complexity of the rooted subtree prune and regraft distance. Annals of Combinatorics 8(4), 409–423 (2005)
7. Rodrigues, E.M., Sagot, M.-F., Wakabayashi, Y.: Some approximation results for the maximum agreement forest problem. In: Goemans, M.X., Jansen, K., Rolim, J.D.P., Trevisan, L. (eds.) RANDOM 2001 and APPROX 2001. LNCS, vol. 2129, pp. 159–169. Springer, Heidelberg (2001)
8. Bonet, M., John, R., Mahindru, R., Amenta, N.: Approximating subtree distances between phylogenies. J. Comput. Biol. 13(8), 1419–1434 (2006)
9. Bordewich, M., McCartin, C., Semple, C.: A 3-approximation algorithm for the subtree distance between phylogenies. J. Discrete Algorithms 6(3), 458–471 (2008)
10. Rodrigues, E., Sagot, M., Wakabayashi, Y.: The maximum agreement forest problem: approximation algorithms and computational experiments. Theoretical Computer Science 374(1-3), 91–110 (2007)
11. Whidden, C., Zeh, N.: A unifying view on approximation and FPT of agreement forests. In: Salzberg, S.L., Warnow, T. (eds.) WABI 2009. LNCS, vol. 5724, pp. 390–402. Springer, Heidelberg (2009)
12. Whidden, C., Beiko, R., Zeh, N.: Fixed-parameter and approximation algorithms for maximum agreement forests. CoRR. abs/1108.2664 (2011)

Oblivious Integral Routing for Minimizing the Quadratic Polynomial Cost[*]

Yangguang Shi[1,2], Fa Zhang[1,3], and Zhiyong Liu[1,4]

[1] Institute of Computing Technology, Chinese Academy of Sciences, China
[2] University of Chinese Academy of Sciences, China
[3] Key Laboratory of Intelligent Information Processing, ICT, CAS, China
[4] Key Laboratory of Computer System and Architecture, ICT, CAS, China
{shiyangguang,zhangfa,zyliu}@ict.ac.cn

Abstract. In this paper, we study the problem of minimizing the cost for a set of multicommodity traffic request $\mathcal{R}$ in an undirected network $G(V, E)$. Motivated by the energy efficiency of communication networks, we will focus on the case where the objective is to minimize $\sum_e (l_e)^2$. Here l_e represents the load on the edge e. For this problem, we propose an oblivious routing algorithm, whose decisions don't rely on the current traffic in the network. This feature enables our algorithm to be implemented efficiently in the high-capacity backbone networks to improve the energy efficiency of the entire network.

The major difference between our work and the related oblivious routing algorithms is that our approach can satisfy the integral constraint, which does not allow splitting a traffic demand into fractional flows. We prove that with this constraint no oblivious routing algorithm can guarantee the competitive ratio bounded by $o(|E|^{\frac{1}{3}})$. By contrast, our approach gives a competitive ratio of $O\left(|E|^{\frac{1}{2}} \log^2 |V| \cdot \log D\right)$, where D is the maximum demand of the traffic requests. This competitive ratio is tight up to $O\left(|E|^{\frac{1}{6}} \log^2 |V| \cdot \log D\right)$.

Keywords: Oblivious Routing, Randomization Algorithm, Hardness of Approximation, Competitive Ratio.

1 Introduction

In a min-cost multicommodity flow problem, we are given a network $G(V, E)$ and a set of traffic requests $\mathcal{R} = \{R_1, R_2, \cdots, R_k, \cdots\}$. Here V and E represent the set of nodes and edges of the network respectively. In this paper we assume that G is an undirected graph, i.e., each edge $e \in E$ is bidirectional. Each traffic request $R_k \in \mathcal{R}$ specifies its source-target pair $(s_k, t_k) \in V \times V$ and the demand (i.e., the volume of traffic needs to be routed) $d_k \in \mathbb{Z}^+$. Let l_e be the load on the edge e incurred by routing the traffic requests along it and $f(l_e)$ be the cost corresponding to the load. The objective is to minimize the overall cost $\sum_e f(l_e)$.

---

[*] This research was supported in part by National Natural Science Foundation of China grant 61020106002, 61161160566 and 61221062.

J. Chen, J.E. Hopcroft, and J. Wang (Eds.): FAW 2014, LNCS 8497, pp. 216–228, 2014.

In particular, we focus on a case that the cost function is a quadratic polynomial $f(l_e) = (l_e)^2$ and the flow is restricted to be integral. In such case we refer to the problem as the *quadratic polynomial cost minimization for multicommodity integral flow* (abbr. QPC-MIF). Correspondingly, the relaxation version of QPC-MIF where fractional flows are allowed will be referred to as QPC-MF. Note that the objective of the QPC-MIF problem and the QPC-MF problem can also be represented by minimizing $\|\vec{l}\|_2^2$, where $\vec{l}$ represents the *load vector* composed of every l_e and $\|\cdot\|_2^2$ represents the square of the L_2-norm.

In this paper, we will consider the *oblivious routing* [6,8–10,12,13] strategy for the QPC-MIF problem. For an oblivious routing algorithm, its routing decisions should be made independently of the current traffic in the network. It implies that the path selection for the traffic requests depends only on the topology of the network G, the source-target pair, and some random bits, while utilizing no information of other traffic requests and actual loads on the current edges. The oblivious routing algorithm can be viewed as a routing "template" precomputed before any traffic requests are known. For each source-target pair (s, t) in the network, a unit flow $u_{s,t}$ is specified by the template. To determine the routing path of each traffic request R_k, the precomputed flow u_{s_k,t_k} is scaled by d_k to meet the request's demand [6,8,13].

To the best of our knowledge, only a few of the existing oblivious routing algorithms can be applied to minimize quadratic polynomial cost for multicommodity traffic, e.g., [5,6,12]. The major difference between our work and the existing ones is that our oblivious routing algorithm can satisfy the *integral constraint*, which requires that each traffic request should be routed along one single path rather than being divided into a set of fractional flows. This constraint is important for many applications in practice [1], especially for the packet switching networks where the routing unit is a "packet" that cannot be split further.

To satisfy the integral constraint in oblivious routing, a conventional approach is the *single-path routing* (SPR) [8,9], which specifies the same fixed path for any traffic requests between a given source-target pair. However, we prove that no SPR approach can give a competitive ratio of $o(|E|)$ due to the superadditivity of the quadratic cost function. Here *competitive ratio* refers to the largest gap between the cost incurred by the oblivious routing algorithm and the cost associated with the optimal possible solution [12,14]. It implies that the SPR approach cannot guarantee a good performance for the QPC-MIF problem.

In this paper, we adopt a different approach to satisfy the integral constraint. For each traffic request, we will select a path from a set of precomputed candidates in a randomized manner. The randomized selection procedure will be performed independently for each traffic request according to a precomputed probability distribution. Our approach will be termed by *randomized oblivious integral routing* (ROIR) algorithm. When being applied to the QPC-MIF problem, ROIR guarantees that the *competitive ratio* can be bounded by $O(|E|^{\frac{1}{2}} \log^2 |V| \cdot \log D)$, where $D = max_k d_k$. This result cannot be improved significantly since we prove that any oblivious routing algorithm which satisfies the integral constraint is unable to guarantee the competitive ratio bounded by $o(|E|^{\frac{1}{3}})$.

Our algorithm is applicable in particular for the energy conservation problem of the high-capacity communication networks. Related research works have presented that some of the electronic network devices can adjust their working speeds dynamically via the *speed scaling* mechanism to save energy [3,16]. In such cases the dynamic CMOS power consumption can be modeled by $f(s) = s^\alpha$, where α is an input parameter no bigger than 3 [11]. Further estimation shows that for real devices, the value of α is close to 2 [16]. Related literatures (e.g., [3]) indicate that for every edge e in the network, we can use $f(l_e)$ to appropriately represent the dynamic power consumption of both the processing devices and transmission devices deployed along e with the load l_e. Thus, our approach can be applied to minimize the energy consumption of the entire network globally.

The reason why we prefer an oblivious algorithm to improve the energy efficiency of high-bandwidth networks is its simplicity in implementation. Since both the candidate paths and the probability distribution can be precomputed and stored in the routing table of every node, our ROIR algorithm can be implemented efficiently in a distributed manner. It is significant especially for the high-capacity network routers, where dynamic traffic requests arrive on a transient timescale of the nanosecond range [15,17]. In this circumstance, it will be time-consuming to evaluate the network traffic pattern to select a proper path for a particular traffic request, which implies that a routing algorithm dependent on the current traffic may be inefficient. By comparison, our ROIR algorithm is able to make the routing decisions timely by simply looking up the routing table. Thus, while the algorithm developed in this paper theoretically can guarantee a competitive ratio of $O(|E|^{\frac{1}{2}} \log^2 |V| \cdot \log D)$, it is practically efficient for routing in high-capacity backbone networks.

1.1 Related Works

Motivated by improving the energy efficiency of the communication networks, some integral routing algorithms (e.g. [2,3]) have been developed for minimizing the cost $\sum_{e \in E} (l_e)^\alpha$. In these works, the entire set of traffic requests are known in advance, and the routing decisions are made offline. Particularly, in [3] Andrews et al. proposed a randomized path selection algorithm which can keep the approximation ratio bounded by $O(\log^{\alpha-1} D \cdot \alpha^\alpha)$. Their works rely on the global fractional optimal solution, which is obtained in a static scenario.

Due to their simplicity, the oblivious routing algorithms have attracted considerable attention. As summarized in [14], most of the existing works are devoted to two categories of objectives, L_1-norm minimization (finding short routing paths) and L_∞-norm minimization (minimizing the congestion) [9,13]. However, only a little attention has been paid to the oblivious routing for superadditive polynomial objectives, which is important for the energy saving of networks. In [10], an oblivious routing algorithm is developed for the case where the objective function is a quadratic polynomial as ours but all the traffic requests are directed to a same target. The competitive ratio of their algorithm can be bounded by $O(\log |V|)$. However, their approach is restricted to the single-target case very much and cannot be generalized to the multicommodity case [6].

In [6], Englert et al. design an oblivious routing algorithm to minimize the L_p-norm of the load vector (abbr. L_p-norm minimization). Their approach is made constructive in [5]. When being applied to minimize the quadratic polynomial cost, their approach can keep the competitive ratio bounded by $O(\log^2 |V|)$. Their result can be obtained only when the fractional flow is allowed. We will show that their approach cannot be extended to the case where integral constraint exists by proving that in such case no oblivious routing algorithm can keep the competitive ratio bounded by $o(|E|^{\frac{1}{3}})$. It implies that the integral constraint makes our problem much more challenging.

In [12], Lawler et al. propose an oblivious routing algorithm to simultaneously minimize all L_p-norms ($1 \leq p \leq \infty$) of the load vector. The competitive ratio of their approach is bounded by $O(\sqrt{|E|})$ for the general topologies, which is proved to be optimal up to some constant. When being applied to optimize the quadratic polynomial cost in the networks with general topologies, it will give a large competitive ratio of $O(|E|)$. Their approach also needs to split the traffic requests into fractional flows and cannot satisfy the integral constraint.

2   Hardness Results

In this part, we will analyse the hardness of approximation of our problem corresponding to the integral constraint. To satisfy the integral constraint, the existing oblivious routing approaches generally fix a unique path for each source-target pair (e.g., [8,9]). However, the following result shows that such a *single-path routing* (SPR) approach is unable to give a tight competitive ratio for our problem due to the superadditivity of our cost function.

Lemma 1. *For any $|E| \geq 4$, there exist input instances such that any single-path oblivious routing algorithm will give a competitive ratio of $\Omega(|E|)$ for the QPC-MIF problem.*

Proof. Our proof is based on the network $G_1(V_1, E_1)$ shown in Fig. 1(a), where $|E_1| = |E|$. There are $\lfloor |E_1|/2 \rfloor$ paths connecting the node pair (u_1, v_1). The length of each path is 2. Furthermore, we add a node w_1 to G_1 and connecting u_1 and w_1 if and only if $2\lfloor |E_1|/2 \rfloor < |E_1|$, i.e., $|E_1|$ is odd. Consider a traffic request set $\mathcal{R}_1 = \{R_1, R_2, \ldots, R_{\lfloor |E_1|/2 \rfloor}\}$. For each $R_k \in \mathcal{R}_1$, $(s_k, t_k) = (u_1, v_1)$ and $d_k = 1$. A single-path oblivious routing algorithm will route all the traffic requests in $\mathcal{R}$ along one of the $\lfloor |E_1|/2 \rfloor$ edge-disjoint paths between u_1 and v_1. The corresponding cost will be $f\left(\left\lfloor \frac{|E_1|}{2} \right\rfloor \cdot 1\right) \cdot 2 = \left\lfloor \frac{|E_1|}{2} \right\rfloor^2 \cdot 2$. By contrast, the optimal solution will route each R_k along a distinct path, whose cost will be $2\lfloor |E_1|/2 \rfloor$. Thus, the competitive ratio will be at least $\lfloor |E_1|/2 \rfloor$.   □

Accordingly, in this paper we develop a routing strategy that selects a path for each traffic request randomly and independently. Nevertheless, the theorem below indicates that it is still difficult for oblivious routing algorithms to give a small competitive ratio even if the randomization is utilized.

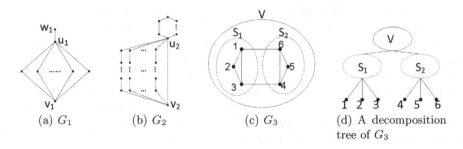

Fig. 1. Topologies of Networks and Decomposition Tree

Theorem 1. *No oblivious integral routing algorithm can guarantee the competitive ratio bounded by $o(|E|^{\frac{1}{3}})$ if it selects paths randomly.*

Proof. Here we consider another network $G_2(V_2, E_2)$ in Fig. 1(b), where $|E_2| = |E|$. It is constructed as follows. The node u_2 and v_2 is connected directly by an edge e_{u_2,v_2}, which will be called the *short path* between u_2 and v_2. In addition, there are τ^2 acyclic paths with length $\tau = \lfloor(|E_2| - 1)^{1/3}\rfloor$ connecting u_2 and v_2. They will be referred to as the *long paths*. Under the case that $(|E_2| - 1)^{\frac{1}{3}} \notin \mathbb{Z}^+$, a ring with $|E_2| - \lfloor(|E_2| - 1)^{\frac{1}{3}}\rfloor^3 - 1$ edges will be attached to the node u_2 to complement the graph. An oblivious routing algorithm Φ will route the traffic requests between u_2 and v_2 along the short path integrally with the probability $\lambda_s \geq 0$. Denote the cost corresponding to Φ by C_Φ and the cost corresponding to the optimal integral solution by OPT_I. Next, we will consider two cases.

1. $\lambda_s \geq \frac{\sqrt{5}-1}{2}$. In such case, we suppose that τ^2 independent traffic requests need to be routed between u_2 and v_2. For each request R_k, let $d_k = 1$. In such case, $\text{E}[C_\Phi] \geq (\lambda_s \cdot \tau^2)^2$ while $\text{OPT}_I \leq \tau^3$. Thus, the competitive ratio will be at least $\frac{(\lambda_s \cdot \tau^2)^2}{\tau^3} = \frac{3-\sqrt{5}}{2}\tau$.

2. $\lambda_s < \frac{\sqrt{5}-1}{2}$. Now we consider a single traffic request R_{large} with $d_{\text{large}} = \tau^2$ between u_2 and v_2. R_{large} will be routed along a long path with probability at least $1 - \lambda_s$. Then we have $\text{Ex}[C_\Phi] \geq (1 - \lambda_s)\tau^5$ and $\text{OPT}_I \leq \tau^4$. The competitive ratio will be at least $\frac{(1-\lambda_s)\tau^5}{\tau^4} = \frac{3-\sqrt{5}}{2}\tau$.

To sum up, the competitive ratio is no less than $\frac{3-\sqrt{5}}{2}\lfloor(|E_2| - 1)^{1/3}\rfloor$.   □

3   Definitions and Procedure

In this part we will introduce some essential definitions and specify the algorithm procedure. We will start by giving a brief overview for the convex combination of decomposition trees [6,13], which will be used by our routing approach.

Given a network $G(V, E)$, a *decomposition tree* T of G is a rooted tree whose nodes are corresponding to the subsets of nodes in V [13]. As an illustration, a decomposition tree of a network G_3 in Fig. 1(c) is given in Fig. 1(d). An embedding of T into G is specified in [6,13], which maps tree nodes to the nodes in V. Particularly, the leaf nodes are mapped to V by a bijection. For each node $u \in G$, we use u^T to represent its corresponding leaf node in T. This embedding also maps each edge $e^T \in T$ to a path P_{e^T} in G. Routing according to T means that a traffic request between the source-target pair (u, v) will be routed along the path uniquely defined by the path connecting u^T and v^T in T, which will be denoted by P_{u^T, v^T}^T. Such a routing approach can be expressed as a $|E| \times \binom{n}{2}$-dimensional binary matrix M_T. For each e and each (s_k, t_k), $M_T(e, k) = 1$ if there exists an $e^T \in T$ such that $e \in P_{e^T}$ and $e_T \in P_{s_k^T, t_k^T}^T$, otherwise $M_T(e, k) = 0$.

In [6,13], Räcke et al. define the *convex combination* of decomposition trees. It is composed of a set of decomposition trees $\{T_1, T_2, \ldots, T_k, \ldots\}$ for G, each of which is associated with a weight λ_i such that $\vec{\lambda} \succeq 0$ and $\|\vec{\lambda}\|_1 = 1$, where $\vec{\lambda}$ represents the vector composed of every λ_i. Based on the convex combination of decomposition trees, Englert et al. propose an oblivious routing strategy in [6] to minimize the L_p-norm of the load vector. According to their routing strategy, each traffic request R_k will be split into a set of fractional flows $fl_1, fl_2, \ldots, fl_i, \ldots$ such that each $fl_i = \lambda_i \cdot d_k$. Each fractional flow fl_i will be routed according to the tree T_i respectively. Such a routing strategy can be represented by a $|E| \times \binom{n}{2}$-dimensional matrix OBL $= \sum_i \lambda_i M_{T_i}$, which is called the *tree-based matrix* [6]. Owing to the split operation on the traffic requests, their routing strategy is unable to satisfy the *integral constraint*.

3.1 Notations

Some notations corresponding to the definitions described above are listed as follows. They will be used to define and analyse our ROIR algorithm.

- $\overline{\text{OBL}}$: For a tree-based matrix OBL, we use $\overline{\text{OBL}}$ to represent the $|E| \times |E|$-dimensional matrix induced from OBL by removing all columns corresponding to node pairs which aren't connected by an edge directly in G.
- $\|\overline{\text{OBL}}\|_p$: For any $p \geq 1$, we use $\|\overline{\text{OBL}}\|_p$ to represent the induced L_p-norm of $\overline{\text{OBL}}$. Formally,

$$\|\overline{\text{OBL}}\|_p = \max_{\|\vec{l}\|_p > 0} \frac{\|\overline{\text{OBL}} \cdot \vec{l}\|_p}{\|\vec{l}\|_p} \tag{1}$$

- $\|\overline{\text{OBL}}\|_p^p$. It represents the p-th power of $\|\overline{\text{OBL}}\|_p$.
- $\vec{l}_{\text{OPT}_F}^{\mathcal{R}}$ and $\vec{l}_{\text{OPT}_I}^{\mathcal{R}}$. Load vectors respectively corresponding to the *fractional* optimal solution and the *integral* optimal solution for the traffic set $\mathcal{R}$.
- $\vec{l}_{\text{OBL}_F}^{\mathcal{R}}$. The load vector incurred by routing $\mathcal{R}$ according to a tree-based matrix OBL *fractionally* in Englert et al.'s manner [6].
- $\vec{l}_{\text{OBL}_I}^{\mathcal{R}}$. The load vector incurred by routing $\mathcal{R}$ via our ROIR algorithm.
- $\text{OPT}_F(\mathcal{R})$. The cost corresponding to $\vec{l}_{\text{OPT}_F}^{\mathcal{R}}$, i.e., $\|\vec{l}_{\text{OPT}_F}^{\mathcal{R}}\|_2^2 = \text{OPT}_F(\mathcal{R})$. The notation $\text{OPT}_I(\mathcal{R})$ is defined in a similar way.

- $\vec{l}(e)$. It represents the element of the load vector $\vec{l}$ which is corresponding to the edge e, i.e., $\vec{l}(e) = l_e$. This notation will be used along with the subscripts and superscripts defined above.

Based on the notations defined above, now we give the formal definition of the competitive ratio. According to [6], it can be defined as $\max_{\mathcal{R}} \left\{ \frac{\|\vec{l}_{OBL_I}^{\mathcal{R}}\|_2^2}{OPT_I(\mathcal{R})} \right\}$.

3.2   Algorithm Procedure

Now we will present the procedure of our randomized oblivious integral routing algorithm (ROIR), which can give a competitive ratio close to the hardness factor given in Theorem 1. Specifically, it consists of two phases:

1. Precomputation Phase. Given a network $G(V, E)$, we precompute a particular convex combination of trees OBL^* of G with Bhaskara et al.'s algorithm [5] such that $\|\overline{OBL^*}\|_2 = O(\log |V|)$. According to [5,6], the time complexity of this step can be bounded by a polynomial of $|E|$. The obtained set of decomposition trees will be denoted by $\mathcal{T}^* = \{T_1^*, T_2^*, \ldots, T_k^*, \ldots\}$, and the weight associated to each tree T_k^* is denoted by λ_k^*, correspondingly. This phase will be accomplished offline before any traffic requests arrive.
2. Rolling Dice Phase. Whenever a traffic request arrive, we will randomly and independently select a decomposition tree $T_k^* \in \mathcal{T}^*$ and route the request according to T_k^*. As every traffic request is routed according to only one tree, we can guarantee that the integral constraint is satisfied. The probability that a tree T_k^* is selected, Pr_k^*, is set to its weight λ_k^*. This step is consistent since the weights satisfies $\vec{0} \leq \vec{\lambda} \leq \vec{1}$ and $\|\vec{\lambda}\|_1 = 1$.

Note that Rolling Dice Phase is different from the conventional techniques termed by the probabilistic approximation of metric spaces with tree metrics [4,7]. These techniques typically requires that the traffic requests should share a common randomness source so that two different requests can utilize the same random tree [9]. By contrast, the randomized decisions in our approach will be made independently, which guarantees that our approach is oblivious to the current state of the system, and can be realized in a decentralized manner. Furthermore, the number of random bits required by our ROIR algorithm to specify each path is bounded by $O(\log |E|)$, which indicates that these randomized decisions are made efficiently.

4   Randomized Path Selection

The competitive ratio of our ROIR algorithm can be decomposed into factors contributed by the Precomputation Phase and the Rolling Dice Phase respectively. In this part, we will study the influence of the Rolling Dice Phase on the competitive ratio in isolation. It will be achieved by assuming that we are given an arbitrary tree-based matrix OBL and analysing the competitive ratio of performing the Rolling Dice Phase according to OBL. It will be demonstrated that in such case the competitive ratio is bounded by $O(\beta \cdot \log D)$, where β is a parameter dependent on OBL. In the next section we will show that the tree-based matrix, OBL*, can guarantee an $O(|E|^{\frac{1}{6}} \log^2 |V|)$-tight bound on β.

To analyse the competitive ratio, a typical approach is to bound the gap between $\|\vec{l}^{\mathcal{R}}_{OBL_I}\|_2^2$ and $\|\vec{l}^{\mathcal{R}}_{OPT_F}\|_2^2$. However, it has be proved in [3] that the gap between the optimal solution of the QPC-MF problem and any solution of the QPC-MIF problem can be as large as $O(|E|)$. To overcome this difficulty, an auxiliary function $g_j(x)$ [3] will be adopted. Formally,

$$g_j(x) = \max\{j \cdot x, \ x^2\} \tag{2}$$

where j is a non-negative parameter. The auxiliary function $g_j(x)$ *linearize* the cost function $f(x) = x^2$ in the interval $[0, j]$. According to [3]:

Lemma 2 ([3]). *Let $X_1, X_2, \ldots, X_k, \ldots$ be a set of independent random variables. For each X_k, it takes value j with probability p_k and value 0 with probability $1 - p_k$. There always exists a constant $c_0 \leq 1 + \lceil 2\lg 6 \rceil \cdot 2^{2(\lceil 2\lg 6 \rceil + 1) + \lg e}$ such that $E[g_j(\sum_k X_k)] \leq c_0 \cdot g_j(E[\sum_k X_k])$. Here e represents Euler's number.*

By replacing the objective from minimizing $\sum_e f(l_e)$ with $\sum_e g_j(l_e)$, a variation of the QPC-MIF problem will be obtained. The variation parameterized by j will be denoted by Ψ_I^j, where the subscript I is used to emphasize the existence of the integral constraint. Correspondingly, the fractional relaxation of Ψ_I^j will be denoted by Ψ_F^j. Given a traffic request set $\mathcal{R}$, the optimal solution of Ψ_I^j and Ψ_F^j will be represented by $OPT_I^j(\mathcal{R})$ and $OPT_F^j(\mathcal{R})$ respectively. An important property of Ψ_I^j is that if the demand d_k of every traffic request R_k is equivalent to j, the QPC-MIF problem and the Ψ_I^j problem will agree on every integral point in the feasible area [3]. A traffic request set with this property will be referred to as the *uniform* request set and denoted by $\mathcal{R}^j$. Thus we have

Lemma 3. $OPT_F^j(\mathcal{R}^j) \leq OPT_I^j(\mathcal{R}^j) = \|\vec{l}^{\mathcal{R}^j}_{OPT_I}\|_2^2 \leq \|\vec{l}^{\mathcal{R}^j}_{OBL_I}\|_2^2.$

Let $\mathcal{R}'$ be a subset of a given traffic request $\mathcal{R}$ (i.e., $\mathcal{R}' \subseteq \mathcal{R}$) and $\vec{l}^{\mathcal{R}}_{OBL_I}(e, \mathcal{R}')$ be the load of the edge e corresponding to the demands in $\mathcal{R}'$ under the case where all the requests in $\mathcal{R}$ are routed integrally according to OBL. In addition, we will use $\vec{l}^{\mathcal{R}'}_{OBL_F}(e)$ to represent the load of e corresponding to the case where only the traffic requests belonging to $\mathcal{R}'$ are routed fractionally according to OBL in Englert' manner. Then we have:

Lemma 4. $E\left[\vec{l}^{\mathcal{R}}_{OBL_I}(e, \mathcal{R}')\right] = \vec{l}^{\mathcal{R}'}_{OBL_F}(e)$. *Here the symbol $E[\cdot]$ represents the expectation of random variables.*

Proof. Let $\delta : V \times V \mapsto \mathbb{N}$ be an arbitrary function that maps each source-target pair to a distinct integer in the interval $[1, \binom{|V|}{2}]$. For each traffic request in $\mathcal{R}'$, we construct a $\binom{|V|}{2}$-dimensional vector $\vec{d}_k$ such that its $\delta(s_k, t_k)$-th element is set to d_k and other elements are set to 0. Obviously, we have $\vec{l}^{\mathcal{R}'}_{OBL_F} = \sum_{R_k \in \mathcal{R}'} OBL \cdot \vec{d}_k$.

For simplicity, here we use the symbol $[\cdot]_e$ to represent the e-th row of a matrix. Then $\mathrm{E}\left[\vec{l}_{\mathrm{OBL}_I}^{\mathcal{R}}(e, \mathcal{R}')\right] = \sum_k d_k \sum_i \lambda_i \cdot M_{T_i}(e, \delta(s_k, t_k)) = \sum_k \left[\sum_i \lambda_i M_{T_i}\right]_e \cdot \vec{d}_k$.
On the other hand, $\vec{l}_{\mathrm{OBL}_F}^{\mathcal{R}'}(e) = \sum_{R_k \in \mathcal{R}'}[\mathrm{OBL}]_e \cdot \vec{d}_k = \sum_{R_k \in \mathcal{R}'}\left[\sum_i \lambda_i M_{T_i}\right]_e \vec{d}_k$.
Thus our proposition follows.   □

Theorem 2. *Suppose that there exists a parameter β such that for any $j \geq 1$,
$\dfrac{\sum_e g_j\left(\vec{l}_{OBL_F}^{\mathcal{R}^j}(e)\right)}{OPT_F^j(\mathcal{R}^j)} \leq \beta$ for any uniform traffic request set $\mathcal{R}^j$. In such case, the competitive ratio can be bounded by $O(\log D \cdot \beta)$, where $D = \max_k d_k$.*

Proof. For any given traffic request set $\mathcal{R}$, we will construct an exponentially discrete request set $\widetilde{\mathcal{R}}$. In particular, for each request $R_k \in \mathcal{R}$, there exists a corresponding request $\widetilde{R}_k \in \widetilde{\mathcal{R}}$ such that $(\tilde{s}_k, \tilde{t}_k) = (s_k, t_k)$ and $\tilde{d}_k = 2^{\lceil \log_2 d_k \rceil}$. The request set $\widetilde{\mathcal{R}}$ can be divided into a sequence of uniform subsets $\widetilde{\mathcal{R}}^1, \ldots, \widetilde{\mathcal{R}}^{2^j}, \ldots$ such that $\widetilde{\mathcal{R}}^{2^j} = \{\widetilde{R}_k \mid \tilde{d}_k = 2^j\}$ for each $j \in [0, \lceil \log_2 D \rceil]$. Correspondingly, the request set $\mathcal{R}$ will also be divided into subsets $\mathcal{R}_1, \ldots, \mathcal{R}_{2^j}, \ldots$ such that $\mathcal{R}_{2^j} = \{R_k \mid \widetilde{R}_k \in \widetilde{\mathcal{R}}^{2^j}\}$. Then we have:

$$\mathrm{E}[(\vec{l}_{\mathrm{OBL}_I}^{\mathcal{R}}(e))^2] \leq \mathrm{E}\left[\left(\sum_{j=0}^{\lceil \log_2 D \rceil} \vec{l}_{\mathrm{OBL}_I}^{\widetilde{\mathcal{R}}}(e, \widetilde{\mathcal{R}}^{2^j})\right)^2\right]$$

$$\leq \mathrm{E}\left[\lceil \log_2 D \rceil \cdot \sum_j \left(\vec{l}_{\mathrm{OBL}_I}^{\widetilde{\mathcal{R}}}(e, \widetilde{\mathcal{R}}^{2^j})\right)^2\right]$$

$$\leq \lceil \log_2 D \rceil \cdot \sum_j c_0 \cdot g_{2^j}\left(\mathrm{E}\left[\vec{l}_{\mathrm{OBL}_I}^{\widetilde{\mathcal{R}}}(e, \widetilde{\mathcal{R}}^{2^j})\right]\right)$$

$$\leq \lceil \log_2 D \rceil \cdot \sum_j c_0 \cdot g_{2^j}\left(\vec{l}_{\mathrm{OBL}_F}^{\widetilde{\mathcal{R}}^{2^j}}(e)\right)$$

The second inequality above is based on the convexity of the cost function [3]. The third one follows from Lemma 2 while fourth one follows from Lemma 4.

$$\mathrm{E}[\|\vec{l}_{\mathrm{OBL}_I}^{\mathcal{R}}\|_2^2] = \sum_e \mathrm{E}[(\vec{l}_{\mathrm{OBL}_I}^{\mathcal{R}}(e))^2] \leq \lceil \log_2 D \rceil \cdot \sum_j c_0 \cdot \sum_{e \in E} g_{2^j}\left(\vec{l}_{\mathrm{OBL}_F}^{\widetilde{\mathcal{R}}^{2^j}}(e)\right)$$

$$\leq \lceil \log_2 D \rceil \cdot c_0 \cdot \beta \sum_j \mathrm{OPT}_F^{2^j}(\widetilde{\mathcal{R}}^{2^j})$$

$$\leq \lceil \log_2 D \rceil \cdot c_0 \cdot \beta \cdot 2^2 \cdot \sum_j \mathrm{OPT}_F^{2^j}(\mathcal{R}_{2^j})$$

$$\leq 4\lceil \log_2 D \rceil c_0 \beta \cdot \sum_j \mathrm{OPT}_I(\mathcal{R}_{2^j})$$

$$\leq 4\lceil \log_2 D \rceil c_0 \beta \cdot \mathrm{OPT}_I(\mathcal{R})$$

The third inequality above follows from the fact that $\tilde{d}_k \leq 2 \cdot d_k$ for each $R_k \in \mathcal{R}$. The fourth inequality follows from Lemma 3. The last one is owing to the superadditivity of $g_j(x)$ [3]. To sum up, the competitive ratio will be $\max_{\mathcal{R}} \frac{E[\|\vec{l}^{\mathcal{R}}_{\mathrm{OBL}_I}\|_2^2]}{\mathrm{OPT}_I(\mathcal{R})} \leq 4\lceil \log_2 D \rceil c_0 \beta$. Thus this proposition holds. $\square$

5 Induced Norm Minimization

In this part, we will consider how to minimize the factor β in the competitive ratio. It will be shown that the tree-based matrix OBL* obtained in the Precomputation Phase can guarantee an $O(|E|^{\frac{1}{6}} \log^2 |V|)$-tight bound on β. We start with the following lemmas:

Lemma 5. *For a tree-based matrix OBL and any $j \geq 1$, if there exists a parameter γ such that $\max\{\|\overline{OBL}\|_1, \|\overline{OBL}\|_2^2\} \leq \gamma$, then $\beta \leq 2 \cdot \gamma$.*

Proof. Let $\vec{l}^{\mathcal{R}}_{\mathrm{OPT}^j_I}$ be the load vector corresponding to $\mathrm{OPT}^j_I(\mathcal{R})$. According to [6], for any tree-based matrix OBL, $\vec{l}^{\mathcal{R}}_{\mathrm{OBL}_F} \preceq \overline{OBL} \cdot \vec{l}^{\mathcal{R}}_{\mathrm{OPT}^j_I}$. Thus for any set $\mathcal{R}$:

$$\sum_{e \in E} g_j\left(\vec{l}^{\mathcal{R}}_{\mathrm{OBL}_F}(e)\right) = \sum_{e \in E} \max\left\{j \cdot \vec{l}^{\mathcal{R}}_{\mathrm{OBL}_F}(e), \left(\vec{l}^{\mathcal{R}}_{\mathrm{OBL}_F}(e)\right)^2\right\}$$
$$\leq j \cdot \|\vec{l}^{\mathcal{R}}_{\mathrm{OBL}_F}\|_1 + \|\vec{l}^{\mathcal{R}}_{\mathrm{OBL}_F}\|_2^2$$
$$\leq j \cdot \|\overline{OBL} \cdot \vec{l}^{\mathcal{R}}_{\mathrm{OPT}^j_I}\|_1 + \|\overline{OBL} \cdot \vec{l}^{\mathcal{R}}_{\mathrm{OPT}^j_I}\|_2^2$$
$$\leq \gamma(j \cdot \|\vec{l}^{\mathcal{R}}_{\mathrm{OPT}^j_I}\|_1 + \|\vec{l}^{\mathcal{R}}_{\mathrm{OPT}^j_I}\|_2^2)$$
$$\leq 2\gamma \cdot \sum_{e \in E} \max\left\{j \cdot \vec{l}^{\mathcal{R}}_{\mathrm{OPT}^j_I}(e), \left(\vec{l}^{\mathcal{R}}_{\mathrm{OPT}^j_I}(e)\right)^2\right\}$$
$$= 2\gamma \cdot \mathrm{OPT}^j_F(\mathcal{R})$$

The second inequality above follows from $\vec{l}^{\mathcal{R}}_{\mathrm{OBL}_F} \preceq \overline{OBL} \cdot \vec{l}^{\mathcal{R}}_{\mathrm{OPT}^j_I}$ [6]. $\square$

This lemma implies that we can find a proper bound on β by simultaneously minimizing $\|\overline{OBL}\|_1$ and $\|\overline{OBL}\|_2^2$ over all tree-based matrices of the network G. The following theorem gives a lower bound on $\max\{\|\overline{OBL}\|_1, \|\overline{OBL}\|_2^2\}$.

Theorem 3. *There exists a network G with $|E|$ edges for which no algorithm can compute a tree-based matrix OBL such that $\max\{\|\overline{OBL}\|_1, \|\overline{OBL}\|_2^2\}$ is bounded by $o(|E|^{\frac{1}{3}})$.*

Proof. Here we consider the network G_2 in Fig. 1(b). Suppose that there exists a tree-based matrix OBL' for G_2 such that $\max\{\|\overline{OBL'}\|_1, \|\overline{OBL'}\|_2^2\} = o(|E_2|^{\frac{1}{3}})$. According to [5,6], in such case by fractionally routing according to OBL' it can be guaranteed that:

$$\max_{p \in \{1,2\}} \max_{\mathcal{R}} \left\{\frac{\|\vec{l}^{\mathcal{R}}_{\mathrm{OBL'}}\|_p^p}{\mathrm{OPT}^p_F(\mathcal{R})}\right\} = o(|E_2|^{\frac{1}{3}}) \tag{3}$$

where $\text{OPT}_F^p(\mathcal{R})$ represents the optimal cost incurred by routing $\mathcal{R}$ fractionally according to OBL′ when the cost function is $f(l_e) = (l_e)^p$.

Following, we will demonstrate that there exists a traffic request set $\mathcal{R}$ on G_2 such that formulation (3) doesn't hold. Let $\mathcal{R} = \{R_1, R_2, \ldots, R_k, \ldots, R_{\tau^2+1}\}$ such that for every $R_k \in \mathcal{R}$, $(s_k, t_k) = (u_2, v_2)$ and $d_k = 1$. In such case, $\text{OPT}_F^1(\mathcal{R}) = \tau^2 + 1$ and $\text{OPT}_F^2(\mathcal{R}) \leq \tau^3 + 1$. Suppose that an arbitrary routing algorithm Φ (not restricted to oblivious routing algorithm) route ε units of flow along the short path between u_2 and v_2, where ε is a real number in the interval $[0, \tau^2 + 1]$. It will make $\|\vec{l}_\Phi^{\mathcal{R}}\|_1 \geq \varepsilon + \tau \cdot (\tau^2 + 1 - \varepsilon)$, while $\|\vec{l}_\Phi^{\mathcal{R}}\|_2^2 \geq \varepsilon^2 + (\frac{\tau^2+1-\varepsilon}{\tau^2})^2$, where $\vec{l}_\Phi^{\mathcal{R}}$ represents the load vector incurred by routing $\mathcal{R}$ with Φ. Plugging in these bounds and fixing the value of τ, $\max\left\{\frac{\|\vec{l}_\Phi^{\mathcal{R}}\|_1}{\text{OPT}_f^1(\mathcal{R})}, \frac{\|\vec{l}_\Phi^{\mathcal{R}}\|_2^2}{\text{OPT}_f^2(\mathcal{R})}\right\}$ can be represented by a function $h(\varepsilon)$. By taking the derivative of $h(\varepsilon)$, we have $h(\varepsilon) \geq \frac{3}{4}\tau$. Since $\tau = \Theta(|E_2|^{\frac{1}{3}})$, this proposition follows.   □

Recall that in the Precomputation Phase, we obtain a tree-based matrix OBL* for which $\|\overline{\text{OBL}^*}\|_2 = O(\log|V|)$. For this matrix we have:

Theorem 4. $\max\left\{\|\overline{OBL^*}\|_1, \|\overline{OBL^*}\|_2^2\right\} = O\left(|E|^{\frac{1}{2}} \log^2 |V|\right).$

Proof. The Precomputation Phase guarantees that $\|\overline{\text{OBL}^*}\|_2^2 = O(\log^2 |V|)$. Thus, here we only need to focus on $\|\overline{\text{OBL}^*}\|_1$. According to the theories on linear algebra, we have $\|\overline{\text{OBL}^*}\|_1 \leq |E|^{\frac{1}{2}} \cdot \|\overline{\text{OBL}^*}\|_2 = O(|E|^{\frac{1}{2}} \cdot \log|V|)$. Thus this proposition follows.   □

According to Theorem 3, the result in Theorem 4 is tight up to $O(E^{\frac{1}{6}} \log^2 |V|)$. By combining Theorem 2 with Lemma 5 and Theorem 4, we have

Theorem 5. *The competitive ratio of our ROIR algorithm can be bounded by* $O(|E|^{\frac{1}{2}} \log^2 |V| \cdot \log D)$.

It can be inferred from Theorem 1 that our competitive ratio is tight up to $O(|E|^{\frac{1}{6}} \log^2 |V| \cdot \log D)$.

6   Conclusion

Motivated by the importance of the energy conservation of the high-capacity backbone networks, we study the multicommodity integral flow problem whose objective is to minimize a quadratic polynomial function. For this problem, we develop an oblivious routing algorithm – ROIR, which can be implemented efficiently in the high-capacity backbone networks in a decentralized manner.

The most important feature of our algorithm, which makes our work different from the existing ones, is that our ROIR algorithm will route every traffic request integrally rather than splitting it into a set of fractional flows. This feature makes our work more practical for the packet switching networks. However, comparing with the case that the fractional routing is allowed, the integral routing increases the hardness of approximation significantly. In particular, we find that no oblivious routing algorithm can give a competitive ratio of $o(|E|^{\frac{1}{3}})$ for QPC-MIF, whereas our ROIR algorithm can guarantee that the competitive ratio is bounded by $O(|E|^{\frac{1}{2}} \log^2 |V| \cdot \log D)$. So the gap between our competitive ratio and hardness result can be bounded by $O(|E|^{\frac{1}{6}} \log^2 |V| \cdot \log D)$.

Although the competitive ratio of ROIR cannot be improved significantly on the networks with general topologies, it may be possible to make improvement for the networks with some special topologies. Whether ROIR can give a tight competitive ratio on the special topologies is still unknown. This problem will be studied in our future work.

References

1. Ahuja, R., Magnanti, T., Orlin, J.: Multicommodity Flows. In: Network Flows: Theory, Algorithms, and Applications, pp. 649–694. Prentice Hall (1993)
2. Andrews, M., Antonakopoulos, S., Zhang, L.: Minimum-cost network design with (dis)economies of scale. In: 2010 51st Annual IEEE Symposium on Foundations of Computer Science (FOCS), pp. 585–592 (October 2010)
3. Andrews, M., Anta, A.F., Zhang, L., Zhao, W.: Routing for power minimization in the speed scaling model. IEEE/ACM Transactions on Networking 20(1), 285–294 (2012)
4. Bartal, Y.: Probabilistic approximation of metric spaces and its algorithmic applications. In: Proceedings of the 37th Annual Symposium on Foundations of Computer Science, FOCS 1996. IEEE Computer Society (1996)
5. Bhaskara, A., Vijayaraghavan, A.: Approximating matrix p-norms. In: Proceedings of the Twenty-Second Annual ACM-SIAM Symposium on Discrete Algorithms, SODA 2011, pp. 497–511. SIAM (2011)
6. Englert, M., Räcke, H.: Oblivious routing for the lp-norm. In: Proceedings of the 2009 50th Annual IEEE Symposium on Foundations of Computer Science, FOCS 2009, pp. 32–40. IEEE Computer Society, Washington, DC (2009)
7. Fakcharoenphol, J., Rao, S., Talwar, K.: A tight bound on approximating arbitrary metrics by tree metrics. In: Proceedings of the Thirty-fifth Annual ACM Symposium on Theory of Computing, STOC 2003, pp. 448–455. ACM (2003)
8. Goyal, N., Olver, N., Shepherd, F.: Dynamic vs. oblivious routing in network design. Algorithmica 61(1), 161–173 (2011)
9. Gupta, A., Hajiaghayi, M.T., Räcke, H.: Oblivious network design. In: Proceedings of the Seventeenth Annual ACM-SIAM Symposium on Discrete Algorithm, SODA 2006, pp. 970–979. ACM, New York (2006)
10. Harsha, P., Hayes, T.P., Narayanan, H., Räcke, H., Radhakrishnan, J.: Minimizing average latency in oblivious routing. In: Proceedings of the Nineteenth Annual ACM-SIAM Symposium on Discrete Algorithms, SODA 2008, pp. 200–207. Society for Industrial and Applied Mathematics (2008)
11. Intel: Enhanced intel speedstep technology for the intel pentium m processor. Intel White Paper 301170-001 (2004)
12. Lawler, G., Narayanan, H.: Mixing times and lp bounds for oblivious routing. In: Proceedings of the 5th Workshop on Analytic Algorithmics and Combinatorics (ANALCO), pp. 66–74 (2009)

13. Räcke, H.: Optimal hierarchical decompositions for congestion minimization in networks. In: Proceedings of the 40th Annual ACM Symposium on Theory of Computing, STOC 2008, pp. 255–264. ACM (2008)
14. Räcke, H.: Survey on oblivious routing strategies. In: Ambos-Spies, K., Löwe, B., Merkle, W. (eds.) CiE 2009. LNCS, vol. 5635, pp. 419–429. Springer, Heidelberg (2009)
15. Tucker, R.S.: The role of optics and electronics in high-capacity routers. Journal of Lightwave Technology 24(12), 4655–4673 (2006)
16. Wierman, A., Andrew, L., Tang, A.: Power-aware speed scaling in processor sharing systems. In: IEEE INFOCOM 2009, pp. 2007–2015 (April 2009)
17. Zervas, G., De Leenheer, M., Sadeghioon, L., Klonidis, D., Qin, Y., Nejabati, R., Simeonidou, D., Develder, C., Dhoedt, B., Demeester, P.: Multi-granular optical cross-connect: Design, analysis, and demonstration. IEEE/OSA Journal of Optical Communications and Networking 1(1), 69–84 (2009)

Finding Simple Paths on Given Points in a Polygonal Region[*]

Xuehou Tan[1,2] and Bo Jiang[1]

[1] Dalian Maritime University, Linghai Road 1, Dalian, China
[2] Tokai University, 4-1-1 Kitakaname, Hiratsuka 259-1292, Japan
tan@wing.ncc.u-tokai.ac.jp

Abstract. Given a set X of points inside a polygonal region P, two distinguished points $s, t \in X$, we study the problem of finding the simple polygonal paths that turn only at the points of X and avoid the boundary of P, from s to t. We present an $O((n^2 + m) \log m)$ time, $O(n^2 + m)$ space algorithm for computing a simple path or reporting no such path exists, where n is the number of points of X and m is the number of vertices of P. This gives a significant improvement upon the previously known $O(m^2 n^2)$ time and space algorithm, and $O(n^3 \log m + mn)$ time, $O(n^3 + m)$ space algorithm.

An important result of this paper, termed the *Shortest-path Dependence Theorem*, is to characterize the simple paths of the minimum link distance, in terms of the shortest paths between the points of X inside P. It finally turns out that the visibility graph of X, together with an implicit representation of the shortest paths between all pairs of the points of X, is sufficient to compute a simple path from s to t or report no simple paths exist. The Shortest-path Dependence Theorem is of interest in its own right, and might be used to solve other problems concerning simple paths or polygons.

Keywords: Computational geometry, Visibility graph, Simple paths, Shortest paths, Dijkstra paradigm.

1 Introduction

Motivated by the work of generating random polygons with given vertices [1,10], the following problem has been studied in the literature [3,6]: Given a simple polygon P with m vertices, a set X of n points inside P, we want to find a *simple polygonal path* (without self-intersections) from a start point $s \in X$ to an end point $t \in X$, which turns only at the points of X and lies in the interior of the polygon P. Without loss of generality, assume that the line segment between s and t intersects the boundary of P. (Note that the points of X may be on the boundary of P).

[*] This work was partially supported by the Grant-in-Aid (MEXT/JSPS KAKENHI 23500024) for Scientific Research from Japan Society for the Promotion of Science and the National Natural Science Foundation of China under grant 61173034.

J. Chen, J.E. Hopcroft, and J. Wang (Eds.): FAW 2014, LNCS 8497, pp. 229–239, 2014.
© Springer International Publishing Switzerland 2014

As pointed out by Cheng et al. [3], the problem of finding arbitrary paths that avoid the boundary of P, from s to t, can simply be solved using the visibility graph of X inside P. The *visibility graph* of X inside P is a graph whose arc set records all point pairs (x, y), $x, y \in X$, such that x is visible from y in P. However, it becomes harder when the requirement that the found paths not be self-intersected is added. Fig. 1 shows an example of such a simple path from s to t (in dotted line) inside a polygon.

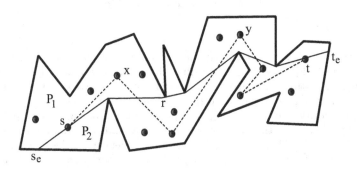

Fig. 1. An example of a simple path (in dotted line) from s to t

A related, more difficult problem is to find a simple path that avoids the boundary of P and uses *all* points of X, not just a subset. Clearly, it is equivalent to computing *simple Hamiltonian paths* from s to t. When P is convex, a simple Hamiltonian path can simply be found in $O(n \log n)$ time [3]. However, when P is an arbitrary polygon, no polynomial-time solution is known. This work finds applications in constructing a non-crossing railroad that connects all digging sites in a mine.

Previous Work. Cheng et al. were the first to study the problem of finding simple paths that turn on the points of X and avoid the boundary of P, from s to t. A simple path R from s to t is *shortcut-free* if no turning points can be deleted from R so as to obtain another simple path. They investigated the following "monotone" property of simple shortcut-free paths (Lemma 4 of [3]). For any simple shortcut-free path R with k points of X, there exists a sequence of k corresponding vertices of P such that the line segment connecting a turning point of R and its corresponding vertex can be extended to partition P into two disjoint subpolygons, one containing s and the other containing t, and moreover, any subpolygon containing s is completely contained in the one immediately after it (in the order from s to t). An $O(m^2n^2)$ time, $O(m^2n^2)$ space algorithm is then developed using dynamic programming. Whether a more efficient algorithm (say, with running time $O(mn(m + n))$) can be given is left as an open problem [3]. (They also showed that finding the simple paths among arbitrary obstacles is NP-complete.)

Later, Daescu and Luo studied the "3-shortcut-free" paths R, in which no shortcuts can be made among any three consecutive points of R, and developed a series of properties and lemmas on 3-shortcut-free simple paths. Let G denote the visibility graph of X inside P. A so-called *pyramid graph* G' is then constructed from G, in which a node of G' corresponds to an arc of G and an arc of G' corresponds to a particular pair of two (connected) arcs of G that may appear in a simple 3-shortcut-free path oriented from s to t. Thus, the pyramid graph has $O(n^2)$ nodes and $O(n^3)$ arcs. They finally claimed an $O(n^3 \log m + mn)$ time, $O(n^3 + m)$ space algorithm to find $O(n^2)$ simple paths from s to t or report no simple paths exist [6].

Our Result. We present a new approach to finding the simple paths from s to t, which is based on the structure of the shortest paths from s and t to all other points of X. Let $\pi(a, b)$ denote the shortest path between two points a and b inside P, which does not go across the boundary of P. The point set X can be partitioned into two subsets X_1 and X_2 by the path $\pi(s, t)$ as well as the extensions of its first and last segments inside P. If there are simple paths from s to t, then there exists a simple shortcut-free path R such that for the points $x_0(= s), x_1, x_2, \ldots, x_k, x_{k+1}(= t)$ of X_i ($i = 1$ or 2), appearing on R in this order, any two paths $\pi(s, x_j)$ and $\pi(x_{j+1}, t)$ $(1 \le j \le k - 1)$ are completely disjoint. This result, termed the *Shortest-path Dependence Theorem*, is then used to check the simplicity of a path from s to t. Furthermore, we show that the visibility graph of X, together with an implicit representation of the shortest paths between all pairs of the points of X, is sufficient to compute a simple path from s to t. Our algorithm runs in $O((n^2 + m) \log m)$ time and $O(n^2 + m)$ space. This gives a significant improvement upon the previously known $O(m^2 n^2)$ time and space algorithm, and $O(n^3 \log m + mn)$ time, $O(n^3 + m)$ space algorithm, and answers an open question posed by Cheng et al. [3].

The rest of this paper is organized as follows. We give basic definitions in Section 2, and present the Shortest-path Dependence Theorem in Section 3. The algorithm of finding a simple path from s to t is described in Section 4, which is a combined application of the Dijkstra paradigm and Shortest-path Dependence Theorem to the visibility graph of X inside P. Concluding remarks are given in Section 5.

2   Definitions

Let P be a simple polygon with m edges. Two points p, $q \in P$ are said to be mutually *visible* if the line segment connecting them, denoted by pq, is entirely contained in P.

Suppose that a set X of n points, and two distinguished points $s, t \in X$ are given in P. Assume that s and t are not mutually visible. A polygonal path R from s to t is called the (s, X, t)-*path* if it turns only at the points of X and avoids going across the boundary of P. Thus, two consecutive points on a (s, X, t)-path are mutually visible. The path R is *simple* if it does not have self-intersections. We refer to the points of X, which are used in R, as the *vertices* of R. Let x, y

be the two points (not necessarily vertices) on the path R such that x is closer to s than y on R. We denote by $R[x, y]$ ($R(x, y)$) the closed (open) portion of R from x to y.

A (s, X, t)-path with k vertices (including s and t) is said to have the *link distance* $k - 1$. A (s, X, t)-path R is *shortcut-free* if any two *non-consecutive* vertices of R are not mutually visible. Thus, the link distance of a shortcut-free path can never be decreased. We call a simple (s, X, t)-path, whose link distance is the minimum among all simple (s, X, t)-paths, the simple *minimum-link* (s, X, t)-path. Clearly, any simple minimum-link (s, X, t)-path is shortcut-free, and its link distance is at most $n - 1$. In the following, we use BN to denote a *big number*, which is larger than n.

For two arbitrary points a, b inside the polygon P (a or/and b may belong to X), denote by $\pi(a, b)$ the shortest path between a and b, which does not cross the boundary of P. Assume that the path $\pi(a, b)$ is oriented from a to b. Let us extend the first and last segments of $\pi(a, b)$, from a and b, until they first intersect the boundary of P. Denote by a_e and b_e two such intersection points, respectively. (Note that $\pi(a, b)$ is a portion of $\pi(a_e, b_e)$.) See Fig. 1 for an example, where the paths $\pi(s, t)$ and $\pi(s_e, t_e)$ are shown in solid line. The path $\pi(s_e, t_e)$ clearly separates the interior of P into two subpolygons P_1 and P_2. Denote by X_i, $i = 1, 2$, the subset of the points $x \in X \cap P_i$. For ease of presentation, assume that no other points of X, excluding s and t, are on the path $\pi(s_e, t_e)$. Thus, $X_1 \cap X_2 = \{s, t\}$ and $X_1 \cup X_2 = X$.

3   Shortest-Path Dependence Theorem

Two vertices x, y of a (s, X, t)-path are said to be X_i-*consecutive* (with respect to the (s, X, t)-path), $i = 1$ or 2, if both x and y belong to X_i and there is no other vertex $z \in X_i$ between x and y in the (s, X, t)-path (see also Fig. 1). In this section, we prove the following *Shortest-path Dependence Theorem*: A shortcut-free (s, X, t)-path R is simple if and only if for any two X_i-consecutive vertices x and y ($i = 1$ or 2), with respect to the (s, X, t)-path, the paths $\pi(s, x)$ and $\pi(y, t)$ are completely disjoint.

Lemma 1. *(The Consecutive Intersection Property [3].) Let R be a simple shortcut-free (s, X, t)-path, and $a, b \in P$ be two points that see each other in P. Let $i_1, i_2, \ldots, i_k$ be the intersection points of the path R with the line segment ab, indexed in the order in which they appear on R. Then, $i_1, i_2, \ldots, i_k$ are consecutive on ab, that is, i_j is between i_{j-1} and i_{j+1} for all $j = 2, \ldots, k - 1$.*

The Consecutive Intersection Property also holds for a path $\pi(s, x)$ (or $\pi(x, t)$), where x is a vertex of the path R.

Lemma 2. *Assume that R is a simple shortcut-free (s, X, t)-path and x ($x \neq s$ and $x \neq t$) is a vertex of R. Let $i_1, i_2, \ldots, i_k$ be the intersection points of R with the path $\pi(s, x)$, indexed in the order in which they appear on R. Then, $i_1, i_2, \ldots, i_k$ are consecutive on $\pi(s, x)$, that is, i_j is between i_{j-1} and i_{j+1} for all $j = 2, \ldots, k - 1$.*

Proof. First, no segment of R can intersect two segments of $\pi(s,x)$; otherwise, $\pi(s,x)$ is not the shortest path between s and x in P, a contradiction. Thus, if a segment of R intersects $\pi(s,x)$, only one intersection point occurs. Let us now apply the Consecutive Intersection Property (Lemma 1) to the segments of $\pi(s,x)$, which *intersect* R, one by one in the order that they appear on $\pi(s,x)$. Since both R and $\pi(s,x)$ start at s, the claim stated in the lemma thus follows.   □

Lemma 3. *(See [3]). If there exists a simple (s,X,t)-path in P, then there is a simple shortcut-free (s,X,t)-path R.*

Lemma 4. *Assume that R is a shortcut-free (s,X,t)-path in P. Let x, y be two X_i-consecutive vertices, $i = 1$ or 2, such that x precedes y on R. If two paths $\pi(s,x)$ and $\pi(y,t)$ share at least one common point, then R is not simple.*

Proof. The proof is given by contradiction. Assume that the shortcut-free path R is simple, but there are two X_i-consecutive vertices x and y in $R(s,t)$ such that $\pi(s,x)$ and $\pi(y,t)$ share some points. Let us extend the first segment and the last segment of the path $\pi(x,y)$ until they intersect the boundary of P, and denote by x_e, y_e two such intersection points. See Fig. 2.

Assume without loss of generality that (x,y) is the *first* pair of the X_i-consecutive vertices in R such that $\pi(s,x)$ and $\pi(y,t)$ intersect, in the sense that no vertex before x can contribute to such a vertex pair. See Fig. 2. So, $R[s,x)$ cannot go across $\pi(x_e,y_e)$; otherwise, there exists a vertex z in $R[s,x)$ such that $\pi(s,z)$ and $\pi(y,t)$ intersect, or y is visible from some vertex of $R[s,x)$, a contradiction in either case. If $R(y,t]$ does not go across the path $\pi(x_e,y_e)$ either, the last segment of $R[s,x]$ intersects $R[y,t]$ (Fig. 2(a)), contradicting the assumption that R is simple.

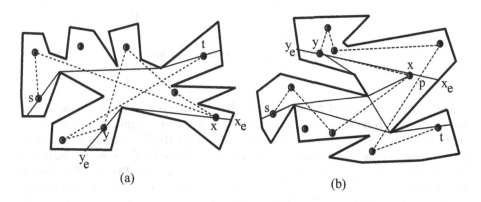

(a)   (b)

Fig. 2. Illustrating the proof of Lemma 4

Consider now the situation in which $R(y,t]$ goes across $\pi(x_e,y_e)$. Following from Lemma 2, all the intersection points of $\pi(s,x)$ with the simple shortcut-free path R are in the same order as they appear on R. Since x is a vertex of

R before y, $R(y, t]$ cannot intersect $\pi(s, x)$. This implies that $R(y, t]$ needn't go across the path $\pi(x, y_e)$. So, $R(y, t]$ goes across the line segment xx_e. Denote by p the intersection point of $R[y, t]$ and xx_e, see Fig. 2(b). Clearly, all the vertices of $R[y, p]$ form a simple polygon. At least one vertex of $R(y, p)$ is then visible from x (Fig. 2(b)), contradicting the shortcut-free property of R. The proof is complete.   □

Theorem 1. *(The Shortest-path Dependence Theorem.) Assume that R is a shortcut-free (s, X, t)-path in P. The path R is simple if and only if for all pairs of the X_i-consecutive vertices x and y, $i = 1$ or 2, such that x precedes y on R, two paths $\pi(s, x)$ and $\pi(y, t)$ are disjoint.*

Proof. The necessity follows from Lemma 4. We prove below sufficiency by induction on the number of vertices of the shortcut-free path R. It is clearly true if R contains three or four vertices. Assume below that the number of vertices of the path R is at least five, and for any two X_i-consecutive vertices x and y of $R(s, t)$, the paths $\pi(s, x)$ and $\pi(y, t)$ are disjoint.

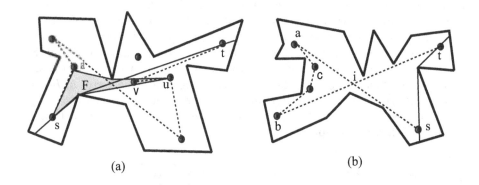

Fig. 3. Illustrating the proof of Theorem 1

Let a be the vertex of R, which immediately succeeds s. We first claim that the sufficiency condition holds for the path $R[a, t]$, too. Again, we prove it by contradiction. Assume that (u, v) is the first pair of the X_i-consecutive vertices in $R(a, t)$ such that $\pi(a, u)$ and $\pi(v, t)$ intersect. Denote by F the region bounded by the line segment sa and two paths $\pi(s, u)$, $\pi(a, u)$. See Fig. 3. (The region F is usually called a *funnel*.) From the assumption that $\pi(s, u)$ and $\pi(v, t)$ are disjoint, v is contained in F and thus visible from at least one vertex of $R[a, u)$. See Fig. 3(a). This contradicts the shortcut-free property of R, and our claim is proved. From the induction assumption, the shortcut-free path $R[a, t]$ is simple. Analogously, the shortcut-free path $R[s, b]$ is simpe, where b denotes the vertex of R immediately preceding t.

For the simplicity of the path R, the remaining job is to show that two segments sa and bt cannot intersect. Again, assume by contradiction that sa and bt

intersect, say, at a point i (Fig. 3(b)). Since both $R[a, t]$ and $R[s, b]$ are shortcut-free and simple, all the vertices of $R(a, b)$ are contained in the geodesic triangle, which is bounded by ai, ib and $\pi(a, b)$. See Fig. 3(b). Thus, all the vertices of $R[a, b]$ belong to a same subset, say, X_1. Let c be the vertex of R immediately succeeding a. Clearly, two paths $\pi(s, a)$ and $\pi(c, t)$ intersect, contradicting the sufficiency condition. Therefore, the theorem is proved. □

Before closing this section, we give a simple method to determine whether two paths $\pi(s, x)$ and $\pi(y, t)$ share a common point. For any two points a and b on the boundary of a polygon P_i ($i = 1$ or 2), if a is encountered strictly before b in the scan of the boundary of P_i, starting from s and then s_e, we write $a \prec b$.

Lemma 5. *Assume that R is a shortcut-free (s, X, t)-path in P. Let (x, y) be the first pair of the X_i-consecutive vertices, $i = 1$ or 2, such that x precedes y on R and two paths $\pi(s, x)$ and $\pi(y, t)$ share a common point. Then, $y_e \prec x_e$, where the points x_e and y_e are computed from $\pi(x, y)$.*

Proof. It follows easily from the fact that (x, y) is the first pair of the X_i-consecutive vertices such that $\pi(s, x)$ and $\pi(y, t)$ share a common point (see also Fig. 2). □

4   Algorithm

Let G denote the visibility graph of X inside P. That is, the node set $V(G)$ is the same as the point set X, and the arc set $E(G)$ records all the point pairs (x, y), $x, y \in X$, such that x is visible from y in P. We will apply the Dijkstra paradigm to the graph G, with the modification that the Shortest-path Dependence Theorem applies to all the found paths. If there exist simple (s, X, t)-paths in P, our algorithm can report a simple (s, X, t)-path of the minimum link distance.

Let L denote a simple shortcut-free (s, X, v)-path in P, $v \neq t$. The path consisting of the (s, X, v)-path and the line segment vt, without considering the visibility between v and t, is called the *pseudo-simple path* of L. Then, we have the following result.

Lemma 6. *Suppose that L is a simple shortcut-free (s, X, v)-path in P, $v \neq t$. If the Shortest-path Dependence Theorem applies to the pseudo-simple path of L, then L may appear as a portion of a simple shortcut-free (s, X, t)-path.*

We now describe how to apply the Dijkstra paradigm (i.e., the single-source shortest-paths algorithm) to the graph G. For a point $x \in X_i$ ($i = 1$ or 2), we call X_i the *belonging set* of x. To apply the Shortest-path Dependence Theorem, we maintain, for each node $v \in V(G)$, a variable v' to indicate the last point on the found (s, X, v)-path, whose belonging set differs from that of v. Suppose that a simple minimum-link (s, X, v)-path is found, and u is the vertex immediately preceding v. If the belonging sets of u and v are the same, then $v' = u'$; otherwise, $v' = u$. For a simple minimum-link (s, X, v)-path, there may exist multiple points

v'; in this case, our algorithm always selects the vertex $v1'$ such that $v1'_e \prec v2'_e$, where $v1'$ and $v2'$ are two arbitrary candidates of v' and the points $v1'_e$ and $v2'_e$ are computed from the path $\pi(v1', v2')$ (Lemma 5).

Let W be the weight function W on the arcs of $E(G)$, which is defined as follows: For any two nodes $x, y \in V(G)$, if (x, y) is an arc of the visibility graph G, then the weight $W(x, y)$ is set to one; otherwise, $W(x, y)$ is equal to BN. For each $v \in V(G)$, a minimum link distance $D[v]$ is then computed if there exists a simple (s, X, v)-path of link distance $D[v]$ and the Shortest-path Dependence Theorem applies to its pseudo-simple path (Lemma 6). Otherwise, $D[v]$ is set to BN. Initially, the value $D[s]$ is zero, and all others $D[x]$ are set to BN.

The following modifications are made to the Dijkstra paradigm: Whenever a minimum link distance $D[v]$ as well as its simple, minimum-link (s, X, v)-path is found, the variable v' is accordingly updated to u or u', where u is the vertex immediately preceding v on the (s, X, v)-path. For the points w whose simple, minimum-link (s, X, w)-paths have not yet been found, we update its value $D[w]$ if (i) $D[v] + W(v, w) \le D[w]$ and (ii) $\pi(s, v)$ is disjoint from $\pi(w, t)$, or $\pi(s, v')$ is disjoint from $\pi(w, t)$, depending on whether the belonging sets of v and w are the same. If $D[v] + W(v, w) < D[w]$, then w' is accordingly set to v or v'. In the case that $D[v] + W(v, w) = D[w]$, if the belonging sets of v and the current w' are same and $v_e \prec w'_e$ (computed from $\pi(v, w')$), then let $w' \leftarrow v$. After the Dijkstra paradigm stops, if $D[t]$ is equal to BN, report "no simple (s, X, t)-paths exist"; otherwise, we output a simple, minimum-link (s, X, t)-path.

We give below the pseudo-code of our algorithm.

Algorithm ModifiedDijkstraParadigm

Input. A graph G of n nodes, two distinguished points $s, t \in V(G)$, and a function W from the arcs to non-negative integers. For any two nodes $x, y \in V(G)$, if (x, y) is an arc of the graph G, then the weight $W(x, y)$ is set to one; otherwise, $W(x, y)$ is equal to BN.

Output. For each $v \in V(G)$, a minimum link distance $D[v]$ if there exists a simple (s, X, v)-path of link distance $D[v]$ and the Shortest-path Dependence Theorem applies to its pseudo-simple path; otherwise, $D[v]$ is equal to BN.

1. Let $T \leftarrow V(G) - \{s\}$, and let $D[s] \leftarrow 0$ and $s' \leftarrow s$.
2. For all points $x \in T$, let $D[x] \leftarrow W(s, x)$.
3. While T is not empty, do the followings:

   (a) Choose a point v from T such that the link distance $D[v]$ is a minimum.
   (b) Let u be the vertex immediately preceding v on the found (s, X, v)-paths. If the belonging sets of v and u are the same, $v' \leftarrow u'$. Otherwise, $v' \leftarrow u$.
   (c) Let $T \leftarrow T - \{v\}$.
   (d) For each point w of T, let $D[w] \leftarrow D[v] + W(v, w)$ if $D[v] + W(v, w) \le D[w]$ and $\pi(s, v)$ is disjoint from $\pi(w, t)$, or $\pi(s, v')$ is disjoint from $\pi(w, t)$, depending on whether the belonging sets of v and w are the same. If $D[v] + W(v, w) < D[w]$, then w' is accordingly set to v or v'. In the case that $D[v] + W(v, w) = D[w]$, if the belonging sets of v and the current w' are same and $v_e \prec w'_e$, let $w' \leftarrow v$. Finally, if the value

of $D[w]$ or w' is ever *changed*, we record v as the vertex immediately preceding w.

4. If $D[t]$ is not equal to BN, then output a simple, minimum-link (s, X, t)-path; otherwise, report "no simple (s, X, t)-paths exist".

Fig. 4 shows an example of the graph G, and the values of all elements of D after our algorithm terminates. The simple (s, X, t)-path of minimum link distance reported by our algorithm is s, e, c, f, d and t. Although s, g and c also form a simple path, the simple (s, X, c)-path found in Step 3(d) is s, e and c, because $e_e \prec g_e$ (Fig. 4(b)). Although there exists the simple (s, X, a)-path (resp. (s, X, b)-path), the Shortest-path Dependence Theorem does not hold for any pseudo-simple path towards a (resp. b), and the final value of $D[a]$ (resp. $D[b]$) is thus equal to BN, see Fig. 4(b).

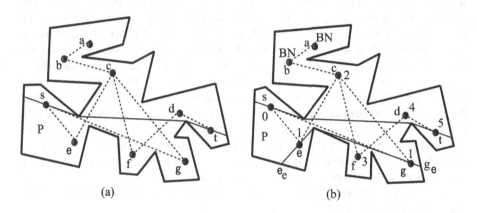

Fig. 4. An example of the graph G and the running result of our algorithm

Lemma 7. *All points x_e and y_e of the paths $\pi(x, y)$, $x, y \in X_i$, $i = 1$ or 2, can be computed in $O((n^2 + m) \log m)$ time.*

Proof. We preprocess in $O(m)$ time the polygon P so that a shortest path query between two given points can be answered in logarithmic time plus the time required to report the path itself [8,9]. Also, we preprocess in $O(m \log m)$ time the polygon P so that a ray-shooting query can be answered in $O(\log m)$ time [4]. For two points $x, y \in X_i$, $i = 1$ or 2, after the first (resp. last) turning point of the path $\pi(x, y)$ is computed in $O(\log m)$, the point x_e (resp. y_e) can be found in $O(\log m)$ time by a ray-shooting query. Since the total number of point pairs (x, y) is $O(n^2)$, the lemma follows.   □

By now, we can conclude the main result of this paper.

Theorem 2. *The problem of finding a simple (s, X, t)-path inside a simple polygon P can be solved in $O((n^2 + m) \log m)$ time and $O(n^2 + m)$ space, where n is the number of points of X and m is the number of vertices of P.*

Proof. All the simple paths of link distance one, from the point s, can clearly be found in the graph G. If there are multiple simple (s, X, v)-paths of link distance $k \geq 1$, we find the (s, X, v)-path with the minimum point v'_e, among those points of all possible candidates v'. Such a simple, minimum-link (s, X, v)-path is computed in Step 3(a). The simple (s, X, w)-paths of the minimum link distance $k+1$ are then computed in Step 3(d). For a point $w \in T$, its value $D[w]$ is updated if $D[v] + W(v, w) \leq D[w]$ and the Shortest-path Dependence Theorem applies to the pseudo-simple path (s, X, w) (Lemma 6), and moreover, the variable w' is accordingly updated. Hence, the correctness of **ModifiedDijkstraParadigm** follows.

Consider the running time of our algorithm. It takes $O(n^2 + n \log n \log(mn) + m)$ time to compute the visibility graph of X inside P [2]. Note that the term $O(n \log n \log(mn))$ can simply be replaced by $O(n^2 \log m)$, no matter which of m and n is larger. The weighted graph G is then constructed in $O(n^2)$ time. For any two points $x, y \in X_i$, $i = 1$ or 2, we compute the points x_e and y_e of the path $\pi(x, y)$ (Lemma 7). Finally, applying the modified Dijkstra's algorithm to the graph G finds a simple minimum-link (s, X, t)-path or reports no such path exists. In Step 3(d), we need to determine whether two paths $\pi(w, t)$ and $\pi(s, v)$ or $\pi(s, v')$ are disjoint. Since all the points x_e and y_e for the point pairs (x, y) have beeen precomputed, it can be verified in constant time (Lemma 5). Step 3(d) can thus be performed in $O(n)$ time. In Step 4, a simple (s, X, t)-path (if it exists) can be reported by starting from $w = t$ and tracing back by using its preceding vertex (recorded in Step 3(d)), until s is reached. Hence, the algorithm **ModifiedDijkstraParadigm** itself takes $O(n^2)$ time [5]. In summary, the time complexity of our algorithm is $O((n^2 + m) \log m)$.

Finally, since the size of the graph G is $O(n^2)$, the space requirement of our algorithm is $O(n^2 + m)$.   □

5   Concluding Remarks

In this paper, we study the problem of computing the simple polygonal paths that turn only at the points of the given set X and avoid the boundary of the given polygon P, from a start point s to an end point t. We have presented an $O((n^2 + m) \log m)$ time, $O(n^2 + m)$ space algorithm for outputting a simple path or reporting no such path exists, where n is the number of points of X and m is the number of vertices of P. This gives a significant improvement upon the previously known $O(m^2 n^2)$ time, $O(m^2 n^2)$ space algorithm, and $O(n^3 \log m + mn)$ time, $O(n^3 + m)$ space algorithm [3,6]. Moreover, since our algorithm makes use of the standard data structures and paradigms, it is much simpler than the previous ones.

We pose several open questions for further research. Analogous to the Dijkstra paradigm, it is desirable to develop an algorithm for computing the simple minimum-link (s, X, x)-paths for all points $x \in X$, $x \neq s$. A simple solution can be given by invoking our algorithm $n - 1$ times. Whether a more efficient algorithm can be developed is open. Also, it is an interesting work to find the application of the Shortest-path Dependence Theorem to other problems concerning simple paths or simple polygons [1,10]. Finally, it remains open to find a polynomial-time solution to the problem of computing the simple Hamiltonian path from s to t, which avoids the boundary of P and uses all points of X [3]. Note that even the problem of finding arbitrary Hamiltonian path from s to t is still open.

Acknowledgements. The authors would like to thank Dr. Jun Luo of Shenzhen Institute of Advanced Technology for his valuable comments on a preliminary version of the paper.

References

1. Auer, T., Held, M.: Heuristics for the generation of random polygons. In: Proc. 8th Candian Conf. on Comput. Geom., pp. 38–44 (1996)
2. Ben-Moshe, B., Hall-Holt, O., Katz, M.J., Mitchell, J.S.B.: Computing the visibility graph of points within a polygon. In: Proc. ACM Sympos. Comput. Geom., pp. 27–35 (2004)
3. Cheng, Q., Chrobak, M., Sundaram, G.: Computing simple paths among obstacles. Comput. Geom. 16, 223–233 (2000)
4. Chazelle, B., Guibas, L.: Visibility and intersection problem in plane geometry. Discrete Comput. Geom. 4, 551–581 (1989)
5. Corman, T.H., Leiserson, C.E., Rivest, R.L., Stein, C.: Introduction to algorithms, 3nd edn. The MIT Press (2009)
6. Daescu, O., Luo, J.: Computing simple paths on points in simple polygons. In: Ito, H., Kano, M., Katoh, N., Uno, Y. (eds.) KyotoCGGT 2007. LNCS, vol. 4535, pp. 41–55. Springer, Heidelberg (2008)
7. Guibas, L., Hershberger, J., Leven, D., Sharir, M., Tarjan, R.: Linear time algorithms for visibility and shortest path problems inside triangulated simple polygons. Algorithmica 2, 209–233 (1987)
8. Guibas, L.J., Hershberger, J.: Optimal shortest path queries in a simple polygon. J. Comput. Syst. Sci. 39, 126–152 (1989)
9. Hershberger, J.: A new data structure for shortest path queries in a simple polygon. Inform. Process. Lett. 38, 231–235 (1991)
10. Zhou, C., Sundaram, G., Snoeyink, J., Mitchell, J.S.B.: Generating random polygons with given vertices. Comput. Geom. 6, 277–290 (1996)

Near Optimal Algorithms
for Online Maximum Weighted b-Matching

Hingfung Ting and Xiangzhong Xiang

Department of Computer Science
The University of Hong Kong
{hfting,xzxiang}@cs.hku.hk

Abstract. We study the online maximum weighted b-matching problem, in which the input is a bipartite graph $G = (L, R, E, w)$. Vertices in R arrive online and each vertex in L can be matched to at most b vertices in R. Assume that the edge weights in G are no more than $w_{\max}$, which may not be known ahead of time. We show that a randomized algorithm GREEDY-RT which has competitive ratio $\Omega(\frac{1}{\prod_{j=1}^{\log^* w_{\max}-1} \log^{(j)} w_{\max}})$. We can improve the competitive ratio to $\Omega(\frac{1}{\log w_{\max}})$ if $w_{\max}$ is known to the algorithm when it starts. We also derive an upper bound $O(\frac{1}{\log w_{\max}})$ suggesting that GREEDY-RT is near optimal. Deterministic algorithms are also considered and we present a near optimal algorithm GREEDY-D which is $\frac{1}{1+2\xi(w_{max}+1)^{\frac{1}{\xi}}}$-competitive, where $\xi = \min\{b, \lceil \ln(1 + w_{\max}) \rceil\}$. We propose a variant of the problem called online two-sided vertex-weighted matching problem, and give a modification of the randomized algorithm GREEDY-RT called GREEDY-vRT specially for this variant. We show that GREEDY-vRT is also near optimal.

1   Introduction

This paper studies the following *online (bipartite) maximum weighted b-matching problem*: The input is a weighted bipartite graph $G = (L, R, E, w)$. The vertex set L is known ahead of time, and the vertices in R arrive online. When a vertex in $r \in R$ arrives, all the edges $(\ell, r) \in E$ incident to r, as well as their weights $w((\ell, r))$, are revealed and we must either (i) pick one of r's unmatched neighbours $\ell \in L$ and match r to it; or (ii) leave r unmatched forever. Each vertex in L can be matched to at most b vertices in R. The goal is to find such a matching M with the maximum weight $\sum_{e \in M} w(e)$. One direct application of our problem is Internet advertising in which L is the set of advertisers and R is the set of websites, and when a website r is available, those advertisers ℓ who are interested in displaying their advertisements on the site will submit their bids $w((\ell, r))$. Each advertiser can post their advertisements on at most b different websites. Our objective is to accept the bids so as to maximize the revenue gained from the advertisers.

Previous Works. In [12], Karp, Vazirani and Vazirani introduced the *online maximum matching problem*, which is a special case of our problem in which

J. Chen, J.E. Hopcroft, and J. Wang (Eds.): FAW 2014, LNCS 8497, pp. 240–251, 2014.
© Springer International Publishing Switzerland 2014

$b = 1$ and all edges have weight 1. They showed that the simple greedy algorithm GREEDY has competitive ratio $\inf_G |\text{GREEDY}(G))|/|O(G)| = 1/2$ where GREEDY(G) and $O(G)$ are the matchings returned by GREEDY and the optimal algorithm for input G, respectively. Then, they gave a randomized algorithm RANKING for the problem, and proved that the algorithm has the optimal competitive ratio $1 - \frac{1}{e} \approx 0.632$. Since [12], there have been many interesting results for the online maximum matching problem. For example, Goel and Mehta [7], Birnbaum and Mathieu [3], and Devanur, Jain and Kleinberg [5] gave substantially simpler and insightful proofs that RANKING is $(1 - \frac{1}{e})$-competitive. Goel and Metha [7] also proposed to study the problem under the random order model, in which the vertices in R arrive in some random order. For this random order model, they showed that the simple greedy algorithm has competitive ratio $1 - \frac{1}{e}$. Later, Karande *et al.* [11] showed that the algorithm RANKING has competitive ratio at least 0.653 for this model, and independently Mahdian and Yan [14] showed that the bound should be at least 0.696. There are also studies on the problem with other assumptions on the input [2,4,6,8,15].

The online unweighted b-matching problem is studied by Kalyanasundaram and Pruhs [10]. They assumed that all edges are unweighted and presented a deterministic algorithm BALANCE which achieves an optimal competitive ratio of $1 - \frac{1}{(1+\frac{1}{b})^b}$.

Recently, Aggarwal, Goel, Karande and Mehta [1] proposed to study the *online vertex-weighted maximum matching problem*, which is also a special case of our problem in which $b = 1$, and all edges incident to the same vertex $\ell \in L$ have the same weight v_ℓ (i.e., for every edge $(\ell, y) \in E$, $w((\ell, y)) = v_\ell$). They gave an algorithm called PERTURBED-GREEDY which achieves an optimal competitive ratio of $1 - \frac{1}{e}$. They also showed that their solution effectively solves the well-known *online budgeted allocations* proposed in [16].

In [9], Kalyanasundaram and Pruhs studied the online maximum weighted matching problem, which is again a special case of our problem in which $b = 1$. By making the assumption that the vertices are points on some metric space, they proved that a greedy algorithm has competitive ratio at least $1/3$. We note that this metric space assumption is rather restrictive: it is easy to prove that without this assumption, no algorithm for the problem can have constant competitive ratio (see Subsection 4.2). In [13], Korula and Pál studied the problem with a different assumption, namely the size of R is known and its vertices arrive in random order. These assumptions allow them to apply the powerful "sample-and-price" technique to design an algorithm as follows: Based on $|R|$ choose a suitable sample size k. Then, read the first k vertices of R arriving online without doing any real matching. Find a maximum weight matching of the subgraph of G restricted to these k vertices, and use this information to assign a threshold $price(\ell)$ for every vertex $\ell \in L$. Then, the algorithm will ignore all edges with weight smaller than the thresholds, or more precisely, when a new vertex r arrives, the algorithm will match it to the unmatched vertex ℓ only if the edge

$e = (\ell, r)$ has weight $w(e) \geq price(\ell)$. By using a greedy strategy for choosing ℓ, they got an algorithm for the problem with competitive ratio at least $1/8$.

Our Results. Since we do not have any constraint on the edge weights, the techniques used in the design and analysis of competitive algorithms for the online maximum matching and the online vertex-weighted maximum matching problems are not applicable to solve our problem. We are interested in the sample-and-price technique that has been applied so successfully by Korula and Pál [13] for the more general maximum weighted matching problem with the assumption that the size of R is known and its vertices arrive in random order. We now do not have this random ordering assumption and thus sampling the set of the first k vertices no longer helps because an adversary can always give us the worst possible set to fool us. However, it is not obvious whether pricing, i.e., the threshold idea helps or not. In particular, it would be interesting to find out how bad a solution can be if we just pick a threshold randomly, and then use a greedy algorithm to find a maximum weighted matching among all edges whose weights above this threshold. A major result of ours is a non-trivial proof of the surprising fact that this algorithm, which we call GREEDY-RT, is almost optimal for online maximum weighted b-matching. We summarize below our results. For the sake of simplicity, we assume that the smallest edge weight is 1 (after some normalization) and the largest edge weight is no more than $w_{\max}$, which may not be known ahead of time.

- We show that by choosing a random threshold according to some carefully chosen probability distribution, GREEDY-RT has competitive ratio $\Omega(\frac{1}{\log w_{\max} \log^{(2)} w_{\max} \dots \log^{(\Delta)} w_{\max}})$ where $\Delta = \log^* w_{\max} - 1$. Here, $\log^{(1)} w_{\max} = \log w_{\max}$, $\log^{(j)} w_{\max} = \log(\log^{(j-1)} w_{\max})$ for $j \geq 2$, and $\log^* w_{\max}$ is the iterated logarithm of $w_{\max}$, which is the number of times the logarithm function must be iteratively applied before the result is less than or equal to 1.
- If we know the value of $w_{\max}$ ahead of time, then by choosing a random threshold uniformly in $\{e^0, e^1, e^2, \dots, e^{\lceil \ln w_{max}+1 \rceil -1}\}$, GREEDY-RT has competitive ratio $\Omega(\frac{1}{\log w_{\max}})$.
- No randomized algorithm can do better than $O(\frac{1}{\log w_{\max}})$-competitive.

We also consider deterministic algorithms for online maximum weighted b-matching. We show that no deterministic algorithm can achieve a competitive ratio greater than $\min\{\frac{2}{\lceil \log_2 w_{max} \rceil}, \frac{1}{(w_{max})^{\frac{1}{b}}}\}$. When $w_{\max}$ is not known ahead of time, no deterministic algorithm can be competitive. With the assumption that $w_{\max}$ is known ahead of time, we present a deterministic algorithm GREEDY-D which is $\frac{1}{1+2\xi(w_{max}+1)^{\frac{1}{\xi}}}$-competitive, where $\xi = \min\{b, \lceil \ln(1 + w_{\max}) \rceil\}$ [1].

We also present a near optimal algorithm for a special case of our problem, namely the online two-side vertex-weighted b-matching problem, in which every

---

[1] If b is extremely large compared with w_{max}, then $\xi = \lceil \ln(1 + w_{\max}) \rceil$, the competitive ratio of GREEDY-D is $\frac{1}{1+2e\lceil \ln(1+w_{\max}) \rceil}$ while the upper bound is $\frac{2}{\lceil \log_2 w_{max} \rceil}$.

vertex $\ell \in L$ is associated with a weight v_ℓ and every vertex $r \in R$ is associated with a weight c_r, and every edge $(\ell, r) \in E$ has weight $v_\ell c_r$. Note that this problem is a generalization of the online vertex-weighted maximum matching problem of Aggarwal *et al.* [1], and when applying to Internet advertising, it corresponds to the case when the bids are separable. We present a competitive randomized algorithm GREEDY-vRT for this problem; it has competitive ratio $\Omega(\frac{1}{\log c_{max}})$ where c_{max} is the maximum weight associated with some vertex $r \in R$.

2 Problem Definitions

The online maximum weighted b-matching problem is defined formally as follows: The input of the problem is a bipartite graph $G = (L, R, E, w)$ where $L = \{\ell_1, \ell_2, \ldots, \ell_n\}$ and $R = \{r_1, r_2, \ldots r_m\}$ are the vertex sets and $E \subseteq L \times R$ is the edge set in which every edge e is associated with a weight $w(e)$. For the sake of simplicity, we assume that the smallest edge weight is 1 (after some normalization) and the largest edge weight is no more than $w_{\max}$, which may not be known ahead of time. An algorithm for the problem only knows the set L when it starts, and the sequence of vertices $r_1, r_2, \ldots r_m$ arrives online one by one. When a vertex $r_i \in R$ arrives, all edges incident to r_i, as well as their weights, are revealed, and the algorithm should immediately match r_i to one of its neighbor in L, or leave r_i unmatched forever. Each vertex ℓ_i in L can be matched to at most b vertices in R. The goal is to find a matching M of G with the maximum weight $\sum_{(\ell,r) \in M} w((\ell, r))$.

This paper also studies a special case of the online maximum weighted b-matching problem, namely the online two-side vertex-weighted b-matching problem, which has the following weight structure: each edge (ℓ, r) has weight $v_\ell c_r$ where v_ℓ and c_r are constants assigned to vertices ℓ and r, respectively. For all $\ell \in L$, the value of v_ℓ is known to the algorithm when it starts, but c_r is revealed only when r arrives. We still assume that the smallest weight of vertices in R is 1 (after some normalization, i.e., $\min_{r \in R} c_r = 1$) and largest weight of vertices in R is no more than $c_{\max}$, which may be unknown ahead of time. Note that when $b = 1$ and all c_r's are equal to 1, this problem is just the online vertex-weighted matching problem introduced in [1].

Let A be a randomized algorithm for the online maximum weighted matching problem. For any input bipartite graph G, let $A_{G,\sigma}$ be the matching returned by A for G based on the random choice σ, and let $w(A_{G,\sigma})$ be its weight. Let $E(w(A_{G,\sigma}))$ be the expected weight of the matching, over all possible random choices, returned by A for G. We say that A has competitive ratio c, and is c-competitive if for all input G, we have $E(w(A_{G,\sigma})) \geq c \cdot w(O_G)$ where $w(O_G)$ is the weight of the maximum weight matching O_G of G. Similarly, a deterministic algorithm A is c-competitive if $w(A_G) \geq c \cdot w(O_G)$, where A_G is the matching returned by A for graph G.

The rest of the paper is structured as follows: In Section 3, we discuss randomized algorithms for online maximum weighted b-matching. In Section 4, we discuss deterministic algorithms. In Section 5, we consider the online two-side vertex-weighted b-matching problem.

3   Randomized Algorithms for Online Maximum Weighted b-Matching

In this section, we show a competitive randomized algorithm GREEDY-RT for one special case of the online maximum weighed b-matching problem, in which $b = 1$. GREEDY-RT is proved to be $\Omega(\frac{1}{\prod_{j=1}^{\log^* w_{max}-1} \log^{(j)} w_{max}})$-competitive. If w_{max} is known ahead of time, its competitive ratio can be increased to $\Omega(\frac{1}{\log w_{max}})$. Then we show that the general online maximum weighted b-matching problem can be reduced to the previous special case. We also show an upper bound $O(\frac{1}{\log w_{max}})$ on the competitive ratio of any randomized algorithm.

3.1   Randomized Algorithm GREEDY-RT for Special Case: $b = 1$

In the special case, each vertex in L can be matched to only one vertex in R. To find the maximum weight matching, our strategy is to "guess" a random threshold and set it as the lowest weight of edges that could be included in the returned matching. Surprisingly, by a choosing proper threshold, the simple strategy is near optimal. The randomized algorithm GREEDY-RT is shown below. Here, e is the base of the natural logarithm. We will specify the probability distribution p_i later.

Choose an integer κ randomly from the set $N = \{0, 1, 2, \ldots\}$ with probability $\Pr[\kappa = i] = p_i$;
Set $\tau = e^\kappa$;
while *a new vertex $r \in R$ arrives* **do**
    $S = \{\ell \mid \ell$ is r's unmatched neighbor in L and $w((\ell, r)) \geq \tau\}$;
    if $S = \emptyset$ **then**
        | leave r unmatched forever;
    else
        | match r to an arbitrary vertex in S;
    end
end

We now prove some useful properties on GREEDY-RT. For any subset of edges F of the input graph $G = (L, R, E)$, let $L(F) = \{\ell \in L \mid \exists r$ s.t. $(\ell, r) \in F\}$ be the set of left ends of the edges in F. Let $w(F) = \sum_{(\ell,r)\in F} w((\ell, r))$ where $w((\ell, r))$ is the weight of (ℓ, r). Let M be the random variable which denotes the matching returned by GREEDY-RT for G. For any $i \geq 0$, let $M_{\geq e^i}$ be the matching returned by GREEDY-RT if the threshold τ is e^i, or equivalently, when $\kappa = i$. We are interested in the expected value of $w(M)$, which is defined to be

$$E(w(M)) = \sum_{0 \leq i \leq \infty} w(M_{\geq e^i}) \Pr(\kappa = i) = \sum_{0 \leq i \leq \infty} w(M_{\geq e^i}) p_i. \qquad (1)$$

Let O be the maximum weight matching of G. Define

$$O_{[e^i, e^{i+1})} = \{x \in O \mid w(x) \in [e^i, e^{i+1})\}$$

to be the subset of edges in O whose weight are in the interval $[e^i, e^{i+1})$.

Lemma 1. *For any $i \geq 0$, we have $|L(O_{[e^i, e^{i+1})}) - L(M_{\geq e^i})| \leq |L(M_{\geq e^i})|$*

Proof. Consider any vertex $\ell \in L(O_{[e^i, e^{i+1})}) - L(M_{\geq e^i})$. Since $\ell \in L(O_{[e^i, e^{i+1})})$, there is a unique $r \in R$ such that $(\ell, r) \in O_{[e^i, e^{i+1})}$, and since $w((\ell, r)) \geq e^i$, GREEDY-RT would have considered it when constructing $M_{\geq e^i}$. Note that $\ell \notin L(M_{\geq e^i})$ and is unmatched, and the fact that GREEDY-RT did not match r to ℓ implies that r is matched to another $\ell' \in L(M_{\geq e^i})$. In conclusion, every vertex $\ell \in L(O_{[e^i, e^{i+1})}) - L(M_{\geq e^i})$ can be mapped to a unique vertex $\ell' \in L(M_{\geq e^i})$; the lemma follows. □

Based on Lemma 1, we can prove the following lemma.

Lemma 2. *For any $i \geq 0$, $w(M_{\geq e^i}) \geq \frac{1}{2e} w(O_{[e^i, e^{i+1})})$.*

Proof. By Lemma 1, we have, for any $i \geq 0$,

$$|O_{[e^i, e^{i+1})}| = |L(O_{[e^i, e^{i+1})})| = |L(O_{[e^i, e^{i+1})}) \cap L(M_{\geq e^i})| +$$
$$|L(O_{[e^i, e^{i+1})}) - L(M_{\geq e^i})| \leq 2|L(M_{\geq e^i})| = 2|M_{\geq e^i}|,$$

and

$$w(M_{\geq e^i}) \geq e^i |M_{\geq e^i}| \geq \frac{e^i}{2} |O_{[e^i, e^{i+1})}| = \frac{e^{i+1}}{2e} |O_{[e^i, e^{i+1})}| \geq \frac{1}{2e} w(O_{[e^i, e^{i+1})}),$$

the lemma follows. □

We are now ready to describe how to choose the threshold. The following theorem shows that if $w_{\max}$ is known ahead of time, then selecting a random κ uniformly will make GREEDY-RT competitive.

Theorem 1. *Let $g = \lceil \ln(1 + w_{\max}) \rceil$. By choosing $\kappa \in \{0, 1, \ldots, g - 1\}$ uniformly, i.e., set $p_i = 1/g$ for every $0 \leq i \leq g - 1$, and $p_i = 0$ for all other i, GREEDY-RT achieves a competitive ratio of at least $\frac{1}{2e\lceil \ln(1 + w_{\max}) \rceil}$.*

Proof. Since $g = \lceil \ln(1 + w_{\max}) \rceil$ and all edges in O have weight no greater than $w_{\max}$, we have $\sum_{0 \leq i \leq g-1} w(O[e^i, e^{i+1})) = w(O)$. Then, by Equation (1) and Lemma 2, $E(w(M)) = \sum_{0 \leq i \leq g-1} w(M_{\geq e^i}) p_i \geq \frac{1}{2e} \sum_{0 \leq i \leq g-1} w(O_{[e^i, e^{i+1})}) \frac{1}{g} = \frac{1}{2eg} w(O)$; the theorem follows. □

We now consider the case when $w_{\max}$ is not known ahead of time. The previous probability distribution does not work as it depends on the value of $w_{\max}$. Consider a more elaborate probability distribution: $p_i = \frac{1}{\alpha' f(i+1)}$ where $f(x) = x \cdot \prod_{j=1}^{\log_2^*(x)-1} \log_2^{(j)} x$ and $\alpha' = \sum_{0 \leq i \leq \infty} \frac{1}{f(i+1)}$. Note that the sum $\sum_{0 \leq i \leq \infty} \frac{1}{f(i+1)}$ converges as the integral $\int_1^{\infty} \frac{dx}{f(x)}$ converges.

Theorem 2. *By setting* $p_i = \frac{1}{\alpha' \cdot (i+1) \prod_{j=1}^{\log^*(i+1)-1} \log^{(j)}(i+1)}$, GREEDY-RT *has competitive ratio* $\Omega(\frac{1}{\prod_{j=1}^{\log^* w_{\max}-1} \log^{(j)} w_{\max}})$.

Proof. Since no edge has weight greater than $w_{\max}$, $M_{\geq e^i}$ and $O_{[e^i, e^{i+1})}$ is empty for all $i \geq \lceil \ln(1 + w_{\max}) \rceil$. Then by Equation (1) and Lemma 2,

$$E(w(M)) = \sum_{0 \leq i \leq \lceil \ln(1+w_{\max}) \rceil - 1} w(M_{\geq e^i}) \, p_i$$

$$\geq \frac{1}{2e} \sum_{0 \leq i \leq \lceil \ln(1+w_{\max}) \rceil - 1} w(O_{[e^i, e^{i+1})}) \frac{1}{\alpha' \cdot (i+1) \prod_{j=1}^{\log^*(i+1)-1} \log^{(j)}(i+1)}$$

$$\geq \frac{1}{2e\alpha' \cdot \lceil \ln(1+w_{\max}) \rceil \cdot \prod_{j=1}^{\log^*(\lceil \ln(1+w_{\max}) \rceil)-1} \log^{(j)}(\lceil \ln(1+w_{\max}) \rceil)} w(O).$$

The theorem follows.   □

3.2 Randomized Algorithms for General Cases

We can show that online maximum weighted b-matching problem can be reduced to the special case where $b = 1$. For simplicity, we name the special case *online weighted matching problem*. Given an instance G of the online maximum weighted b-matching problem, we could construct an input instance G' of the online weighted matching problem by simply making b copies of each vertex in L. If ℓ' (one copy of $\ell \in L$) is matched to r in the online weighted matching problem, then ℓ would be matched to r in the online maximum weighted b-matching problem. So we could use the randomized algorithm GREEDY-RT to solve the online maximum weighted b-matching problem. From Theorem 1 and 2, we have:

Theorem 3. *For the online maximum weighted b-matching problem,* GREEDY-RT *has competitive ratio* $\Omega(\frac{1}{\prod_{j=1}^{\log^* w_{\max}-1} \log^{(j)} w_{\max}})$ *when* w_{max} *is unknown and has competitive ratio* $\frac{1}{2e\lceil \ln(1+w_{\max}) \rceil}$ *when* w_{max} *is known ahead of time.*

3.3 Upper Bound of Randomized Algorithms

We show an upper bound on the competitive ratio of any randomized algorithm, using a technique inspired by Yao's Lemma [17].

Theorem 4. *For the online maximum weighted b-matching problem, no randomized algorithm can do better than* $\frac{2}{\lceil \log_2(w_{\max}+1) \rceil + 1}$*-competitive.*

Proof. Let c be the competitive ratio of any randomized algorithm A. First, we construct some bipartite graphs G_i's in which the maximum edge weight is no more than $w_{\max}$ as follows. Let $g = \lceil \log_2(w_{\max} + 1) \rceil$. For $0 \leq i \leq g - 1$, $G_i = (L_i, R_i, E_i, w)$. The set L_i has only one vertex ℓ, R_i has $b(i + 1)$ vertices

$\underbrace{r_0, \ldots, r_0}_{b} \underbrace{r_1, \ldots, r_1}_{b} \cdots \underbrace{r_i, \ldots, r_i}_{b}$ and there are edges between ℓ and all vertices in R_i. For any $0 \leq j \leq i$, the edge (ℓ, r_j) has weight 2^j.

Now, we consider the following input distribution on $\mathcal{G} = \{G_0, G_1, \ldots, G_{g-1}\}$. The probability of having G_i is given as follows:

- For any $0 \leq i \leq g - 2$, the input is G_i with probability $q_i = \frac{1}{2^{i+1}}$.
- The input is G_{g-1} with probability $q_{g-1} = \frac{1}{2^{g-1}}$.

For any $G_i \in \mathcal{G}$, let O_{G_i} be the matching returned by the offline optimal algorithm on input G_i, and $A_{G_i,\sigma}$ is the one returned by A on input G_i and random choice σ.

Since $c \cdot w(O_{G_i}) \leq E_\sigma[w(A_{G_i,\sigma})]$, we have:

$$c \cdot E_{G_i}[w(O_{G_i})] \leq E_{G_i}[E_\sigma[w(A_{G_i,\sigma})]] \leq E_\sigma[E_{G_i}[w(A_{G_i,\sigma})]] \leq \max_\sigma E_{G_i}[w(A_{G_i,\sigma})]. \tag{2}$$

Note that

$$E_{G_i}[w(O_{G_i})] = \sum_{0 \leq i \leq g-1} q_i \cdot 2^i \cdot b = \sum_{0 \leq i \leq g-2} \frac{1}{2^{i+1}} \cdot 2^i \cdot b + \frac{1}{2^{g-1}} \cdot 2^{g-1} \cdot b = \frac{(g+1)}{2} \cdot b. \tag{3}$$

Below, we prove that for any fixed σ, $E_{G_i}[w(A_{G_i,\sigma})] \leq b$. Then by (2) and (3), we conclude that $c \leq \frac{2}{g+1}$ and the theorem follows.

Consider any deterministic algorithm D. Let D_{G_i} be the total weight returned by D on input G_i. Suppose that G_{g-1} is given to D as input. Assume that in D, for each $0 \leq j \leq g - 1$, the number of r_j's which are matched to ℓ is m_j. By the deterministic nature of D, we conclude that when the input is G_i, for any $0 \leq j \leq i$, there are still m_j r_j's matched to ℓ. Thus

$$E_{G_i}[D_{G_i}] = \sum_{0 \leq i \leq g-1} q_i \cdot \sum_{0 \leq j \leq i} 2^j m_j = \sum_{0 \leq j \leq g-1} 2^j m_j \sum_{j \leq i \leq g-1} q_i$$
$$= \sum_{0 \leq j \leq g-1} 2^j m_j \frac{1}{2^j} = \sum_{0 \leq j \leq g-1} m_j \leq b.$$

Therefore, we conclude, $E_{G_i}(D_{G_i}) \leq b$. Note that when σ is fixed, A is essentially a deterministic algorithms and therefore $E_{G_i}[w(A_{G_i,\sigma})] \leq b$, as required. $\square$

4 Deterministic Algorithms for Online Maximum Weighted b-Matching

In this section, we consider deterministic algorithms. With the assumption that $w_{\max}$ is known ahead of time, we present a deterministic algorithm GREEDY-D which is $\frac{1}{1+2\xi(w_{max}+1)^{\frac{1}{\xi}}}$-competitive, where $\xi = \min\{b, \lceil \ln(1 + w_{\max}) \rceil\}$. We also show an upper bound $\min\{\frac{2}{\lceil \log_2 w_{max} \rceil}, \frac{1}{(w_{max})^{\frac{1}{b}}}\}$ for any deterministic algorithm.

4.1 A Competitive Deterministic Algorithm: GREEDY-D

If w_{max} is not known ahead of time, no deterministic algorithm could be competitive as we cannot know what the value of w_{max} is or when the edge with the maximum weight will be revealed. We concentrate on the case that w_{max} is known ahead of time in this subsection.

Recall that each $\ell \in L$ can be matched to at most b vertices in R. Consider the scenario that one algorithm has already matched vertex ℓ to b vertices in R and all the edges between them have low weights. Then a new vertex $r \in R$ arrives, ℓ is the only neighbor of r and $w((\ell, r)) = w_{max}$. Then r cannot be matched and the algorithm has a poor performance. Inspired by this example, we should avoid the following case: one vertex $\ell \in L$ is matched to b vertices in R and all the edges between them have low weights.

We design a competitive deterministic greedy algorithm, GREEDY-D. Let ξ be $\min\{b, \lceil \ln(1 + w_{\max}) \rceil\}$. To understand GREEDY-D better, we could represent each vertex $\ell_i \in L$ by ξ "nodes", $\{\ell_{i,0}, \ldots, \ell_{i,\xi-1}\}$ and each node could be matched to at most $\lfloor \frac{b}{\xi} \rfloor$ vertices in R. (Note that all the ξ nodes of ℓ_i will be matched to no more than $\lfloor \frac{b}{\xi} \rfloor \cdot \xi \leq b$ vertices in total. As $b \geq \xi$, we get that $\lfloor \frac{b}{\xi} \rfloor$ is at least 1.) For any k, $w((\ell_{i,k}, r)) = w((\ell_i, r))$. In the algorithm, we will make sure that node $\ell_{i,k}$ can be matched to a vertex r only if $w((\ell_i, r)) \in [(w_{max} + 1)^{\frac{k}{\xi}}, (w_{max} + 1)^{\frac{k+1}{\xi}})$. In this way, it will not happen that ℓ_i is matched to b vertices and all the edges between them have low weights.

GREEDY-D is shown below. Variable $x_{i,k}$ is used to record the number of vertices in R that the node $\ell_{i,k}$ has been matched to. When r arrives, we try to match it to its neighbor greedily (in other words, edges with higher weights are preferred). Before matching r to ℓ_i, we compute which node of ℓ_i should be used. Only if the specific node $\ell_{i,k}$ has not been matched by as many as $\lfloor \frac{b}{\xi} \rfloor$ vertices, would we match r to ℓ_i formally.

Theorem 5. *If w_{max} is known ahead of time, GREEDY-D achieves competitive ratio* $\dfrac{1}{1 + 2\xi(w_{max}+1)^{\frac{1}{\xi}}}$ *where* $\xi = \min\{b, \lceil \ln(1 + w_{\max}) \rceil\}$.

Proof. Let M be the matching produced by GREEDY-D and O be the optimal matching. Note that in GREEDY-D, each $\ell \in L$ can be represented by ξ "nodes" and each "node" is matched to at most $\lfloor \frac{b}{\xi} \rfloor$ vertices in R. Let M' be the matching from "nodes" to vertices in R. Then: $w(M) = w(M') = \sum_{(\ell_{i,k}, r_j) \in M'} w((\ell_{i,k}, r_j))$ and $w(O) = \sum_{(\ell_i, r_j) \in O} w((\ell_i, r_j))$. For simplicity, we say one node is "full" if it has been matched to as many as $\lfloor \frac{b}{\xi} \rfloor$ vertices in R.

We would map each edge in O to one vertex or node matched in M'. Assume that $(\ell_i, r_j) \in O$ and $w((\ell_i, r_j)) \in [(w_{max} + 1)^{\frac{k}{\xi}}, (w_{max} + 1)^{\frac{k+1}{\xi}})$. In GREEDY-D, when r_j arrives, there are two cases:

I. Node $\ell_{i,k}$ is not "full". Then r_j will be matched by node $\ell_{i',k'}$ (which may be just $\ell_{i,k}$). We map edge (ℓ_i, r_j) to vertex r_j. Note that as $\ell_{i,k}$ is available for r_j, it must be true that $w((\ell_i, r_j)) \leq w((\ell_{i',k'}, r_j))$.

Set $\xi := \min\{b, \lceil \ln(1 + w_{\max}) \rceil\}$;
For each $\ell_i \in L$, define variables $(x_{i,0}, \ldots, x_{i,\xi-1}) := 0$;
while *a new vertex* $r \in R$ *arrives* **do**
   $t := 1$;
   while $t \leq |L|$ **do**
      Let ℓ_i be neighbor of r such that the weight of edge between them is the
      t-th highest among that of all edges adjacent to r; if not exist, break;
      Choose k such that $w((\ell_i, r)) \in [(w_{max}+1)^{\frac{k}{\xi}}, (w_{max}+1)^{\frac{k+1}{\xi}})$;
      if $x_{i,k} < \lfloor \frac{b}{\xi} \rfloor$ **then**
         match r to ℓ_i and $x_{i,k} := x_{i,k} + 1$;
         break;
      else
         $t := t+1$;
      end
   end
end

II. Node $\ell_{i,k}$ is "full". We map edge (ℓ_i, r_j) to node $\ell_{i,k}$. Let $M'_{(\ell_{i,k},*)} = \{(\ell_{i,k}, r) | (\ell_{i,k}, r) \in M'\}$. Note that there are $\lfloor \frac{b}{\xi} \rfloor$ edges in $M'_{(\ell_{i,k},*)}$ and each edge has weight no less than $(w_{max}+1)^{\frac{k}{\xi}}$. Then $w(M'_{(\ell_{i,k},*)}) \geq \lfloor \frac{b}{\xi} \rfloor (w_{max}+1)^{\frac{k}{\xi}} \geq \lfloor \frac{b}{\xi} \rfloor (w_{max}+1)^{-\frac{1}{\xi}} w((\ell_i, r_j))$.

We divide all edges in O into two parts O_1 and O_2: O_1 includes edges which falls into case (I) and $O_2 = O \setminus O_1$. In case (I), each edges $(\ell_i, r_j) \in O_1$ is mapped to vertex r_j matched in M'; it is impossible that two edges in O_1 are mapped to the same vertex matched in M'. As $w((\ell_i, r_j)) \leq w((\ell_{i'}, k', r_j))$, we have $w(O_1) \leq w(M')$. In case (II), each edge $(\ell_i, r_j) \in O_2$ is mapped to one "full" node $\ell_{i,k}$ matched in M'. As ℓ_i can match at most b vertices, there are at most b edges in O_2 which are mapped to the same node $\ell_{i,k}$ matched in M'. As $w((\ell_i, r_j)) \leq (\lfloor \frac{b}{\xi} \rfloor)^{-1} (w_{max}+1)^{\frac{1}{\xi}} w(M'_{(\ell_{i,k},*)})$ and $w(M') = \sum_{i,k} w(M'_{(\ell_{i,k},*)})$, we have $w(O_2) \leq b(\lfloor \frac{b}{\xi} \rfloor)^{-1} (w_{max}+1)^{\frac{1}{\xi}} w(M') \leq 2\xi (w_{max}+1)^{\frac{1}{\xi}} w(M')$ (note that the last inequality is true as: let $b = \eta_1 + \eta_2$, where $\eta_1 = \lfloor \frac{b}{\xi} \rfloor$, $0 \leq \eta_2 < \xi$; then $b(\lfloor \frac{b}{\xi} \rfloor)^{-1} = b/\eta_1 \leq (\eta_1 \xi + \xi)/\eta_1 \leq (1 + 1/\eta_1)\xi \leq 2\xi$). As $w(O) = w(O_1) + w(O_2)$ and $w(M') = w(M)$, we conclude that $w(O) \leq \left(1 + 2\xi (w_{max}+1)^{\frac{1}{\xi}}\right) w(M)$ and the theorem follows.   □

4.2 Upper Bound of Deterministic Algorithms

Theorem 6. *For online maximum weighted b-matching, no deterministic algorithm can have competitive ratio larger than* $\min\{\frac{2}{\lceil \log_2 w_{max} \rceil}, \frac{1}{(w_{max})^{\frac{1}{b}}}\}$.

Proof. Because of page limit, we move the proof into the full version of this paper (http://www.cs.hku.hk/~xzxiang/papers/bmatching.pdf).   □

Remarks. Recall that we have shown the lower bound. When b is extremely large compared with $w_{\max}$, then $\frac{1}{1+2e\lceil \ln(1+w_{\max})\rceil} \le c \le \frac{2}{\lceil \log_2 w_{max} \rceil}$; when b is very small compared with $w_{\max}$, then $\frac{1}{1+2b(1+w_{max})^{\frac{1}{b}}} \le c \le \frac{1}{(w_{max})^{\frac{1}{b}}}$. The gap between the lower bound and the upper bound is not large.

5   Online Two-Side Vertex-Weighted b-Matching

In this section, we consider randomized algorithms for online two-side vertex-weighted b-matching. We show a competitive randomized algorithm GREEDY-vRT for one special case of the problem, in which $b = 1$. GREEDY-vRT is a simple modification of GREEDY-RT. Note that every edge $(\ell, r) \in E$ has weight $w((\ell, r)) = v_\ell c_r$ and the highest edge weight w_{max} may be much greater than the highest vertex weight c_{max}. In the algorithm GREEDY-RT, we set a random threshold for the weight of any edge. However, in GREEDY-vRT, we set a random threshold for the weight of any vertex in R. GREEDY-vRT is proved to be $\Omega(\frac{1}{\prod_{j=1}^{\log^* c_{\max}-1} \log^{(j)} c_{\max}})$-competitive. If $c_{\max}$ is known ahead of time, its competitive ratio can be increased to $\Omega(\frac{1}{\log c_{\max}})$. Then we show that the general online two-side vertex-weighted b-matching problem can be reduced to the previous special case. We also show an upper bound $O(\frac{1}{\log c_{\max}})$ on the competitive ratio of any randomized algorithm. Because of page limit, we put the details in the full version of this paper (http://www.cs.hku.hk/~xzxiang/papers/bmatching.pdf).

References

1. Aggarwal, G., Goel, G., Karande, C., Mehta, A.: Online vertex-weighted bipartite matching and single-bid budgeted allocations. In: Proceedings of the Twenty-Second Annual ACM-SIAM Symposium on Discrete Algorithms, pp. 1253–1264. SIAM (2011)
2. Bahmani, B., Kapralov, M.: Improved bounds for online stochastic matching. In: de Berg, M., Meyer, U. (eds.) ESA 2010, Part I. LNCS, vol. 6346, pp. 170–181. Springer, Heidelberg (2010)
3. Birnbaum, B., Mathieu, C.: On-line bipartite matching made simple. ACM SIGACT News 39(1), 80–87 (2008)
4. Hubert Chan, T.-H., Chen, F., Wu, X., Zhao, Z.: Ranking on arbitrary graphs: Rematch via continuous lp with monotone and boundary condition constraints. In: SODA, pp. 1112–1122 (2014)
5. Devanur, N.R., Jain, K., Kleinberg, R.D.: Randomized primal-dual analysis of ranking for online bipartite matching. In: SODA, pp. 101–107. SIAM (2013)
6. Feldman, J., Mehta, A., Mirrokni, V., Muthukrishnan, S.: Online stochastic matching: Beating 1-1/e. In: 50th Annual IEEE Symposium on Foundations of Computer Science, FOCS 2009, pp. 117–126. IEEE (2009)
7. Goel, G., Mehta, A.: Online budgeted matching in random input models with applications to adwords. In: Proceedings of the Nineteenth Annual ACM-SIAM Symposium on Discrete Algorithms, pp. 982–991. Society for Industrial and Applied Mathematics (2008)

8. Haeupler, B., Mirrokni, V.S., Zadimoghaddam, M.: Online stochastic weighted matching: Improved approximation algorithms. In: Chen, N., Elkind, E., Koutsoupias, E. (eds.) Internet and Network Economics. LNCS, vol. 7090, pp. 170–181. Springer, Heidelberg (2011)
9. Kalyanasundaram, B., Pruhs, K.: Online weighted matching. J. Algorithms 14(3), 478–488 (1993)
10. Kalyanasundaram, B., Pruhs, K.R.: An optimal deterministic algorithm for online b-matching. Theoretical Computer Science 233(1), 319–325 (2000)
11. Karande, C., Mehta, A., Tripathi, P.: Online bipartite matching with unknown distributions. In: Proceedings of the 43rd Annual ACM Symposium on Theory of Computing, pp. 587–596. ACM (2011)
12. Karp, R.M., Vazirani, U.V., Vazirani, V.V.: An optimal algorithm for on-line bipartite matching. In: Proceedings of the Twenty-second Annual ACM Symposium on Theory of Computing, pp. 352–358. ACM (1990)
13. Korula, N., Pál, M.: Algorithms for secretary problems on graphs and hypergraphs. In: Albers, S., Marchetti-Spaccamela, A., Matias, Y., Nikoletseas, S., Thomas, W. (eds.) ICALP 2009, Part II. LNCS, vol. 5556, pp. 508–520. Springer, Heidelberg (2009)
14. Mahdian, M., Yan, Q.: Online bipartite matching with random arrivals: an approach based on strongly factor-revealing lps. In: Proceedings of the 43rd Annual ACM Symposium on Theory of Computing, pp. 597–606. ACM (2011)
15. Manshadi, V.H., Gharan, S.O., Saberi, A.: Online stochastic matching: Online actions based on offline statistics. Mathematics of Operations Research 37(4), 559–573 (2012)
16. Mehta, A., Saberi, A., Vazirani, U., Vazirani, V.: Adwords and generalized online matching. Journal of the ACM (JACM) 54(5), 22 (2007)
17. Yao, A.C.-C.: Probabilistic computations: Toward a unified measure of complexity. In: 18th Annual Symposium on Foundations of Computer Science, pp. 222–227 (October 1977)

Tree Convex Bipartite Graphs: $\mathcal{NP}$-Complete Domination, Hamiltonicity and Treewidth[*]

Chaoyi Wang[1], Hao Chen[1], Zihan Lei[2], Ziyang Tang[1], Tian Liu[1,**],
and Ke Xu[3,**]

[1] Key Laboratory of High Confidence Software Technologies, Ministry of Education,
Institute of Software, School of Electronic Engineering and Computer Science,
Peking University, Beijing 100871, China
lt@pku.edu.cn
[2] School of Mathematical Science, Peking University, Beijing 100871, China
[3] National Lab of Software Development Environment,
Beihang University, Beijing 100191, China
kexu@nlsde.buaa.edu.cn

Abstract. There are seven graph problems grouped into three classes of domination, Hamiltonicity and treewidth, which are known to be $\mathcal{NP}$-complete for *bipartite* graphs, but tractable for *convex bipartite* graphs. We show these problems to remain $\mathcal{NP}$-complete for *tree convex bipartite* graphs, even when the associated trees are *stars* or *combs* respectively. Tree convex bipartite graphs generalize convex bipartite graphs by associating a tree, instead of a path, on one set of the vertices, such that for every vertex in another set, the neighborhood of this vertex is a subtree.

Keywords: $\mathcal{NP}$-complete, domination, Hamiltonianicity, treewidth, tree convex bipartite graphs.

1 Introduction

Some graph problems are $\mathcal{NP}$-complete for *bipartite* graphs, but tractable for *restricted* bipartite graphs. For example, treewidth is $\mathcal{NP}$-complete for bipartite graphs, but tractable for *chordal bipartite* graphs [14], while domination and Hamiltonian circuit are $\mathcal{NP}$-complete for chordal bipartite graphs, but tractable for *convex bipartite* graphs [3,17,18]. *Tree convex bipartite* graphs [9] generalizes both chordal bipartite graphs [5] and convex bipartite graphs [6] by associating a tree, instead of a path as in convex bipartite graphs, on one set of vertices, such that for every vertex in another set, the neighborhood of this vertex induces a subtree. In chordal bipartite graphs, each cycle of length at least six has a chord. Chordal bipartite graphs are sandwiched between convex bipartite graphs and tree convex bipartite graphs [28,10]. Then, a natural question is

– *Are these graph problems also tractable for tree convex bipartite graphs?*

---

[*] Partially supported by National 973 Program of China (Grant No. 2010CB328103) and Natural Science Foundation of China (Grant Nos. 61370052 and 61370156).
[**] Corresponding authors.

J. Chen, J.E. Hopcroft, and J. Wang (Eds.): FAW 2014, LNCS 8497, pp. 252–263, 2014.

In this paper, we show some of them are $\mathcal{NP}$-complete for tree convex bipartite graphs, even when the associated trees are *stars* or *combs* respectively, which are then called *star convex bipartite* or *comb convex bipartite* graphs respectively. The seven problems under consideration are grouped into three classes of domination, Hamiltonicity and treewidth (TW). For domination (DS), we also consider its variants connected/paired/total domination (CDS/PDS/TDS). For Hamiltonicity, we consider both Hamiltonian circuit (HC) and Hamiltonian path (HP). Recently, similar results are shown for feedback vertex set (FVS) and independent domination (IDS) [9,28,10,27]. See Table 1 for a summary.

Table 1. Complexity results for various bipartite graphs

Graph classes	FVS	DS	CDS	IDS	PDS	TDS	HC	HP	TW
Bipartite	N[29]	N[4]	N[26]	N[8]	N[2]	N[26]	N[17]	N[17]	N[1,14]
Star conv. b.	N[9,10]	N[⋆]	N[⋆]	N[27]	N[⋆]	N[⋆]	N[⋆]	N[⋆]	N[⋆]
Comb conv. b.	N[28,10]	N[⋆]	N[⋆]	N[27]	N[⋆]	N[⋆]	N[⋆]	N[⋆]	O
Chordal b.	P[15]	N[24]	N[24]	N[3]	N[25]	P[3]	N[23]	N[23]	P[14]
Convex b.	P[19]	P[3]	P[3]	P[3]	P[3,7]	P[3]	P[18]	O	P[14]

(N: $\mathcal{NP}$-complete, P: Polynomial time, O: Open, ⋆: This paper)

Our interests on tree convex bipartite graphs are mainly from a theoretical perspective. No application is known for these graphs yet. The most interesting fact about tree convex bipartite graphs is that the associated trees have effects on complexity of some graph problems. For example, FVS, CDS and IDS have been shown to be $\mathcal{NP}$-complete for star convex bipartite and comb convex bipartite graphs [9,28,10,27], but tractable for *triad convex bipartite* graphs [11,10,27,22,20], where a triad is three paths with a common endpoint. Since these graphs are proper subclasses of tree convex bipartite graphs and triad convex bipartite graphs are also sandwiched between convex bipartite graphs and tree convex bipartite graphs, the following interesting question arises naturally

– *Where is the boundary between tractability and intractability of these problems for tree convex bipartite graphs?*

At this moment, a complexity classification of these problems for different restricted tree convex bipartite graphs is largely unknown. In this paper, we focus on the intractability results for two subclasses of tree convex bipartite graphs, namely star convex bipartite graphs and comb convex bipartite graphs.

This paper is structured as follows. After introducing basic notions and facts (Section 2), $\mathcal{NP}$-completeness for tree convex bipartite graphs is shown to domination (Section 3), Hamiltonicity (Section 4), and Treewidth (Section 5) respectively, and finally are some concluding remarks (Section 6).

2 Preliminaries

A graph $G = (V, E)$ has vertex set V and edge set E. Each edge has two ends in V, they are *adjacent*. The neighborhood of a vertex x, denoted $N_G(x)$, is the set of all adjacent vertices to x. A *bipartite* graph $G = (A, B, E)$ has a bipartition $A \cup B = V$ with no adjacent vertices in A (B, respectively). A *path* is a vertex sequence with every two consecutive vertices adjacent. A *connected* graph has every two vertices on a path. A *circuit* or a *cycle* is a path with the same start and end vertices. A *tree* is a connected cycle-free graph. Given a cycle, a *chord* is an edge whose two endpoints are not consecutive on the cycle. A *star* is a group of edges with a common end (called *central* vertex, other vertices are *leaves*). A *matching* is a set of edges with no common end for any two edges in it. A *perfect matching* is a matching with every vertex on it. A *comb* is a tree $T = (A, F)$ where $A = \{a_1, a_2, \cdots, a_{2p}\}$ and $F = \{(a_i, a_{p+i}) \mid 1 \le i \le p\} \cup \{(a_i, a_{i+1}) \mid 1 \le i < p\}$. The vertex set $\{a_i \mid 1 \le i < p\}$ or the path $a_1 a_2 \cdots a_p$ is *backbone* of the comb, and $a_{p+1}, a_{p+2}, \cdots, a_{2p}$ are *teeth* of the comb.

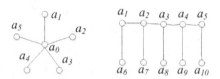

Fig. 1. A star and a comb

A *dominating set* is a subset of vertices with each vertex outside having a neighbor in. A dominating set is *connected, independent, paired, total* respectively, if the set induces a connected subgraph, induces a subgraph with no edge, admits a perfect matching inside it, induces a subgraph with no isolated vertex respectively. A *Hamiltonian circuit* (*path*, respectively) is a circuit (path, respectively) visiting each vertex exactly once.

A *tree decomposition* of a graph is a tree of bags and each bag is a subset of vertices, such that each edge of the graph has its two ends in a bag, and for each vertex of the graph, all bags containing this vertex induce a subtree. The *width* of a tree decomposition is the maximum size of bags minus one. The *treewidth* of a graph is the minimum width over all tree decompositions of the graph. For more on treewidth and other graph notions, see e.g. [13,16].

We will consider the following seven graph problems related to Domination, Hamiltonicity, and Treewidth: (i) Domination (DS). *Input*: a graph G and a positive integer k. *Output*: whether G admits a dominating set of size at most k. (ii) Connected Domination (CDS). *Input*: a graph G and a positive integer k. *Output*: whether G admits a connected dominating set of size at most k. (iii) Paired Domination (PDS). *Input*: a graph G and a positive integer k. *Output*: whether G admits a paired dominating set of size at most k. (iv) Total Domination (TDS). *Input*: a graph G and a positive integer k. *Output*: whether G admits a total dominating set of size at most k. (v) Hamiltonian Circuit (HC). *Input*: a graph G. *Output*: whether G admits a Hamiltonian circuit. (vi) Hamiltonian Path (HP). *Input*: a graph G. *Output*: whether G admits a Hamiltonian path. (vii) Treewidth (TW). *Input*: a graph G and a positive integer k. *Output*: whether G admits a tree decomposition of width at most k.

Definition 1. *A bipartite graph $G = (A, B, E)$ is called* tree convex bipartite *[9,10] (*circular convex bipartite *[18], respectively), if there is an associated tree $T = (A, F)$ (a circular ordering on A, respectively), such that for each vertex b in B, its neighborhood $N_G(b)$ induces a subtree of T (a circular arc, respectively). When T is a star (comb, path, respectively), G is called* star convex bipartite *[9] (*comb convex bipartite *[27,28,10], convex bipartite *[6], *respectively).*

There is a standard way (called *canonical transformation*) to transform any bipartite graph into a star (comb) convex bipartite graph as follows.

Lemma 1. *For any bipartite graph $G = (A, B, E)$, where $A = \{a_1, a_2, \cdots, a_{|A|}\}$, $B = \{b_1, b_2, \cdots, b_{|B|}\}$, the graph $G' = (A', B, E')$ is star convex bipartite, where $A' = A \cup \{a_0\}$ and $E' = E \cup \{(a_0, b)|b \in B\}$, and the graph $G'' = (A'', B, E'')$ is comb convex bipartite, where $A'' = A \cup \{a_{|A|+1}, a_{|A|+2}, \cdots, a_{2|A|}\}$ and $E'' = E \cup \{(a_i, b)||A| + 1 \leq i \leq 2|A|\}$.*

Proof. We add a new vertex a_0 adjacent to every vertex in B to A'. The star T' on A' has central vertex a_0 and leaves $a_1, \cdots, a_{|A|}$. For any vertex b in G', its neighborhood $N_{G'}(b) = N_G(b) \cup \{a_0\}$ is a subtree of T'. For a star, any subset of leaves and the central vertex form a subtree. See Figure 2.

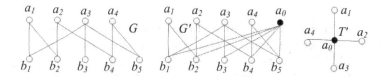

Fig. 2. Canonical transformation to a star convex bipartite graph

We add $|A|$ new vertices $a_{|A|+1}, a_{|A|+2}, \cdots, a_{2|A|}$ each adjacent to every vertex in B to A''. The comb T'' on A'' has backbone $\{a_{|A|+i} \mid 1 \leq i \leq |A|\}$ and teeth $a_1, a_2, \cdots, a_{|A|}$. For any vertex b in B, its neighborhood $N_{G''}(b) = N_G(b) \cup \{a_{|A|+i}|1 \leq i \leq |A|\}$ is a subtree on T''. For a comb, any subset of teeth and the backbone form a subtree. See Figure 3. $\qquad\square$

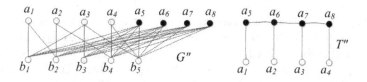

Fig. 3. Canonical transformation to a comb convex bipartite graph

3 Domination

Theorem 1. *DS, CDS, PDS and TDS are $\mathcal{NP}$-complete for star convex bipartite graphs and comb convex bipartite graphs.*

Proof. These problems are well known in $\mathcal{NP}$. We reduce from vertex cover (VC) which is $\mathcal{NP}$-complete for general graphs [12]. Given a graph G and a positive integer k, VC asks whether there is at most k vertices, such that each edge has at least one end in them. We assume that G has no isolated vertices which are useless for a vertex cover. The following two constructions work for all.

Reduction 1. *Input*: A graph $G = (V, E)$ and a positive integer k, where $V = \{v_1, v_2, \ldots, v_n\}$, $E = \{e_1, e_2, \ldots, e_m\}$. *Output*: A bipartite graph $G' = (A', B', E')$ and a positive integer $f(k)$ (to be specified later), where $A' = \{a_0, a_1, \cdots, a_{2m}\}$, $B' = \{b_0, b_1, \cdots, b_n\}$, $E' = \{(b_{i_s}, a_{2s-1}), (b_{i_s}, a_{2s}), (b_{j_s}, a_{2s-1}), (b_{j_s}, a_{2s})|e_s = (v_{i_s}, v_{j_s}), 1 \le s \le m\} \cup \{(a_0, b)|b \in B'\}$.

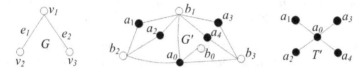

Fig. 4. An example of Reduction 1

Reduction 2. *Input*: The same as in Reduction 1. *Output*: A graph $G'' = (A'', B'', E'')$ and a positive integer $g(k)$ (to be specified later), where $A'' = \{a_1, a_2, \cdots, a_{4m}\}$, $B'' = \{b_0, b_1, \cdots, b_n\}$, $E'' = \{(b_{i_s}, a_{2s-1}), (b_{i_s}, a_{2s}), (b_{j_s}, a_{2s-1}), (b_{j_s}, a_{2s})|e_s = (v_{i_s}, v_{j_s}), 1 \le s \le m\} \cup \{(a_{2m+i}, b)|1 \le i \le 2m, b \in B''\}$.

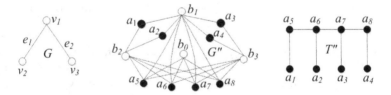

Fig. 5. An example of Reduction 2

Intuitively, we replace each vertex v_i of G by a new vertex b_i and add a new vertex b_0 to form B' (B'' respectively), add two vertices a_{2s-1}, a_{2s} to A' (A'' respectively) for each edge $e_s = (v_{i_s}, v_{j_s})$ of G, add four edges between bipartition $\{b_{i_s}, b_{j_s}\} \cup \{a_{2s-1}, a_{2s}\}$ to E' (E'' respectively). Then for G', add to A' a new vertex a_0 adjacent to all vertices in B'. For G'', add $2m$ new vertices a_{2m+i} each adjacent to all vertices in B''. Clearly, for any specific linear functions $f(k)$ and $g(k)$, Reduction 1 and Reduction 2 are computable in polynomial time. The following lemmas show correctness of Reduction 1 and Reduction 2.

Lemma 2. G' (G'', respectively) is star (comb, respectively) convex bipartite.

Proof. Every edge of G' has ends in A' and B' respectively, and A' and B' are disjoint. Every vertex in B' is adjacent to a_0. Similarly, every edge of G'' has ends in A'' and B'' respectively, and A'' and B'' are disjoint. Every vertex in B'' is adjacent to all vertices a_{2m+i} for $1 \leq i \leq 2m$. By Lemma 1, G' (G'', respectively) is star (comb, respectively) convex bipartite. $\square$

Lemma 3. G has a VC of size at most k if and only if G' (G'', respectively) has a DS of size at most $f(k) = k + 1$ ($g(k) = k + 1$, respectively).

Proof. (*Only if*) Assume that $k' \leq k$ and $C = \{v_{i_j} | 1 \leq j \leq k'\}$ is a VC for G of size k'. Then $D' = \{b_{i_j} | 1 \leq j \leq k'\} \cup \{a_0\}$ is a DS for G' of size $k' + 1$. Indeed, any edge $e_s = (v_{i_s}, v_{j_s})$ of G has an end, say v_{i_s}, in C. Then a_{2s-1} and a_{2s} are adjacent to b_{i_s} in D'. So every vertex in $\{a_i | 1 \leq i \leq 2m\}$ is adjacent to at least one vertex from $\{b_{i_j} | 1 \leq j \leq k'\}$, and every vertex in B' is adjacent to a_0.

Similarly, $D'' = \{b_{i_j} | 1 \leq j \leq k'\} \cup \{a_{2m+1}\}$ is a DS for G'' of size $k'+1$. Indeed, every vertex in $\{a_{2m+i} | 1 \leq i \leq 2m\}$ is adjacent to all b_{i_j} for $1 \leq j \leq k'$, every vertex in $\{a_i | 1 \leq i \leq 2m\}$ is adjacent to at least one vertex from $\{b_{i_j} | 1 \leq j \leq k'\}$, and every vertex in B'' is adjacent to a_{2m+1}.

(*If*) Assume that $k' \leq k$ and D' is a *minimum* dominating set for G' of size $k' + 1$. Such a minimum D' always exists by the existence of a dominating set of size at most $k + 1$. If D' does not contain vertex a_0, then it must contain vertex b_0, since b_0 is only adjacent to a_0. Then we can replace b_0 by a_0 in D' to get a dominating set of the same size. Thus, without loss of generality, we may assume that D' contains a_0 but not b_0. Similarly, assume that for an edge $e_s = (v_{i_s}, v_{j_s})$ in G, D' contains only one vertex from $\{a_{2s-1}, a_{2s}\}$, say a_{2s-1}. Then D' contain at most one vertex from $\{b_{i_s}, b_{j_s}\}$, since otherwise a_{2s-1} is redundant in D, a contradiction to the minimality of D'. Then D' must contain exact one vertex from $\{b_{i_s}, b_{j_s}\}$, say b_{i_s}, to dominate a_{2s}, since a_{2s} is only adjacent to b_{i_s} and b_{j_s}. Then we can replace a_{2s-1} by b_{j_s} in D' to get a dominating set of the same size. If D' contains both two vertices in $\{a_{2s-1}, a_{2s}\}$, then none vertex in $\{b_{i_s}, b_{j_s}\}$ is in D' due to the minimality of D'. Then we can replace $\{a_{2s-1}, a_{2s}\}$ by $\{b_{i_s}, b_{j_s}\}$ in D' to get a dominating set of the same size. Thus, without loss of generality, we may assume that $D' = \{b_{i_j} | 1 \leq j \leq k'\} \cup \{a_0\}$. Then we know that $\{b_{i_j} | 1 \leq j \leq k'\}$ dominates all vertex in $\{a_i | 1 \leq i \leq 2m\}$. This is possible only when $C = \{v_{i_j} | 1 \leq j \leq k'\}$ is a vertex cover for G of size k'.

Similarly, assume that $k' \leq k$ and a *minimum* dominating set D'' of size k' for G'' contains none vertex a_{2m+i} for $1 \leq i \leq 2m$, then D'' must contain the vertex b_0, since b_0 is only adjacent to all a_{2m+i} for $1 \leq i \leq 2m$. Now we can replace b_0 by a_{2m+1} in D'' to get a dominating set of the same size. Thus, without loss of generality, we may assume that D'' contains a_{2m+1}. Then similarly to the above discussion on D', without loss of generality, we may assume that $D'' = \{b_{i_j} | 1 \leq j \leq k'\} \cup \{a_{2m+1}\}$. Then similarly $C = \{v_{i_j} | 1 \leq j \leq k'\}$ is a vertex cover for G of size k'. $\square$

Lemma 4. *G has a VC of size at most k if and only if G' (G", respectively) has a CDS (TDS, respectively) of size at most $f(k) = k + 1$ $(g(k) = k + 1$, respectively).*

Proof. (*Only if*) This part is the same as in proof of Lemma 3, by noticing that the set D' (D'', respectively) in the *only if* part in proof of Lemma 3 is also connected (has no isolated vertex, respectively).

(*If*) This part is a direct consequence of Lemma 3, since a connected (total, respectively) dominating set is already a dominating set.   □

Lemma 5. *G has a VC of size at most k if and only if G' (G", respectively) has a PDS of size at most $f(k) = 2k$ $(g(k) = 2k$, respectively).*

Proof. (*Only if*) Assume that $k' \leq k$ and $C = \{v_{i_j} | 1 \leq j \leq k'\}$ is a VC for G of size k'. For each vertex v_{i_j} in C, assume that v_{i_j} is on an edge e_s (recall that we assume that G has no isolated vertex). Then we pick a unique vertex from $\{a_{2s}, a_{2s-1}\}$ and denote it by $a_{\mu(i_j)}$. Such a *unique* vertex $a_{\mu(i_j)}$ always exists for each $1 \leq i_j \leq k'$. For example, if $e_s = (v_{i_s}, v_{j_s})$, then we can assign a_{2s-1}, a_{2s} *uniquely* to $a_{\mu(i_s)}, a_{\mu(j_s)}$ respectively. Then a paired dominating set of size $2k'$ for G' is $D' = \{b_{i_j}, a_{\mu(i_j)} | 1 \leq j \leq k' - 1\} \cup \{a_0, b_{i_{k'}}\}$. A perfect matching of D' is $\{(b_{i_j}, a_{\mu(i_j)}) | 1 \leq j \leq k' - 1\} \cup \{(b_{i_{k'}}, a_0)\}$. This is assured by the uniqueness of $\mu(i_j)$ for each i_j.

Similarly, $D'' = \{b_{i_j}, a_{\mu(i_j)} | 1 \leq j \leq k' - 1\} \cup \{a_{2m+1}, b_{i_{k'}}\}$ is a paired dominating set of size $2k'$ for G'', with a perfect matching $\{(b_{i_j}, a_{\mu(i_j)}) | 1 \leq j \leq k' - 1\}$ $\cup \{(b_{i_{k'}}, a_{2m+1})\}$ of D''.

(*If*) Assume that $k' \leq k$ and D' is a paired dominating set for G' of size $2k'$. Then D' has k' vertices from A' and k' vertices from B' respectively, and D' has to contain a_0 to dominate b_0 or to do matching with b_0, since b_0 is only adjacent to a_0. If $D' = \{b_{i_j}, a_{\mu(i_j)} | 1 \leq j \leq k' - 1\} \cup \{a_0, b_0\}$ with a perfect matching $\{(b_{i_j}, a_{\mu(i_j)}) | 1 \leq j \leq k' - 1\} \cup \{(b_0, a_0)\}$, where $1 \leq \mu(i_j) \leq 2m$ for all i_j, then we can find a vertex $b_{i_{k'}}$ not in D' to replace b_0 in D'. Such a $b_{i_{k'}}$ must exist, since $k' - 1 < n$ and a_0 is adjacent to every vertex in B'. So we can assume that $D' = \{b_{i_j}, a_{\mu(i_j)} | 1 \leq j \leq k' - 1\} \cup \{a_0, b_{i_{k'}}\}$, with a perfect matching $\{(b_{i_j}, a_{\mu(i_j)}) | 1 \leq j \leq k' - 1\} \cup \{(b_{i_{k'}}, a_0)\}$. Then $C = \{v_{i_j} | 1 \leq j \leq k'\}$ must be a VC of size k' for G. Otherwise, assume that edge $e_s = (v_{i_s}, v_{j_s})$ is not covered by C. Then none of b_{i_s}, b_{j_s} is in D', so D' cannot be a paired dominating set of G, since none of a_{2s-1}, a_{2s} can have a matching vertex in D'.

Similarly, assume that $k' \leq k$ and D'' is a paired dominating set for G'' of size $2k'$. Then we can assume that $D'' = \{b_{i_j}, a_{\mu(i_j)} | 1 \leq j \leq k' - 1\} \cup \{a_{2m+1}, b_{i_{k'}}\}$, with a perfect matching $\{(b_{i_j}, a_{\mu(i_j)}) | 1 \leq j \leq k' - 1\} \cup \{(b_{i_{k'}}, a_{2m+1})\}$. Here, a_{2m+1} in D'' takes the role of a_0 in D', and $\mu(i_j)$ can take values between 1 and $4m$ except the value $2m + 1$. Then $C = \{v_{i_j} | 1 \leq j \leq k'\}$ must be a VC of size k' for G by the same reason as above.   □

The proof of theorem 1 is finished.   □

It is interesting to note that Reduction 1 and Reduction 2 both work for the four problems DS/CDS/PDS/TDS simultaneously.

4   Hamiltonicity

Theorem 2. *Hamiltonian Circuit and Hamiltonian Path are $\mathcal{NP}$-complete for star (comb, respectively) convex bipartite graphs.*

Proof. We reduce from HP which is $\mathcal{NP}$-complete for *bipartite* graphs [17]. [1]
A necessary condition for a bipartite graph $G = (A, B, E)$ to have a Hamiltonian path is $|A| = |B| + \epsilon$ where $\epsilon \in \{-1, 0, 1\}$. If $\epsilon = -1$ and $A = \{a_1, a_2, \cdots, a_{n-1}\}$, $B = \{b_1, b_2, \cdots, b_n\}$, let $G' = (A \cup \{a_n, a_0\}, B \cup \{b_0\}, E')$, $E' = E \cup \{(a_n, b)|b \in B'\} \cup \{(a_0, b_0)\}$. See Figure 6 (left). If $\epsilon = 0$ and $A = \{a_1, a_2, \cdots, a_n\}$, $B = \{b_1, b_2, \cdots, b_n\}$, let $G'' = (A \cup \{a_0\}, B \cup \{b_0\}, E'')$, $E'' = E \cup \{(b_0, a)|a \in A'\}$. See Figure 6 (right). If $\epsilon = 1$, first exchange A and B and second do as above. Then G has a Hamiltonian path if and only if G' (G'', respectively) has a Hamiltonian path. Moreover, every Hamiltonian path of G' (G'', respectively) must start at a_0 and end in B. In rest of this section, we assume this property for inputs.

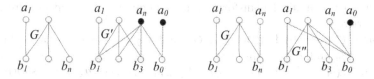

Fig. 6. How to make $|A| = |B|$ and a_0 an end on every Hamiltonian path

Reduction 3. *Input*: A bipartite graph $G = (A, B, E)$, $A = \{a_1, a_2, \cdots, a_n\}$, $B = \{b_1, b_2, \cdots, b_n\}$, and every Hamiltonian path of G has ends at a_1 and in B respectively. *Output*: A bipartite graph $G_3 = (A_3, B_3, E_3)$, $A_3 = A \cup \{a_0\}$, $B_3 = B \cup \{b_0\}$, $E_3 = E \cup \{(a_0, b)|b \in B'\}$. See Figure 7 (left).

Reduction 4. *Input*: The same as in Reduction 3. *Output*: $G_4 = (A_4, B_4, E_4)$, where $A_4 = A_3$, $B_4 = B_3$, $E_4 = E_3 \cup \{(b_0, a_1)\}$. See Figure 7 (right).

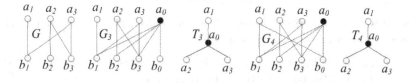

Fig. 7. An example of Reduction 3 and Reduction 4

Reduction 5. *Input*: As in Reduction 3. *Output*: $G_5 = (A_5, B_5, E_5)$, $A_5 = A \cup \{a_{n+1}, a_{n+2}, \cdots, a_{2n}\}$, $B_5 = B \cup \{b_{n+1}, b_{n+2}, \cdots, b_{2n}\}$, $E_5 = E \cup \{(a_i, b)|n+1 \le i \le 2n, b \in B\} \cup \{(a_{n+i}, b_{n+i})|1 \le i \le n\} \cup \{(a_{n+i}, b_{n+i+1})|1 \le i < n\}$.

---

[1] [17] only claims HC but the proof holds for HP too. Both HC and HP are in $\mathcal{NP}$.

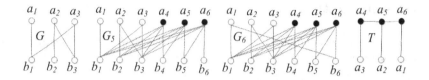

Fig. 8. An example of Reduction 5 and Reduction 6

Reduction 6. *Input*: As in Reduction 3. *Output*: $G_6 = (A_6, B_6, E_6)$, $A_6 = A_5$, $B_6 = B_5$, $E_6 = E_5 \cup \{(b_{2n}, a_1)\}$.

Clearly, Reductions 3, 4, 5, 6 are computable in polynomial time. The following lemmas show correctness of Reductions 3, 4, 5, 6 respectively.

Lemma 6. G_3 *and* G_4 *are star convex bipartite,* G_5 *and* G_6 *are comb convex bipartite.*

Proof. Vertex a_0 in G_3 (G_4, respectively) is adjacent to all vertices in B_3 (B_4, respectively). By canonical transformation in Lemma 1, G_3 and G_4 are star convex bipartite.

The comb T associated with both G_5 and G_6 is consists of a backbone $a_{n+1}a_{n+2} \cdots a_{2n}$ and a matching between a_i with a_{2n-i} for $1 \leq i \leq n$. Then for $1 \leq i \leq n$, $N_{G_5}(b_i) = N_{G_6}(b_i) = N_G(b_i) \cup \{a_{n+1}, a_{n+2}, \cdots, a_{2n}\}$ is a subtree of T, as in proof of Lemma 1. For $n + 1 \leq i < 2n$, $N_{G_5}(b_i) = N_{G_6}(b_i) = \{a_i, a_{i+1}\}$ is a subtree of T, since (a_i, a_{i+1}) is an edge of T. $N_{G_5}(b_{2n}) = \{a_{2n}\}$ and $N_{G_6}(b_{2n}) = \{a_1, a_{2n}\}$ are also subtrees of T, since (a_1, a_{2n}) is an edge of T. □

Lemma 7. G *has a Hamiltonian path if and only if* G_3 (G_4, *respectively*) *has a Hamiltonian path (circuit, respectively).*

Proof. Vertex b_0 in G_3 is of degree one, all Hamiltonian path of G_3 end at b_0. Then P is a Hamiltonian path of G ending in B if and only if Pa_0b_0 is a Hamiltonian path of G_3.

Vertex b_0 in G_3 is of degree two, all Hamiltonian circuit of G_4 contain path $a_0b_0a_1$. Then P is a Hamiltonian path of G starting at a_1 and ending in B if and only if $Pa_0b_0a_1$ is a Hamiltonian circuit of G_4. □

Lemma 8. G *has a Hamiltonian path if and only if* G_5 (G_6, *respectively*) *has a Hamiltonian path (circuit, respectively).*

Proof. Each vertex in $\{b_{n+1}, b_{n+2}, \cdots, b_{2n-1}\}$ in G_5 is of degree two, so every Hamilton path of G_5 contains the path $a_{n+1}b_{n+1}a_{n+2}b_{n+2} \cdots b_{2n}$. Then P is a Hamiltonian path of G ending in B if and only if $Pa_{n+1}b_{n+1}a_{n+2}b_{n+2} \cdots b_{2n}$ is a Hamiltonian path of G_5.

Each vertex in $\{b_{n+1}, b_{n+2}, \cdots, b_{2n}\}$ in G_6 is of degree two, so every Hamilton circuit of G_6 contains the path $a_{n+1}b_{n+1}a_{n+2}b_{n+2} \cdots b_{2n}a_1$. Then P is a Hamiltonian path of G starting at a_1 and ending in B if and only if $Pa_{n+1}b_{n+1}a_{n+2}b_{n+2} \cdots b_{2n}a_1$ is a Hamiltonian circuit of G_6. □

The proof of Theorem 2 is finished. □

5   Treewidth

Theorem 3. *Treewidth is $\mathcal{NP}$-complete for star convex bipartite graphs.*

Proof. Treewidth is well known in $\mathcal{NP}$ [1,13]. We reduce from Treewidth which is $\mathcal{NP}$-complete for general graphs [1].

Reduction 7. *Input:* A graph $G = (V, E)$ and a positive integer k, where $V = \{v_1, v_2, \ldots, v_n\}$ and $E = \{e_1, e_2, \ldots, e_m\}$. *Output:* A bipartite graph $G' = (A, B, E')$ and a positive integer $k+1$, where $A = \{a_0, a_1, \cdots, a_m\}$, $B = \{b_1, b_2, \cdots, b_n\}$, $E' = \{(a_s, b_i), (a_s, b_j) | e_s = (v_i, v_j) \text{ is in } E\} \cup \{(a_0, b) | b \in B\}$.

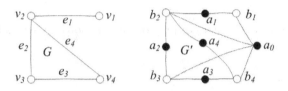

Fig. 9. An example of Reduction 7

Clearly, Reduction 7 is computable in polynomial time. By canonical transformation in Lemma 1, the graph G' is star convex bipartite. The following lemma helps to show the correctness of Reduction 7.

Lemma 9. *Adding a new vertex adjacent to all existing vertices in a graph will increase its treewidth exactly by one.*

Proof. On the one hand, for any tree decomposition of the original graph, putting the new vertex into each bag of the decomposition will increase the size of each bag by one. The result is still a tree decomposition of the new graph, thus the treewidth will increase at most by one.

On the other hand, treewidth is also defined by *elimination order* which is a permutation on vertices. For two adjacent vertices in the graph, we call the later one under the permutation a *higher neighbor* to the previous one. A *fill-in* procedure is from the first to the last vertex under the permutation to make each vertex's higher neighbors a clique in the graph. The *width* of an elimination order is the maximum number of higher neighbors over each vertex after fill-in. The *treewidth* is the minimum width over all elimination orders of the graph.

For any elimination order of the new graph, if we move the new vertex to the end, the width of the elimination order will not increase. If we remove the new vertex from the end of the elimination order, the width will decrease by one. The result is still an elimination order of the original graph. Thus the treewidth of the original graph is at most the treewidth of the new graph minus one.   □

Now, Reduction 7 is correct because treewidth of G is t if and only if treewidth of G' is $t + 1$. Indeed, starting with the graph G, first adding a_0 to G with a_0 adjacent to all vertices of G will increase treewidth exactly by one by Lemma 9, and then adding a_s on e_s for $s = 1, \cdots, m$ will unchange treewidth.   □

6   Conclusions

We have shown $\mathcal{NP}$-completeness of Domination, Connected Domination, Paired Domination, Total Domination, Hamiltonian Path, and Hamiltonian Circuit for star (comb, respectively) convex bipartite graphs, and $\mathcal{NP}$-completeness of Treewidth for star convex bipartite graphs. It remains open to show $\mathcal{NP}$-completeness of Treewidth for comb convex bipartite graphs, and tractability of Hamiltonian Path for convex bipartite graphs (see Table 1).

Recently, similar results are shown on Set Cover, Set Packing and Hitting Set for tree convex set systems and tree-like set systems [21].

Acknowledgments. The help of previous anonymous reviewers has improved our presentation greatly.

References

1. Arnborg, S., Corneil, D.G., Proskurowski, A.: Complexity of finding embeddings in a k-tree. SIAM J. Algebraic Discrete Methods 8, 277–284 (1987)
2. Chen, L., Lu, C., Zeng, Z.: Labelling algorithms for paired-domination problems in block and interval graphs. J. Comb. Optim. 19, 457–470 (2010)
3. Damaschke, P., Muller, H., Kratsch, D.: Domination in Convex and Chordal Bipartite Graphs. Inform. Proc. Lett. 36, 231–236 (1990)
4. Garey, M.R., Johnson, D.S.: Computers and Intractability, A Guide to the Theory of NP-Completeness. W.H. Freeman and Company (1979)
5. Golumbic, M.C., Goss, C.F.: Perfect elimination and chordal bipartite graphs. J. Graph Theory 2, 155–163 (1978)
6. Grover, F.: Maximum matching in a convex bipartite graph. Nav. Res. Logist. Q. 14, 313–316 (1967)
7. Hung, R.-W.: Linear-time algorithm for the paired-domination problem in convex bipartite graphs. Theory Comput. Syst. 50, 721–738 (2012)
8. Irving, W.: On approximating the minimum independent dominating set. Inform. Process. Lett. 37, 197–200 (1991)
9. Jiang, W., Liu, T., Ren, T., Xu, K.: Two hardness results on feedback vertex sets. In: Atallah, M., Li, X.-Y., Zhu, B. (eds.) FAW-AAIM 2011. LNCS, vol. 6681, pp. 233–243. Springer, Heidelberg (2011)
10. Jiang, W., Liu, T., Wang, C., Xu, K.: Feedback vertex sets on restricted bipartite graphs. Theor. Comput. Sci. (2013) (in press), doi: 10.1016/j.tcs.2012.12.021
11. Jiang, W., Liu, T., Xu, K.: Tractable feedback vertex sets in restricted bipartite graphs. In: Wang, W., Zhu, X., Du, D.-Z. (eds.) COCOA 2011. LNCS, vol. 6831, pp. 424–434. Springer, Heidelberg (2011)
12. Karp, R.: Reducibility among combinatorial problems. In: Complexity of Computer Computations, pp. 85–103. Plenum Press, New York (1972)
13. Kloks, T.: Treewidth: Computations and Approximations. Springer (1994)
14. Kloks, T., Kratsch, D.: Treewidth of chordal bipartite graphs. J. Algorithms 19, 266–281 (1995)
15. Kloks, T., Liu, C.H., Pon, S.H.: Feedback vertex set on chordal bipartite graphs. arXiv:1104.3915 (2011)

16. Kloks, T., Wang, Y.L.: Advances in Graph Algorithms (2013) (manuscript)
17. Krishnamoorthy, M.S.: An NP-hard problem in bipartite graphs. SIGACT News 7(1), 26 (1975)
18. Liang, Y.D., Blum, N.: Circular convex bipartite graphs: maximum matching and Hamiltonian circuits. Inf. Process. Lett. 56, 215–219 (1995)
19. Liang, Y.D., Chang, M.S.: Minimum feedback vertex sets in cocomparability graphs and convex bipartite graphs. Acta Informatica 34, 337–346 (1997)
20. Lu, M., Liu, T., Xu, K.: Independent domination: Reductions from circular- and triad-convex bipartite graphs to convex bipartite graphs. In: Fellows, M., Tan, X., Zhu, B. (eds.) FAW-AAIM 2013. LNCS, vol. 7924, pp. 142–152. Springer, Heidelberg (2013)
21. Lu, M., Liu, T., Tong, W., Lin, G., Xu, K.: Set cover, set packing and hitting set for tree convex and tree-like set systems. In: Gopal, T.V., Agrawal, M., Li, A., Cooper, S.B. (eds.) TAMC 2014. LNCS, vol. 8402, pp. 248–258. Springer, Heidelberg (2014)
22. Lu, Z., Liu, T., Xu, K.: Tractable connected domination for restricted bipartite graphs (Extended abstract). In: Du, D.-Z., Zhang, G. (eds.) COCOON 2013. LNCS, vol. 7936, pp. 721–728. Springer, Heidelberg (2013)
23. Müller, H.: Hamiltonian circuits in chordal bipartite graphs. Disc. Math. 156(1-3), 291–298 (1996)
24. Müller, H., Brandstät, A.: The NP-completeness of steiner tree and dominating set for chordal bipartite graphs. Theor. Comput. Sci. 53(2-3), 257–265 (1987)
25. Panda, B.S., Prahan, D.: Minimum paired-dominating set in chordal graphs and perfect elimination bipartite graphs. J. Comb. Optim. (2012) (in press), doi:10.1007/s10878-012-9483-x
26. Pfaff, J., Laskar, R., Hedetniemi, S.T.: NP-completeness of total and connected domination, and irredundance for bipartite graphs. Technical Report 428, Dept. Mathematical Sciences, Clemenson Univ. (1983)
27. Song, Y., Liu, T., Xu, K.: Independent domination on tree convex bipartite graphs. In: Snoeyink, J., Lu, P., Su, K., Wang, L. (eds.) AAIM 2012 and FAW 2012. LNCS, vol. 7285, pp. 129–138. Springer, Heidelberg (2012)
28. Wang, C., Liu, T., Jiang, W., Xu, K.: Feedback vertex sets on tree convex bipartite graphs. In: Lin, G. (ed.) COCOA 2012. LNCS, vol. 7402, pp. 95–102. Springer, Heidelberg (2012)
29. Yannakakis, M.: Node-deletion problem on bipartite graphs. SIAM J. Comput. 10, 310–327 (1981)

Zero-Sum Flow Numbers of Triangular Grids

Tao-Ming Wang[1,*], Shih-Wei Hu[2], and Guang-Hui Zhang[3]

[1] Department of Applied Mathematics
Tunghai University, Taichung, Taiwan, ROC
[2] Institute of Information Science
Academia Sinica, Taipei, Taiwan, ROC
[3] Department of Applied Mathematics
National Chung Hsing University, Taichung, Taiwan, ROC

Abstract. As an analogous concept of a nowhere-zero flow for directed graphs, we consider zero-sum flows for undirected graphs in this article. For an undirected graph G, a **zero-sum flow** is an assignment of non-zero integers to the edges such that the sum of the values of all edges incident with each vertex is zero, and we call it a **zero-sum k-flow** if the values of edges are less than k. Note that from algebraic point of view finding such zero-sum flows is the same as finding nowhere zero vectors in the null space of the incidence matrix of the graph. We consider in more details a combinatorial optimization problem, by defining the **zero-sum flow number** of G as the least integer k for which G admitting a zero-sum k-flow. It is well known that grids are extremely useful in all areas of computer science. Previously we studied flow numbers over hexagonal grids and obtained the optimal upper bound. In this paper, with new techniques we give completely zero-sum flow numbers for certain classes of triangular grid graphs, namely, regular triangular grids, triangular belts, fans, and wheels, among other results. Open problems are listed in the last section.

1 Background and Motivation

Let G be a directed graph. A **nowhere-zero flow** on G is an assignment of non-zero integers to each edge such that for every vertex the Kirchhoff current law holds, that is, the sum of the values of incoming edges is equal to the sum of the values of outgoing edges. A **nowhere-zero k-flow** is a nowhere-zero flow using edge labels with maximum absolute value $k - 1$. Note that for a directed graph, admitting nowhere-zero flows is independent of the choice of the orientation, therefore one may consider such concept over the underlying undirected graph. A celebrated conjecture of Tutte in 1954 says that every bridgeless graph has a nowhere-zero 5-flow. F. Jaeger showed in 1979 that every bridgeless graph has a nowhere-zero 8-flow[6], and P. Seymour proved that every bridgeless graph has a nowhere-zero-6-flow[9] in 1981. However the original Tutte's conjecture remains open.

---

[*] The corresponding author whose research is partially supported by the National Science Council of Taiwan under project NSC-102-2115-M-029-003.

J. Chen, J.E. Hopcroft, and J. Wang (Eds.): FAW 2014, LNCS 8497, pp. 264–275, 2014.

There is an analogous and more general concept of a nowhere-zero flow that uses bidirected edges instead of directed ones, first systematically developed by Bouchet[4] in 1983. Bouchet raised the conjecture that every bidirected graph with a nowhere-zero integer flow has a nowhere-zero 6-flow, which is still unsettled. Recently another related nowhere-zero flow concept has been studied, as a special case of bi-directed one, over the undirected graphs by S. Akbari et al.[2] in 2009.

Definition 1. *For an undirected graph G, a* **zero-sum flow** *is an assignment of non-zero integers to the edges such that the sum of the values of all edges incident with each vertex is zero. A* **zero-sum k-flow** *is a zero-sum flow whose values are integers with absolute value less than k.*

Note that from algebraic point of view finding such zero-sum flows is the same as finding nowhere zero vectors in the null space of the incidence matrix of the graph. S. Akbari et al. raised a conjecture for zero-sum flows similar to the Tutte's 5-flow Conjecture for nowhere-zero flows as follows:

Conjecture. (Zero-Sum 6-Flow Conjecture) If G is a graph with a zero-sum flow, then G admits a zero-sum 6-flow.

It was proved in 2010 by Akbari et al. [1] that the above Zero-Sum 6-Flow Conjecture is equivalent to the Bouchet's 6-Flow Conjecture for bidirected graphs. In literatures a more general concept **flow number**, which is defined as the least integer k for which a graph may admit a k-flow, has been studied for both directed graphs and bidirected graphs. We extend the concept in 2011 to the undirected graphs and call it **zero-sum flow numbers**, and also considered general **constant-sum flows** for regular graphs[11].

It is well known that grids are extremely useful in all areas of computer science. One of the main usage, for example, is as the discrete approximation to a continuous domain or surface. Numerous algorithms in computer graphics, numerical analysis, computational geometry, robotics and other fields are based on grid computations.

Also it is known that there are three possible types of regular tessellations, which are tilings made up of squares, equilateral triangles, and hexagons. Formally, a lattice grid, or a **lattice grid graph** is induced by a finite subset of the infinite integer lattice grid $\mathbb{Z} \times \mathbb{Z}$. The vertices of a lattice grid are the lattice points, and the edges connect the points which are at unit distance from each other. The infinite grid $\mathbb{Z} \times \mathbb{Z}$ may be viewed as the set of vertices of a regular tiling of the plane with unit squares.

There are only two other types of plane tiling with regular polygons. One is with equilateral triangles, which defines an infinite triangular grid in the same way. A **triangular grid graph** is a graph induced by a finite subset of the infinite triangular grid. The other type of tiling is with regular hexagons which defines an infinite hexagonal grid. Similarly the graph induced by a finite subset of the infinite hexagonal grid is called a **hexagonal grid graph**. A hexagonal graph is also named a honeycomb graph in literature.

In this paper, we calculate zero-sum flow numbers for some classes of triangular grid graphs. In particular we consider the problem for two classes of well known generalized triangular grid graphs, namely fans and wheels.

2   Zero-Sum Flow Numbers

In the study of nowhere-zero flows of directed graphs(bidirected graphs) one considers a more general concept, namely, the least number of k for which a graph may admit a k-flow. In 2011 [11] we consider similar concepts for zero-sum k-flows:

Definition 2. *Let G be a undirected graph. The **zero-sum flow number** $F(G)$ is defined as the least number of k for which G may admit a zero-sum k-flow. $F(G) = \infty$ if no such k exists.*

Obviously the zero-sum flow numbers can provide with more detailed information regarding zero-sum flows. For example, we may restate the previously mentioned Zero-Sum Conjecture as follow: Suppose a undirected graph G has a zero-sum flow, then $F(G) \leq 6$. In 2012, we showed some general properties of small flow numbers, so that the calculation of zero-sum flow numbers gets easier. It is well known that a graph admits a nowhere-zero 2-flow if and only if it is Eulerian (every vertex has even degree). We obtain the following for zero-sum flows:

Lemma 3 ((T. Wang and S. Hu [12])). *A graph G has zero-sum flow number $F(G) = 2$ if and only if G is Eulerian with even size (even number of edges) in each component.*

Tutte obtained in 1949 that a cubic graph has a nowhere-zero 3-flow if and only if it is bipartite. Similarly for undirected graphs we have that:

Lemma 4 ((T. Wang and S. Hu [12])). *A cubic graph G has zero-sum flow number $F(G) = 3$ if and only if G admits a perfect matching.*

Lemma 5. *Let G be a undirected graph and $G = H_1 \cup H_2$ be an arbitrary union of H_1 and H_2, where flow numbers $F(H_1) = k_1$ and $F(H_2) = k_2$. Then $F(G) \leq k_1 k_2$.*

Proof. Since $F(H_1) = k_1$ and $F(H_2) = k_2$, we have two edge labeling functions $f_1 : E(H_1) \to \{\pm 1, \cdots, \pm(k_1 - 1)\}$ and $f_2 : E(H_2) \to \{\pm 1, \cdots, \pm(k_2 - 1)\}$, which are zero-sum k_i-flow for H_i, $i = 1, 2$. To make an edge labeling for G, we set $f_1^*(e) = f_1(e)$ if $e \in E(H_1)$, and $f_1^*(e) = 0$, otherwise. Also $f_2^*(e) = f_2(e)$ if $e \in E(H_2)$, and $f_2^*(e) = 0$ otherwise, for all $e \in E(G)$. Now, let $f = f_1^* + k_1 f_2^*$ or $k_2 f_1^* + f_2^*$, then it can be seen that the function f forms a zero-sum $k_1 k_2$-flow for G.   □

Then we obtain the following corollaries, which are extremely useful tools for calculating flow numbers in general:

Corollary 6. *Let G be a undirected graph and $G = H_1 \cup H_2$ be an arbitrary union of H_1 and H_2, where flow numbers $F(H_1) = F(H_2) = 2$. Then $F(G) \leq 4$.*

Corollary 7. *Let G be a undirected graph and $G = H_1 \cup H_2$ be an arbitrary union of H_1 and H_2, where flow numbers $F(H_1) = 2$ and $F(H_2) = 3$. Then $F(G) \leq 6$.*

We calculated in 2013 [13] for the zero-sum flow numbers of hexagonal grids. Note that Akbari. et al. showed that in [2] if **Zero-Sum 6-Flow Conjecture** is true for $(2,3)$-graphs (in which every vertex is of degree 2 or 3), then it is true for any graph. Therefore the study can be reduced to $(2,3)$-graphs. It is clear non-trivial hexagonal grid graphs are a special class of $(2,3)$-graphs. In particular we found the optimal upper bound and also provided with infinitely many examples of hexagonal grid graphs with flow number 3 and 4 respectively.

3   Flow Numbers for Regular Triangular Grids

Contrast to the case of hexagonal grids, the upper bound for the flow numbers of an arbitrary triangular grid is somewhat difficult to find. In this section we initiate the study and calculate the exact value of zero-sum flow numbers for the class of regular triangular grids as defined in below.

Definition 8. *Let G be a triangular grid stacked by n layers of triangles with $1, 2, \cdots, n$ triangles in each layer. We call G a **regular triangular grid** and denote it by T_n.*

For example see Figure 1 for T_5.

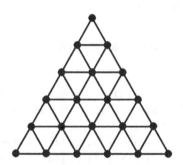

Fig. 1. The regular triangular grid T_5

Theorem 9. *The flow numbers of T_n are as follows:*

$$F(T_n) = \begin{cases} \infty, & n = 1. \\ 2, & n \equiv 3,4 \ (mod \ 4). \\ 3, & n \equiv 1,2 \ (mod \ 4). \\ 4, & n = 2. \end{cases}$$

Proof. Note that T_1 has no zero-sum flow, thus $F(T_1) = \infty$. Then we consider the following for the remaining cases.

Case 1. $n \equiv 3, 4 \ (mod\ 4)$.

Note that in general there is no any odd degree vertex in T_n (the only possible degrees are 2, 4, and 6), and we see $|E(T_n)| = \frac{3n(n+1)}{2}$. So when $n \equiv 3, 4 \ (mod\ 4)$, $|E(T_n)|$ is even. Then by Lemma 3, $F(T_n) = 2$ in this case.

Case 2. $n \equiv 1, 2 \ (mod\ 4)$.

As seen in Case 1, T_n is an even graph. But now when $n \equiv 1, 2 \ (mod\ 4)$, $|E(T_n)|$ is odd. Therefore by Lemma 3, we see $F(T_n) \neq 2$ in this case. Assume first $n \equiv 1 \ (mod\ 4)$, i.e. $n = 4s + 1$ for some s, we find two special even subgraphs that are both of even size. We treat T_{4s+1} as the union of T_{4s} and H as in Figure 2, where the intersection of the edge sets for T_{4s} and H is the triangle $v_1 v_2 v_3$.

We see the size of H equals to $3n + 3 = 12s + 6$, which is even, thus $F(H) = 2$ by Lemma 3. On the other hand, by Case 1 we see $F(T_{4s}) = 2$. Now we just need to focus on the labels of $v_1 v_2, v_2 v_3, v_3 v_1$. We use the labels 1 and -1 as in Figure 2 over two Euler paths on T_{4s} and H respectively, to make the labels of $v_1 v_2, v_2 v_3, v_3 v_1$ in T_{4s} coincide with that of $v_1 v_2, v_2 v_3, v_3 v_1$ in H.

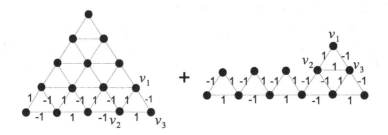

Fig. 2. Two (1, -1)-labeled Euler paths over T_{4s} and H

Then we see that, by extending these labels to the whole T_{4s}, together with H, which give a zero-sum 3-flow for T_{4s+1}, with labels over $v_1 v_2, v_2 v_3, v_3 v_1$ are 2, 2, -2, and 1, -1 everywhere else.

Note that similarly T_{4s+2} equals to $T_{4s+1} \cup H'$, where H' is the edge disjoint union of $4s + 2$ copies of triangles. Now T_{4s+1} part just follow the labels we built previously. Note that the H' part is an even graph with even size, whose flow number is 2. Therefore we give a zero-sum 3-flow for T_{4s+2}.

Case 3. $n = 2$.

T_2 is union of two even graphs with even size as in Figure 3.

Then by Corollary 6, one see $F(T_2) \leq 4$. First by Lemma 3 we have $F(T_2) \neq 2$. We want to show further that $F(T_2) \neq 3$. Suppose a, b, c give a zero-sum 3-flow over T_2 as in Figure 4.

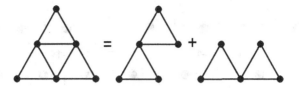

Fig. 3. T_2 as union of two subgraphs

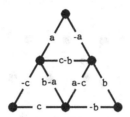

Fig. 4. Labels over T_2

Since it is a zero-sum 3-flow with a, b, c are distinct, there are only 4 possibilities for (a, b, c): $(1, -1, 2)$, $(-1, 1, -2)$, $(1, 2, -2)$, $(-1, -2, 2)$. But all these four cases make $|c - b| = 3$, a contradiction. Hence $F(T_2) = 4$.   □

4   Flow Numbers for Triangular Belts

Definition 10. *A **triangular belt** is exactly the square of paths P_n^2, $n \geq 3$, in which we connect distance two vertices in a path P_n, as in Figure 5.*

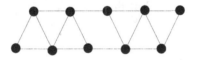

Fig. 5. A triangular belt

Note that $P_3^2 = C_3$ and $P_4^2 = F_3$, and both admit no any zero-sum flow (The fan graph F_3 is defined and seen in Figure 16 next section). We start from P_5^2. A zero-sum 3-flow for P_5^2 is obtained from adding the corresponding edge labels together with two basic labeled parallelograms as in Figure 6.

Theorem 11. *The flow number of the graph square of paths $F(P_n^2) = 3$ when $n \geq 5$, and $F(P_n^2) = \infty$ for $n = 3, 4$.*

Proof. We proceed with induction on $n \geq 5$. Since there are odd degree vertices, $F(P_5^2) = 3$. Suppose that f is a zero-sum 3-flow for P_{n-1}^2, $n \geq 6$, as in Figure 7, which is obtained by using two basic labeled parallelograms as in Figure 6.

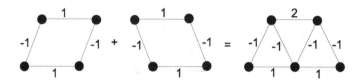

Fig. 6. A zero-sum 3-flow for P_5^2

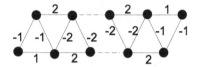

Fig. 7. A zero-sum 3-flow for P_{n-1}^2

Then we consider P_n^2 as the union of a C_4 with the P_{n-1}^2, with intersection over one edge, as shown in Figure 8. Note that $P_n^2 = P_{n-1}^2 \cup C_4$ admits a zero-sum 3-flow and it has odd degree vertices, hence $F(P_n^2) = 3$. □

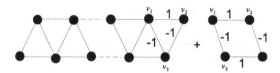

Fig. 8. $P_{n-1}^2 \cup C_4$

5 Flow Numbers for Fans and Wheels

In order to obtain the index sets of fans and wheels, we first describe a *subdivision method* which is commonly used here for the construction of zero-sum flows, in particular fans and wheels.

Triple Subdivision Method

Let G be a graph admitting a zero-sum flow f. Using the triple subdivision method we may obtain a new graph G' with larger order, and a new zero-sum flow f' of G', based upon G and f. We proceed by choosing in G a vertex v and edges e_1, e_2 with $f(e_1) = f(e_2) = x$, which are not incident with v. Then subdivide these two edges by inserting new vertices of degree 3, join them to v respectively. See Figure 9. Now then we may construct a new f' on G' by keeping

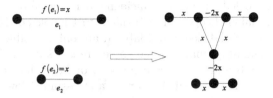

Fig. 9. Triple Subdivision Method

the labels on G unchanged, and labeling x, x, $-2x$ on three newly inserted edges respectively. Note that $x + x - 2x = 0$, then the new labeling f' on G' is still zero-sum flow, and $f'(E(G')) = f(E(G)) \cup \{-2x\}$.

Note that fans and wheels are graphs made from joining one point to a path and a cycle respectively as in Figure 10. Also both graphs consist of triangles, and can be seen as special triangular grids. In below, we use the above triple subdivision method by new edge insertion and induction to justify the results for the flow numbers of fans and wheels.

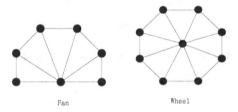

Fan   Wheel

Fig. 10. Fans and Wheels

Theorem 12. *The flow numbers of fans F_n and wheels W_n are as follows:*

$$F(F_n) = \begin{cases} \infty, & n = 1, 2, 3. \\ 3, & n = 3k + 1, \ k \geq 1. \\ 4, & otherwise. \end{cases}$$

$$F(W_n) = \begin{cases} 3, & n = 3k, \ k \geq 1. \\ 5, & n = 5. \\ 4, & otherwise. \end{cases}$$

To prove the above result, we start with the following lemma:

Lemma 13. W_n *does not admit any zero-sum 3-flow for $3 \nmid n$, and F_n does not admit any zero-sum 3-flow for $3 \nmid n - 1$.*

Proof. Assume F_n or W_n admits a zero-sum 3-flow, then only $\pm 1, \pm 2$ can be used for the edge labels. One may see that along the outer cycle of W_n or outer path of F_n, the 1-labeled edge can be incident to 1-labeled or (-2)-labeled edge only. Similarly (-2)-labeled edge can be incident to 1-labeled or (-2)-labeled edge only. By symmetry, (-1)-labeled and 2-labeled are only possible incident edges. Without loss of generality, may assume there are x 1's and y (-2)'s. Then we have that $x - 2y = 0$ and $x + y = n$ for W_n, or have that $x - 2y = 0$ and $x + y = n - 1$ for F_n. Therefore if W_n admits a zero-sum 3-flow, then $3 \mid n$, and if F_n admits a zero-sum 3-flow, then $3 \mid n - 1$. □

By the technique of triple subdivision, we have the following:

Lemma 14. F_{3n+1} and W_{3n} admit zero-sum 3-flows, for $n \geq 1$, respectively.

Proof. By induction from F_{3n+1} to F_{3n+4} for $n \geq 1$, and from W_{3n} to W_{3n+3} for $n \geq 1$ respectively. See the Figure 11 and Figure 12 below, which are triple subdivision (edge insertion) from F_4 to F_7 and from W_3 to W_6 respectively. □

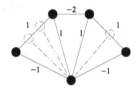

Fig. 11. A zero-sum 3-flow from F_4 to F_7

Fig. 12. A zero-sum 3-flow from W_3 to W_6

Similarly we may obtain by the same triple subdivision that F_{3n+2}, F_{3n+3} and W_{3n+1}, W_{3n+2} admit zero-sum 4-flows, for $n \geq 1$, respectively, with the only exception W_5. See lemmas in below.

Lemma 15. F_{3n+2}, F_{3n+3} and W_{3n+1}, W_{3n+2} admit zero-sum 4-flows, for $n \geq 1$, respectively, with the only exception W_5.

Proof. See Figure 13 and Figure 14, and one may have zero-sum 4-flows by triple subdivision and induction method. □

The following is the only exceptional case W_5:

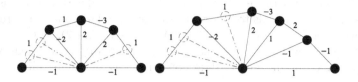

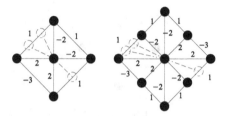

Fig. 13. Zero-sum 4-flows from F_5 and F_6 respectively

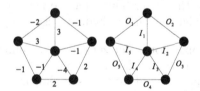

Fig. 14. Zero-sum 4-flows from W_4 and W_8 respectively

Lemma 16. W_5 *does not admit any zero-sum 4-flow.*

Proof. Suppose it does admit a zero-sum 4-flow. See the Figure 15, where we assume the edges of outside cycle are O_i for $1 \leq i \leq 5$, and the edges of inner spokes are O_i for $1 \leq i \leq 5$. Note that only $\pm 1, \pm 2, \pm 3$ can be used to give the 4-flow, and also by adding zero-sums over every vertex one has $\Sigma_{i=1}^5 O_i = 0$. Hence the sum of three consecutively adjacent edge labels must obey $1 \leq |O_i + O_{i+1} + O_{i+2}| \leq 3$ for any i, modulo 5, since $O_i + O_{i+1} + O_{i+2} = -(O_{i+3} + O_{i+4}) = -I_{i+3}$.

Fig. 15. W_5 admits a zero-sum 5-flow

Without loss of generality we need only consider three cases $O_1 = 1$, $O_1 = 2$, or $O_1 = 3$. If $O_1 = 1$, then the only possible labels on two incident edges O_2 and O_5 are $1, 2, -2, -3$. So the possible pairs of (O_2, O_5) by symmetry are $(1, 1)$, $(1, 2)$, $(1, -2)$, $(1, -3)$, $(2, 2)$, $(2, -2)$, $(2, -3)$, $(-2, -2)$, $(-2, -3)$, $(-3, -3)$. We may rule out cases $(1, 2), (1, -2), (2, 2), (2, -3), (-2, -3), (-3, -3)$ by using the condition $1 \leq |O_i + O_{i+1} + O_{i+2}| \leq 3$. On the other hand, we exhausted the discussion over all remaining cases. Say in case the pair $(O_2, O_5) = (1, 1)$, then the pair (O_3, O_4) has to be either $(-1, -2)$ or $(-2, -1)$. However this would end up with some I_i equals to 0, a contradiction. Similarly one would reach the contradiction for the rest of cases $(O_2, O_5) = (1, -3), (2, -2)$, or $(-2, -2)$, and also for the cases when $O_1 = 2$ and $O_1 = 3$. Therefore we are done with all possibilities, and obtain that W_5 does not admit any zero-sum 4-flow.   □

It is clear to see that F_1 has degree 1 vertex and $F_2 \cong C_3$, so $F(F_1)$ and $F(F_2)$ are both infinity. F_3 case are as follows:

Fig. 16. F_3

Suppose F_3 has flow number r, as in Figure 16. Then $x + y + z$ should equal to r. On the other hand, $r - x + z + r - y$ equals to r as well. Thus $r + z = x + y$, but $r = x + y + z$. Therefore $z = 0$, a contradiction. This shows $F(F_3) = \infty$.

Therefore with above Lemmas and the existence of labeling of zero-sum 4-flows and zero-sum 5-flows, we finish calculating the flow numbers of fans and wheels for Theorem 12.

6   Concluding Remark

One may further study various classes of triangular grid graphs. Contrast to the case of hexagonal grids, the upper bound for the flow numbers of an arbitrary triangular grid is somewhat difficult to find. We list two problems as follows:

1. Give the optimal upper bound for zero-sum flow numbers of general triangular grid graphs and classify them;
2. Give the optimal upper bound for zero-sum flow numbers of various classes of planar graphs and classify them.

References

[1] Akbari, S., Daemi, A., Hatami, O., Javanmard, A., Mehrabian, A.: Zero-Sum Flows in Regular Graphs. Graphs and Combinatorics 26, 603–615 (2010)
[2] Akbari, S., Ghareghani, N., Khosrovshahi, G.B., Mahmoody, A.: On zero-sum 6-flows of graphs. Linear Algebra Appl. 430, 3047–3052 (2009)
[3] Akbari, S., et al.: A note on zero-sum 5-flows in regular graphs. The Electronic Journal of Combinatorics 19(2), P7 (2012)
[4] Bouchet, A.: Nowhere-zero integral flows on a bidirected graph. J. Combin. Theory Ser. B 34, 279–292 (1983)

[5] Gallai, T.: On factorisation of grahs. Acta Math. Acad. Sci. Hung. 1, 133–153 (1950)

[6] Jaeger, F.: Flows and generalized coloring theorems in graphs. J. Combin. Theory Ser. B 26(2), 205–216 (1979)

[7] Kano, M.: Factors of regular graph. J. Combin. Theory Ser. B 41, 27–36 (1986)

[8] Petersen, J.: Die Theorie der regularen graphs. Acta Mathematica (15), 193–220 (1891)

[9] Seymour, P.D.: Nowhere-zero 6-flows. J. Combin. Theory Ser. B 30(2), 130–135 (1981)

[10] Tutte, W.T.: A contribution to the theory of chromatic polynomials. Can. J. Math. 6, 80–91 (1954)

[11] Wang, T.-M., Hu, S.-W.: Constant Sum Flows in Regular Graphs. In: Atallah, M., Li, X.-Y., Zhu, B. (eds.) FAW-AAIM 2011. LNCS, vol. 6681, pp. 168–175. Springer, Heidelberg (2011)

[12] Wang, T.-M., Hu, S.-W.: Zero-Sum Flow Numbers of Regular Graphs. In: Snoeyink, J., Lu, P., Su, K., Wang, L. (eds.) AAIM 2012 and FAW 2012. LNCS, vol. 7285, pp. 269–278. Springer, Heidelberg (2012)

[13] Wang, T.-M., Zhang, G.-H.: Zero-Sum Flow Numbers of Hexagonal Grids. In: Fellows, M., Tan, X., Zhu, B. (eds.) FAW-AAIM 2013. LNCS, vol. 7924, pp. 339–349. Springer, Heidelberg (2013)

A Study of Pure Random Walk Algorithms on Constraint Satisfaction Problems with Growing Domains*

Wei Xu[1] and Fuzhou Gong[2]

[1] School of Automation and Electrical Engineering,
University of Science and Technology Beijing, Beijing 100083, China
xuwei_225@163.com
[2] Institute of Applied Mathematics, Academy of Mathematics and System Science,
Chinese Academy of Sciences, Beijing 100190, China
fzgong@amt.ac.cn

Abstract. The performances of two types of pure random walk (PRW) algorithms for a model of constraint satisfaction problems with growing domains (called Model RB) are investigated. Threshold phenomenons appear for both algorithms. In particular, when the constraint density r is smaller than a threshold value r_d, PRW algorithms can solve instances of Model RB efficiently, but when r is bigger than the r_d, they fail. Using a physical method, we find out the threshold values for both algorithms. When the number of variables N is large, the threshold values tend to zero, so generally speaking PRW does not work on Model RB.

Keywords: constraint satisfaction problems, Model RB, random walk, local search algorithms.

1 Introduction

Constraint satisfaction problems (CSPs) arise in a large spectrum of scientific disciplines, such as computer science, information theory, and statistical physics [19, 17, 14]. A typical CSP instance involves a set of variables and a collection of constraints. Variables take values in a finite domain. Constraints contain a few variables and forbid some of their joint values. A solution is an assignment satisfying all the constraints simultaneously. Given a CSP instance, two fundamental scientific questions are to decide the existence of solutions and to find out a solution if it exists. Examples of CSPs are Boolean formula satisfiability (SAT), graph coloring, variants of SAT such as XORSAT, error correction codes, etc.

Random models of CSPs play a significant role in computer science. As instance generators, they provide instances for benchmarking algorithms, help to inform the design of algorithms and heuristics, and provide insight into problem hardness. Classical random CSP models were proposed and denoted

---

* Partially supported by NSFC 61370052 and 61370156.

J. Chen, J.E. Hopcroft, and J. Wang (Eds.): FAW 2014, LNCS 8497, pp. 276–287, 2014.

by A, B, C and D respectively [20, 12], and many alternatives also appeared [1, 27, 25, 11, 9, 10].

Model RB is a typical CSP model with growing domains. It was proposed by Xu and Li [27] to overcome the trivial insolubility of the classical model B, and was proved to have exact satisfiability phase transitions. The instances generated in the phase transition region of Model RB are hard to solve [28, 29] and have been widely used in various kinds of algorithm competitions. Model RB develops a new way to study CSPs, especially CSPs with large domains, thus has gotten considerable attention [14, 32, 15, 31, 18, 3, 13, 26, 16].

Algorithm analysis is a notoriously difficult task. The current rigorous results mostly deal with algorithms that are extremely simple, such as backtrack-free algorithms, which assign variables one by one without backtracking [5, 4]. Pure Random Walk (PRW) algorithm is a process that consists of a succession of random moves. It is relatively simple and has been intensively studied on the k-SAT problem [2, 21, 22, 6–8, 23, 24]. On k-SAT, A frequently studied PRW algorithm (Algorithm 2 in the following) is called Walksat. Another reason for PRW algorithm being studied is that random walk is a part of many local search algorithms [19].

In this paper, we study two types of PRW algorithms on Model RB. By experimental methods, threshold phenomenons on performance of these two PRW algorithms are found, just like that of Walksat on k-SAT. Moreover, by a physical method we locate the thresholds for both algorithms, which are $\frac{1-p}{p}\frac{1}{k\ln N}$, with N being the total number of variables, k the number of variables per constraints, p the portion of forbidden joint values per constraints.

This paper is organized as follows. We first give the definition of Model RB and its main properties in Section 2. In Section 3, we show the threshold behaviors of PRW algorithms by experiments, and also show the different performances before and after the thresholds. In Section 4, we use a physical method to calculate the thresholds for both algorithms. We finally give some concluding remarks in Section 5.

2   Model RB

Both classical and revised models of CSPs can be found in [14]. Here we give the definition of Model RB. Let $k \geq 2$ be an integer. Let $r > 0$, $\alpha > 0$, $0 < p < 1$ be real numbers. Let N be the number of variables and $V = \{\sigma_1, \sigma_2, \cdots, \sigma_N\}$ the set of variables. Each variable takes values from a domain $D = \{1, 2, \cdots, N^\alpha\}$. Each constraint involves k variables and an associated incompatible-set, which is a subset of the Cartesian product D^k. Elements in incompatible-set are called incompatible (forbidden) joint values. Model $RB(N, k, r, \alpha, p)$ is a probability space defined by the following steps to generate its instances.

1. We select with repetition $rN \ln N$ constraints independently at random. Each constraint is formed by selecting without repetition k out of N variables independently at random.

2. For each constraint, we form an incompatible-set by selecting without repetition $pN^{\alpha k}$ elements from D^k independently at random.

A solution is an assignment which satisfies all the constraints. That is to say, the joint values in a solution dose not belong to any incompatible-sets of the constraints. The set of all solutions, denoted by $\mathcal{S}$, is a subset of D^N. Let X be the number of solutions, $X = |\mathcal{S}|$. It is easy to see that in model RB, the expectation of X is

$$\mathbb{E}(X) = N^{\alpha N}(1-p)^{rN \ln N}.$$

Let

$$r_{cr} = -\frac{\alpha}{\ln(1-p)}.$$

If $\alpha > \frac{1}{k}$ and $0 < p < 1$ are two constants, and k and p satisfy the inequality $k \geq \frac{1}{1-p}$, then

$$\lim_{n \to \infty} \Pr(X > 0) = \begin{cases} 1, r < r_{cr}, \\ 0, r > r_{cr}. \end{cases}$$

Thus, Model RB has exact satisfiability phase transitions, see [27, 31].

3   Performance of Pure Random Walk on Model RB

In this section, we study the performance of PRW algorithms on Model RB. By experiments, we find that PRW algorithms exhibit threshold phenomenons, and have different performances before and after the thresholds.

3.1   Pure Random Walk Algorithms

We concentrate on two types of PRW algorithms, called Algorithm 1 and Algorithm 2 respectively. In Algorithm 1, we randomly reassign a variable from conflict set. In algorithm 2, we randomly select an unsat-constraint (unsatisfied constraint), then randomly select one of its variable to reassign it.

Algorithm 1

1. Pick up a random assignment. Set up a maximum number of steps.
2. Let conflict set be the set of all variables that appear in a constraint that is unsatisfied under the current assignment.
   (a) If the conflict set is empty, terminate the algorithm, output the current assignment.
   (b) Otherwise, randomly select a variable in the conflict set, reassign it a value.
3. Repeat step 2, until the maximum number of steps, then output *fail*.

Algorithm 2

1. Pick up a random assignment. Set up a maximum number of steps.

   (a) If current assignment satisfies all constraints, terminate the algorithm, output the current assignment.
   (b) Otherwise, randomly select an unsat-constraint, and randomly select a variable in the constraint, reassign it a value.

2. Repeat step 2, until time of repeating has gotten to the maximum step number, output *fail*.

3.2   Threshold Behavior

Both Algorithm 1 and Algorithm 2 exhibit threshold phenomenons, as shown in Figure 1. The probability of getting a solution drops from 1 to 0 dramatically. Every point in Figure 1 is averaged over 10 runs, and the maximum number of steps is 2000.

The same threshold phenomenon has been found for Walksat on k-SAT problem [7], with a conjectured threshold value $\alpha = 2^k/k$.

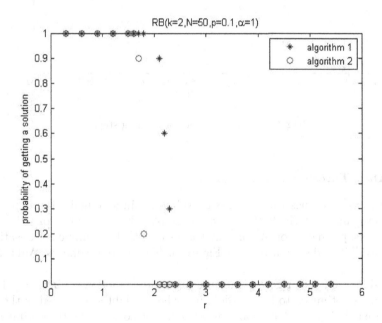

Fig. 1. Probability of getting a solution by PRW algorithms on model RB

3.3 Before Threshold

When $r < r_d$ (r_d is the threshold value) and r is very small, algorithms can find a solution in a short time. If at each step, the number of unsat-constraints decreases by $O(1)$, then the solving time will be $O(N \ln N)$. Figure 2 shows the average number of running steps, each point is averaged over 100 runs. Figure 3 shows the average number of running steps divided by $N \ln N$. So when r is very small, the solving time is in an order of $O(N \ln N)$.

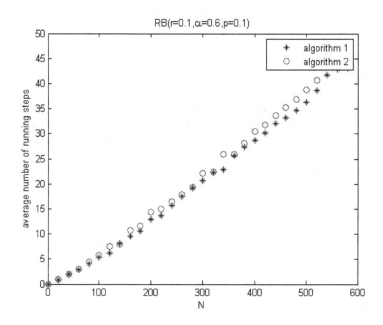

Fig. 2. Average number of running steps

3.4 After Threshold, Algorithm 2

When $r > r_d$, variables will be reassign values again and again, the number of unsat-constraints will fluctuate around some plateau value for a long time, see Figure 4. (Experiments on Algorithm 1 are similar.) The number of unsatisfied clauses exhibit a distribution, see Figure 5. This is the same as Walksat on k-SAT.

Two simple but not rigorous interpretations are as follows. First, the chosen constraint is optimized to be satisfied, but when variables contained in the constraint are reassigned values again for other chosen constraints, the optimization was destroyed. Second, the value of the reassigned variable is optimized, but when variables connected to the reassigned variable are reassigned values, the optimization was destroyed. So when r is big, for example $r > r_d$, optimization (or effect of each reassignment) cannot be retained, and algorithms fail.

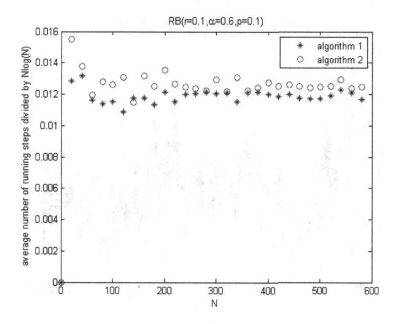

Fig. 3. Average number of running steps divided by $Nlog(N)$

4   Analysis Based on an Approximation

The main method used in this section is from theoretical physics, which has been used on k-SAT and other problems by Semerjian et al [21, 22, 6]. It is not a rigorous method, since an approximation is utilized, but remarkable results have been gotten on k-SAT problem and XORSAT problem with this method. For more background and correctness of this method, we refer to [21].

Approximation. At each step, before a reassignment, we treat the situation at that time as a typical situation, featured by the number of its unsat-constraints. A typical situation featured by M_0 means that, $M = rN \ln N$ constraints are randomly selected (as step 1 of Model RB definition), then M_0 unsat-constraints are randomly chosen from the M constraints. Then the solving process becomes a Markov chain using number of unsat-constraints as its state space, and the transition probability from M_0 to M_0' is the probability that the typical situation featured by M_0 have M_0' unsat-constraints after a step (a reassignment).

First, we should give the transition probability from M_0 to M_0'. We will choose a variable to reassign from typical situation featured by M_0. Let $p(Z_1)$ be the probability that a variable with Z_1 unsat-constraints will be chosen, which depends

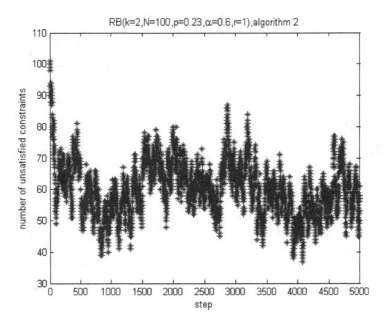

Fig. 4. Number of unsat-constraints, Algorithm 2

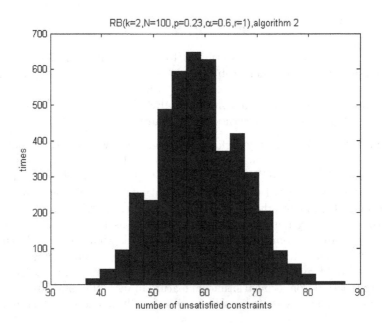

Fig. 5. Histogram of the number of unsat-constraints, where data at the first 500 step were omitted

on algorithms. The probability that Z_2 of Z_1 unsat-constraints become satisfied after reassignment is

$$\binom{Z_1}{Z_2} (1-p)^{Z_2} (p)^{Z_1-Z_2} \triangleq p(Z_2),$$

Z_2 obeys the binomial distribution. When $M_0 > 0$, the probability that Z_3 constraints become unsatisfied from satisfied is

$$\binom{M-M_0}{Z_3} \left(\frac{pk}{N}\right)^{Z_3} \left(1-\frac{pk}{N}\right)^{M-M_0-Z_3} \triangleq p(Z_3), \tag{1}$$

because each of $M - M_0$ feasible satisfied constraints connects to the reassignment variable with probability $\frac{k}{N}$, and each connecting constraint becomes unsatisfied with probability p. When $M_0 = 0$, $Z_3 = 0$ with probability 1.

Then the transition probability from M_0 to M_0' is

$$A_{M_0',M_0} = \sum_{Z_1=0}^{M_0} \sum_{Z_2=0}^{Z_1} \sum_{Z_s=0}^{M-M_0} p(Z_1)p(Z_2)p(Z_3) 1_{M_0'-M_0+Z_2-Z_3}$$

where

$$1_X = \begin{cases} 1, \text{ if } X = 0, \\ 0, \text{ otherwise.} \end{cases}$$

The initial distribution, i.e. the probability that at time 0 the typical situation is featured by M_0, is

$$\Pr[M_0, 0] = \binom{M}{M_0} p^{M_0} (1-p)^{M-M_0}. \tag{2}$$

Iteratively, the probability that at time $T+1$ the typical situation is featured by M_0' is

$$\Pr[M_0', T+1] = \sum_{M_0=0}^{M} A_{M_0'M_0} \Pr[M_0, T].$$

Criterion. If $\Pr[0, T]$ is 0 (almost) in a long time, it is in the phase after the threshold; if $\Pr[0, T]$ becomes positive from 0 in polynomial time, it is in the phase before the threshold. Denote the average fraction of unsat-constraints at time $T = tM$ by $\varphi(t)$,

$$\varphi(t) = \frac{1}{M} \sum_{M_0=0}^{M} M_0 \Pr[M_0, T = tM].$$

In the beginning, $\Pr[0, T] = 0$, but if $\varphi(t)$ become 0 in polynomial time, then $\Pr[0, T]$ will become positive from 0; if $\varphi(t)$ is always positive, then the number of unsat-constraints will fluctuate around some plateau value, $\Pr[0, T]$ will always be 0. So the criterion is whether $\varphi(t)$ becomes 0 in polynomial time when $\Pr[0, T] = 0$.

4.1   Analysis on Algorithm 1

When N is large, we might as well say $\frac{d\varphi}{dt} = (\varphi(t + \frac{1}{M}) - \varphi(t))/(1/M)$, then

$$
\begin{aligned}
\frac{d\varphi}{dt} &= \sum_{M_0'=0}^{M} M_0' \left(\sum_{M_0=0}^{M} A_{M_0' M_0} \Pr[M_0, T] \right) - \sum_{M_0=0}^{M} M_0 \Pr[M_0, T] \\
&= \sum_{M_0=0}^{M} \Pr[M_0, T] \left(\sum_{M_0'=0}^{M} A_{M_0' M_0} (M_0' - M_0) \right) \\
&= \sum_{M_0=0}^{M} \Pr[M_0, T] (\mathbb{E}(Z_3) - \mathbb{E}(Z_2)),
\end{aligned}
\tag{3}
$$

where $\mathbb{E}(Z_3)$ is the average number of constraints becoming unsatisfied from satisfied, referring to (1),

$$
\mathbb{E}(Z_3) = \frac{k}{N}(M - M_0)p;
\tag{4}
$$

$\mathbb{E}(Z_2)$ is the average number of constraints becoming satisfied from unsatisfied,

$$
\mathbb{E}(Z_2) = E(Z_1)(1 - p).
\tag{5}
$$

According to Algorithm 1, we randomly select a variable to reassign from the conflict set. When $M_0 > 0$,

$$
\mathbb{E}(Z_1) = \frac{M_0 k}{N} \beta,
\tag{6}
$$

where $\beta = \mathbb{E}(\frac{1}{X})$; X is the fraction of not empty variables (connecting to at least an unsat-constraint), when we throw M_0 unsat-constraints to N variable. $\mathbb{E}(Z_1)$ is at least 1 and

$$
\mathbb{E}(Z_1) \to 1, \; as \; M_0/M \to 0.
\tag{7}
$$

When $\Pr[0, T] = 0$, from (3)(4)(5)(6), we have

$$
\begin{aligned}
\frac{d\varphi}{dt} &= \frac{krN \ln N}{N}(1 - \varphi)p - \sum_{M_0=1}^{M} \Pr[M_0, T] \frac{k}{N} M_0 \beta(1 - p) \\
&= -(1 - p) + kr \ln N(1 - \varphi)p - \sum_{M_0=1}^{M} \Pr[M_0, T](\frac{k}{N} M_0 \beta - 1)(1 - p).
\end{aligned}
\tag{8}
$$

Let

$$
r_d = \frac{1 - p}{p} \frac{1}{k \ln N}.
$$

For $r < c \cdot r_d$, where $c < 1$ is a constant, from (8) we have

$$
\frac{d\varphi}{dt} < -(1 - c)(1 - p).
$$

From (2), we have $\varphi(t = 0) = p$. So when $r < c \cdot r_d$, φ becomes 0 before $t = \frac{p}{(1-c)(1-p)}$. For $r > c \cdot r_d$, where $c > 1$ is a constant, when φ is near 0, by (7) we can see the last term of (8) is near 0, so $\frac{d\varphi}{dt} > 0$, φ is always positive.

Thus, the threshold value of Algorithm 1 is $r_d = \frac{1-p}{p} \frac{1}{k \ln N}$.

4.2 Analysis on Algorithm 2

For Algorithm 2, when $M_0 > 0$, the probability that a variable with Z_1 unsat-constraints was chosen is

$$p(Z_1) = \binom{M_0 - 1}{Z_1 - 1} \left(\frac{k}{N}\right)^{Z_1 - 1} \left(1 - \frac{k}{N}\right)^{M_0 - Z_1}$$

because we randomly select an unsat-constraint, then each of other $M_0 - 1$ unsat-constraints connects to the variable with probability $\frac{k}{N}$. Therefore

$$\mathbb{E}(Z_1) = 1 + (M_0 - 1)k/N. \tag{9}$$

Similarly, for Algorithm 2, when $\Pr[0, T] = 0$, from (3)(4)(5)(9) we know

$$\frac{d\varphi}{dt} = \sum_{M_0=1}^{M} \Pr[M_0, T] \left(\frac{k}{N}(M - M_0)p - (1 - p)(1 + (M_0 - 1)\frac{k}{N})\right)$$

$$= -(1 - p) + kr \ln Np + \frac{k}{N}(1 - p) - k\varphi r \ln N.$$

Sloving this first-order linear differential equation with the initial condition $\varphi(t = 0) = p$, we get

$$\varphi(t) = p + \frac{1 - p - k(1 - p)/N}{rk \ln N}(e^{-rk \ln Nt} - 1).$$

Sloving equation $\lim_{t \to \infty} \varphi(t) = 0$ of variable r, we have

$$r = (1 - \frac{k}{N})\frac{1 - p}{p}\frac{1}{k \ln N} \triangleq r'_d.$$

For $r < cr'_d$, where $c < 1$ is a constant, $\lim_{t \to \infty} \varphi(t) < 0$. Function $\varphi(t)$ decreases and becomes 0 before $t = \frac{p}{(1-c)(1-p)(1-k/N)}$. For $r > cr'_d$, where $c > 1$ is a constant, $\lim_{t \to \infty} \varphi(t) > 0$.

Thus, the threshold value on Algorithm 2 is

$$r'_d \approx r_d = \frac{1 - p}{p}\frac{1}{k \ln N}.$$

4.3 A Note

The estimates of threshold values from numerical simulations are always larger than the calculated one. Taking RB($k = 2, N = 350, p = 0.2, \alpha = 0.5$) as an example, the calculated value is $r_d = r'_d = 0.34$, but the simulated value is 0.43 for Algorithm 1, and 0.51 for Algorithm 2. However, the simulated values always fall into the region $(r_d, 2r_d)$, so the theoretically calculated values r_d and r'_d reveal the positions of the real threshold values successfully.

5   Conclusion

We have studied performances of pure random walk (PRW) algorithms on a model of random constraint satisfaction problem with growing domains called Model RB. The same threshold behaviors of PRW are shown on Model RB, just like that of Walksat on k-SAT.

From our results, we find that PRW algorithms are more suitable for k-SAT than for Model RB. Taking 3-SAT as an example, Walksat can solve 3-SAT until clause density 2.7, which is not small relative to its satisfiability threshold value of 4.26. But for Model RB, PRW can work until $\frac{1-p}{p}\frac{1}{k\ln N}$, which is very small (tending to 0) relative to its satisfiability threshold value of $-\frac{\alpha}{\ln(1-p)}$ (a constant). This may be due to the fact that the instances of Model RB have large domain size, and a large domain size leads to more constraints and more unsat-constraints, while PRW algorithms can not deal with instances with many unsat-constrains.

In another recent paper, we found out that a backtrack-free algorithm can solve Model RB until a positive constant proportion of $-\frac{\alpha}{\ln(1-p)}$ [30], while it can barely solve k-SAT. Therefore, CSPs with large domain size (such as Model RB) and CSPs with small domain size (such as k-SAT) may have different properties, and different strategies (such as PRW and backtrack-free search) may have different effects on them.

References

1. Achlioptas, D., Kirousis, L., Kranakis, E., Krizanc, D., Molloy, M., Stamatiou, Y.: Random constraint satisfaction: a more accurate picture. In: Smolka, G. (ed.) CP 1997. LNCS, vol. 1330, pp. 107–120. Springer, Heidelberg (1997)
2. Alekhnovich, M., Ben-Sasson, E.: Linear Upper Bounds for Random Walk on Small Density Random 3-cnfs. SIAM J. Comput. 36(5), 1248–1263 (2006)
3. Alphonse, É., Osmani, A.: A model to study phase transition and plateaus in relational learning. In: Železný, F., Lavrač, N. (eds.) ILP 2008. LNCS (LNAI), vol. 5194, pp. 6–23. Springer, Heidelberg (2008)
4. Broder, A.Z., Frieze, A.M., Upfal, E.: On the Satisfiability and Maximum Satisfiability of Random 3-CNF Formulas. In: Proc of SODA, pp. 322–330 (1993)
5. Chao, M., Franco, J.: Probabilistic Analysis of Two Heuristics for the 3-Satisfiability Problem. SIAM J. Comput. 15(4), 1106–1118 (1986)
6. Cocco, S., Monasson, R., Montanari, A., Semerjian, G.: Analyzing search algorithms with physical methods. In: Percus, A., Istrate, G., Moore, C. (eds.) Computational Complexity and Statistical Physics, pp. 63–106. Oxford University Press (2006)
7. Coja-Oghlan, A., Frieze, A.: Analyzing Walksat on random formulas. In: Proc. of ANALCO, pp. 48–55 (2012)
8. Coja-Oghlan, A., Feige, U., Frieze, A., Krivelevich, M., Vilenchik, D.: On smoothed k-CNF formulas and the Walksat algorithm. In: Proc. of SODA, pp. 451–460 (2009)
9. Fan, Y., Shen, J.: On the phase transitions of random k-constraint satisfaction problems. Artif. Intell. 175, 914–927 (2011)
10. Fan, Y., Shen, J., Xu, K.: A general model and thresholds for random constraint satisfaction problems. Artif. Intell. 193, 1–17 (2012)

11. Gao, Y., Culberson, J.: Consistency and random constraint satisfaction problems. J. Artif. Intell. Res. 28, 517–557 (2007)
12. Gent, I., Macintype, E., Prosser, P., Smith, B., Walsh, T.: Random constraint satisfaction: flaws and structure. Constraints 6(4), 345–372 (2001)
13. Jiang, W., Liu, T., Ren, T., Xu, K.: Two hardness results on feedback vertex sets. In: Atallah, M., Li, X.-Y., Zhu, B. (eds.) FAW-AAIM 2011. LNCS, vol. 6681, pp. 233–243. Springer, Heidelberg (2011)
14. Lecoutre, C.: Constraint Networks: Techniques and Algorithms. John Wiley & Sons (2009)
15. Liu, T., Lin, X., Wang, C., Su, K., Xu, K.: Large Hinge Width on Sparse Random Hypergraphs. In: Proc of IJCAI, pp. 611–616 (2011)
16. Liu, T., Wang, C., Xu, K.: Large hypertree width for sparse random hypergraphs. J. Comb. Optim. (2014), doi 10.1007/s10878-013-9704-y
17. Mezard, M., Montanari, A.: Information, Physics and Computation. Oxford University Press (2009)
18. Richter, S., Helmert, M., Gretton, C.: A stochastic local search approach to vertex cover. In: Hertzberg, J., Beetz, M., Englert, R. (eds.) KI 2007. LNCS (LNAI), vol. 4667, pp. 412–426. Springer, Heidelberg (2007)
19. Rossi, F., Van Beek, P., Walsh, T. (eds.): Handbook of Constraint Programming. Elsevier (2006)
20. Smith, B.M., Dyer, M.E.: Locating the Phase Transition in Binary Constraint Satisfaction Problems. Artif. Intell. 81, 155–181 (1996)
21. Semerjian, G., Monasson, R.: Relaxation and Metastability in the Random Walk SAT search procedure. Phys. Rev. E 67, 066103 (2003)
22. Semerjian, G., Monasson, R.: A Study of Pure Random Walk on Random Satisfiability Problems with Physical Methods. In: Giunchiglia, E., Tacchella, A. (eds.) SAT 2003. LNCS, vol. 2919, pp. 120–134. Springer, Heidelberg (2004)
23. Schöning, U.: A probabilistic algorithm for k-SAT based on limited local search and restart. Algorithmica 32, 615–623 (2002)
24. Schöning, U.: A probabilistic algorithm for k-SAT and constraint satisfaction problems. In: Proc. of FOCS, pp. 410–414 (1999)
25. Smith, B.: Constructing an asymptotic phase transition in random binary constraint satisfaction problems. Theoret. Comput. Sci. 265, 265–283 (2001)
26. Wang, C., Liu, T., Cui, P., Xu, K.: A note on treewidth in random graphs. In: Wang, W., Zhu, X., Du, D.-Z. (eds.) COCOA 2011. LNCS, vol. 6831, pp. 491–499. Springer, Heidelberg (2011)
27. Xu, K., Li, W.: Exact Phase Transitions in Random Constraint Satisfaction Problems. J. Artif. Intell. Res. 12, 93–103 (2000)
28. Xu, K., Li, W.: Many Hard Examples in Exact Phase Transitions. Theoret. Comput. Sci. 355, 291–302 (2006)
29. Xu, K., Boussemart, F., Hemery, F., Lecoutre, C.: Random Constraint Satisfaction: Easy Generation of Hard (Satisfiable) Instances. Artif. Intell. 171, 514–534 (2007)
30. Xu, W.: An analysis of backtrack-free algorithm on a constraint satisfaction problem with growing domains (in Chineses). Acta Mathematicae Applicatae Sinica (Chinese Series) (accepted, 2014)
31. Zhao, C., Zheng, Z.: Threshold behaviors of a random constraint satisfaction problem with exact phase transitions. Inform. Process. Lett. 111, 985–988 (2011)
32. Zhao, C., Zhang, P., Zheng, Z., Xu, K.: Analytical and Belief-propagation Studies of Random Constraint Satisfaction Problems with Growing Domains. Phys. Rev. E 85, 016106 (2012)

Calculating the Crossing Probability on the Square Tessellation of a Connection Game with Random Move Order: The Algorithm and Its Complexity*

Yi Yang[1], Shuigeng Zhou[1,2,**], and Jihong Guan[3]

[1] School of Computer Science, Fudan University, Shanghai 200433, China
[2] Shanghai Key Lab of Intelligent Information Processing, Shanghai 200433, China
{yyang1,sgzhou}@fudan.edu.cn
[3] Department of Computer Science & Technology, Tongji University
Shanghai 201804, China
jhguan@tongji.edu.cn

Abstract. This paper presents an algorithm for calculating the crossing probability on the square tessellation of a connection game with random move order. The time complexity of the algorithm is $O(poly(N) \cdot 2.7459...^N)$, where N is the size of the tessellation. We conjecture that the bound is tight within a $poly(N)$ term.

1 Introduction

Connection games [1] is a type of abstract strategy games in which players attempt to complete a specific type of connection with their pieces. In many connection games, the goal is to connect the two opposite sides of the board. Connection games are not necessary played in turn. It can be played in random move order, for example, random-turn hex [2]. It can also be played in bidding manner, for example, bidding hex [3].

Recently, we designed a new connection game Square++ [4], which is played in random move order. To play Square++ well, we need to solve the following problem:

1.1 Problem Statement

To describe the problem, we first draw a square tessellation. Fig. 1 shows an example of square tessellation of size 5, which has 5*5=25 cells. In this tessellation, we mark the 4 sides of the tessellation in 2 colors: black and white, which is shown in Fig. 2. The black color and the white color are at the opposite sides of the tessellation.

---

* This work was supported by National Natural Science Foundation (NSFC) under grant No. 61373036.
** Corresponding author.

J. Chen, J.E. Hopcroft, and J. Wang (Eds.): FAW 2014, LNCS 8497, pp. 288–297, 2014.

Fig. 1. An example of square tessellation of size 5

Fig. 2. An example of square tessellation with marks on its 4 sides

We try to find out: when we randomly and independently color the cells in black with probability p and in white with probability $(1 - p)$, what is the probability that there is a path consisting of black cells that connect the two black sides? Here, a path is a sequence of connected cells. We say two cells are connected if they share a common side. For a path, its first cell connects one black side, and its last cell connects the other black side.

We call the path of black cells that connect the two black sides a *crossing*, and call the probability *crossing probability*. Fig. 3 shows a crossing example and a non-crossing example.

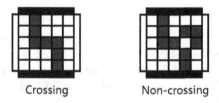

Crossing   Non-crossing

Fig. 3. Examples of *crossing* and *non-crossing*

The problem we want to solve is: given p and the size N of the square tessellation, how to compute the crossing probability?

1.2   Our Result

To solve the problem, we introduce a dynamic programming algorithm. Our result is as follows:

Theorem 1. *Given p and the size N of the square tessellation, there is an algorithm calculates the crossing probability in $O(poly(N) \cdot \lambda^N)$ time, where $\lambda = 2.7459...$ is the real root of $\lambda^5 - 2\lambda^4 - 2\lambda^3 - 1 = 0$.*

To the best of our knowledge, no previous study gives such exact time complexity, and we conjecture that this bound is tight within the $poly(N)$ term.

2   Preliminaries

We first give the definition of *situation* as follows:

Definition 1. *(Situation) A situation S is a mapping from the set of the cells $\{1, 2, ..., N\} \times \{1, 2, ..., N\}$ to the colors $\{\square, \boxtimes, \blacksquare\}$. Here, $\square$ is a white cell, $\blacksquare$ is a black cell, $\boxtimes$ is a cell with unknown color. This cell has p probability being $\blacksquare$, and $(1 - p)$ probability being $\square$. If a situation has no unknown color cells, we call this situation* final situation. *If all the colors of the cells of a situation are unknown, we call this situation* empty situation, *denoted as E.*

So there are 3^{N^2} situations for square tessellation with size N. Each situation has a crossing probability, denoted as $C(S)$. The $C(S)$ can be defined as follows. For the 2^{N^2} final situations, $C(S)$ is either 0 or 1. More specifically, if there is a path with black cells that connect the two opposite black sides, $C(S) = 1$; Otherwise, $C(S) = 0$. For the situations with unknown color cells, we can find the crossing probability recursively as follows.

Lemma 1. *If a situation S has unknown color cells, then $C(S) = p \cdot C(P_i) + q \cdot C(Q_i)$ holds for $i = 1, 2, ..., k$. Where $q = 1 - p$, k is the number of cells with unknown color in S, P_i is the situation S that the i-th unknown color cell is replaced by $\blacksquare$, Q_i is the situation S that the i-th unknown color cell is replaced by $\square$.*

Proof. Since the coloring is randomly and independently, we can specify a color on the i-th $\boxtimes$ first, and then specify the color of the remaining $\boxtimes$ later. Since a $\boxtimes$ cell have p chance to be a $\blacksquare$ cell, forming the situation P_i, and have q chance to be a $\square$ cell, forming the situation Q_i. Thus, we get $C(S) = p \cdot C(P_i) + q \cdot C(Q_i)$ holds for $i = 1, 2, ..., k$.

3   A Naïve Algorithm

After defining the situations, our goal is to calculate $C(E)$. To calculate $C(E)$, we recursively apply Lemma 1 N^2 times, and then we got the following formula.

$$
\begin{aligned}
&C([\boxtimes\,\boxtimes\,...\boxtimes, \boxtimes\,\boxtimes\,...\boxtimes, ..., \boxtimes\,\boxtimes\,...\boxtimes]) \\
&= p \cdot C([\blacksquare\,\boxtimes\,...\boxtimes, \boxtimes\,\boxtimes\,...\boxtimes, ..., \boxtimes\,\boxtimes\,...\boxtimes]) \\
&\quad + q \cdot C([\square\,\boxtimes\,...\boxtimes, \boxtimes\,\boxtimes\,...\boxtimes, ..., \boxtimes\,\boxtimes\,...\boxtimes]) \\
&= p^2 \cdot C([\blacksquare\blacksquare...\boxtimes, \boxtimes\,\boxtimes\,...\boxtimes, ..., \boxtimes\,\boxtimes\,...\boxtimes]) \\
&\quad + pq \cdot C([\blacksquare\square...\boxtimes, \boxtimes\,\boxtimes\,...\boxtimes, ..., \boxtimes\,\boxtimes\,...\boxtimes]) \\
&\quad + qp \cdot C([\square\blacksquare...\boxtimes, \boxtimes\,\boxtimes\,...\boxtimes, ..., \boxtimes\,\boxtimes\,...\boxtimes]) \\
&\quad + q^2 \cdot C([\square\square...\boxtimes, \boxtimes\,\boxtimes\,...\boxtimes, ..., \boxtimes\,\boxtimes\,...\boxtimes]) \\
&= ... \\
&= p^{N^2} \cdot C([\blacksquare\blacksquare...\blacksquare, \blacksquare\blacksquare...\blacksquare, ..., \blacksquare\blacksquare...\blacksquare]) \\
&\quad + p^{N^2-1}q \cdot C([\blacksquare\blacksquare...\blacksquare, \blacksquare\blacksquare...\blacksquare, ..., \blacksquare\blacksquare...\square]) \\
&\quad + ... \\
&\quad + q^{N^2} \cdot C([\square\square...\square, \square\square...\square, ..., \square\square...\square])
\end{aligned}
\tag{1}
$$

In this formula, there are 2^k terms in the k-th step, $k = 0, 1, 2, ..., N^2$. For each term, we are considering a situation that the colors of the first k cells are known, the color of the last $(N^2 - k)$ cells are unknown. We call such a situation *partial filled situation, k-filled situation* or *(i,j)-filled situation*. Where (i,j)-filled situation means the situation that the colors of the first $(i - 1)$ rows and the first j cells of the i-th row are known, other cells are unknown.

In the final step of the formula, we got 2^{N^2} final situations. So we can check them one by one, and sum up the crossing probabilities. Thus this naive algorithm takes $O(N^2 \cdot 2^{N^2})$ time to compute the crossing probability $C(E)$.

4   A Dynamic Programming Algorithm

John Tromp has developed a dynamic programming algorithm to compute the legal positions of Go [5]. Following a similar idea, a dynamic programming algorithm is proposed to compute $C(E)$ with a fixed p.

In Formula (1), there are 2^k k-filled situations to consider in the k-th step. However, most of them can be combined, since they have the same crossing probability. That is, for s different k-filled situations $S_1, S_2, ..., S_s$, if we are sure that $C(S_1) = C(S_2) = ... = C(S_s)$, then $c_1 C(S_1) + c_2 C(S_2) + ... + c_s C(S_s)$ can be rewritten as $(c_1 + c_2 + ... + c_s)C(S_1)$, where $c_1, ..., c_s$ are the polynomials of p and q. After combining, we only need to compute $C(S_1)$ by further recursion, and we do not need to consider the recursion of $C(S_2), ..., C(S_s)$ any more, which will make the computation faster. Figure 4 shows 6 (2,2)-filled situations with $N = 3$ that have the same crossing probability. That is, $C(S_1) = C(S_2) = ... = C(S_6) = p^2$.

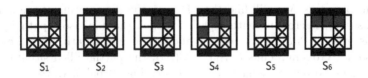

S_1 S_2 S_3 S_4 S_5 S_6

Fig. 4. (2,2)-filled situations with $N = 3$ that have the same crossing probability

To specify which partial filled situations have the same crossing probability, we define *border state*.

Definition 2. *(Border State) A border state extracts the following data of a partial filled situation:*

- $N, (i, j)$
- *the color of the border cells $(i, 1)$, ..., (i, j), $(i - 1, j + 1)$, ..., $(i - 1, N)$,*
- *the information of the border ■ pieces that can be specified which ■ pieces connect the top of the board, which ■ pieces are in the same connection component.*

Where a connection component is a maximum set of ■ *pieces that connect with each other.*

Figure 5 shows a partial filled situation and its border state. In the border state, $N = 5, i = 4, j = 2$, the color of the border cells are ■, ■, ■, □, ■. The first 2 ■s connects the top of the board, the last 2 ■s are in the same connection component.

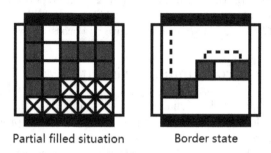

| Partial filled situation | Border state |

Fig. 5. A partial filled situation and its border state

Theorem 2. *The partial filled situations that have the same border state have the same crossing probability.*

Proof. Let S_1 and S_2 be two partial filled situations with the same border state. Then they have the same N and (i, j), and they have the same number of cells with unknown color, and the color of the cells in the border and the connections of the ■ cells in the border are the same. Now we enumerate the colors of the unknown color cells. We will find that, for any specific coloring of the unknown color cells, if S_1 is crossing with a path P_1, then there is also a path P_2 in S_2, thus S_2 is crossing; if S_1 is not crossing and has no such a path, we will find that S_2 is also not crossing. Thus $C(S_1) = C(S_2)$, the partial filled situations that have the same border state have the same crossing probability.

Definition 3. *(Valid Border State) We say a border state is valid if it satisfies all the following conditions:*

- *Adjacent* ■ *pieces are marked to be the same connection component.*
- *if 3* ■ *pieces are ordered horizontally as a, b and c, with a and c connecting with each other, and b connecting the top of the board, then they must all connect the top of the board.*
- *if 4* ■ *pieces are ordered horizontally as a, b, c and d, with a and c connecting with each other, b and d connecting with each other, then they must all connect with each other.*

Theorem 3. *All the partial filled situations have a valid border state.*

Proof. The first property is obvious. The second and the third property can be seen from the planarity of 2D boards, which have been shown in Figure 6 and Figure 7.

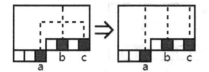 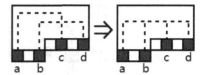

Fig. 6. If a and c are connected, b connects the top, they must all connect the top

Fig. 7. If a and c are connected, b and d are connected, they must be all connected

In the dynamic programming algorithm, we still compute $C(E)$ according to Formula (1). But at every step, we combine the partial filled situations which are sure to have the same crossing probability. As Theorem 2 says, the partial filled situations that have the same border state have the same crossing probability, we only need to consider the recursions of the situations that have different border state. As Theorem 3 says, all the partial filled situations have a valid border state, So, to evaluate the time complexity of the algorithm, it comes up the problem that we need to count the number of valid border states of (i, j)-filled situations of $1 \le i \le N, 1 \le j \le N$. Let $T_{i,j}$ be the number of valid border states of (i, j)-filled situations, then the computation time is $T = T_{1,1} + T_{1,2} + ... + T_{N,N}$.

So, if we want to reduce the time complexity of calculating $C(E)$, it is important to reduce the computation time of considering different valid border states $T_{i,j}$.

We do have a method to reduce the computation time of considering different valid border states $T_{i,j}$. We next define *simplified border state*. To simplify the discussion, we consider (i, N)-filled situations only. This simplification technique can be apply to the general (i, j)-filled situation as well, but with some details needs to be handled.

Definition 4. *(simplified border state) A simplified border state is a valid border state that satisfies the following conditions.*
- *There is no single border ■ cell without outer connections.*
- *There is no two adjacent border ■ cell without outer connections.*
- *There is no three consecutive border ■ cell without outer connections.*

Theorem 4. *All the partial filled situations have a similar crossing probability to that of a simplified border state.*

Proof. Given a partial filled situation, we check its border state. If there is a single border ■ cell that without outer connections, we can replace this single border ■ to □, then the border state of this partial filled situation is changed, but the crossing probability still keeps the same, as shown in Figure 8.

Given a partial filled situation, we check its border state. If there are two adjacent border ■ cells that without outer connections, we can replace this two border ■s to two □s, then the border state of this partial filled situation is changed, but the crossing probability still keeps the same, as shown in Figure 9.

Given a partial filled situation, we check its border state. If there are three consecutive border ■ cells that without outer connections, we can replace the

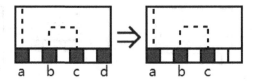

Fig. 8. A single border ■ cell d without outer connections can be replaced by a □ cell

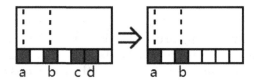

Fig. 9. Two adjacent border ■ cells c, d without outer connections can be replaced by two □ cells

middle border ■ to □, and mark the left ■ piece and the right ■ piece to be the same connection component, as shown in Figure 10. Then, for the situation that there is no ■ path connects the top and the bottom, there is also no ■ path connects the top and the bottom after the replacement. For the situation that there is a ■ path connects the top and the bottom, let P be a shortest ■ path that connects the top and the bottom. If P does not contain three consecutive border ■ cells, then there is a ■ path connects the top and the bottom after the replacement, as shown in Figure 11. If P contains three consecutive border ■ cells, then there is also a ■ path connects the top and the bottom after the replacement, as shown in Figure 12. Thus, the crossing probability keeps the same after the replacement.

We check the border state of the given partial filled situation and modify them repeatedly according to the above 3 operations, then the border state will changed, but the crossing probability remains the same, and the border state will finally satisfied the 3 rules of the simplified border states.

It is obvious that the number of simplified border states is much less than the number of valid border states. So to calculate $C(E)$ according to Formula (1),

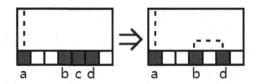

Fig. 10. Three adjacent border ■ cells b, c, d without outer connections can be replaced by the connected ■ b and ■ d

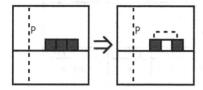
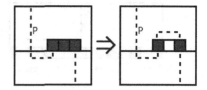

Fig. 11. If P does not contain three consecutive border ■ cells, then there is a ■ path connects the top and the bottom after the replacement

Fig. 12. If P contains three consecutive border ■ cells, then there is also a ■ path connects the top and the bottom after the replacement

we only need to consider the recursion of the simplified border states. Let $R_{i,j}$ be the number of the simplified border states of (i,j)-filled situations, then the time complexity calculating $C(E)$ becomes $R = R_{1,1} + R_{1,2} + ... + R_{N,N}$.

5   Complexity Analysis

To analysis the time complexity of calculating $C(E)$, we need to find the number of simplified border states of (i,j)-filled situations. According to Theorem 3, we can represent the simplified border states to the parenthesis sequences. The parenthesis sequences has 8 symbols " ", "[", "]", "][", "[]", "(", ")", ")(", where:

- " " represents a □ piece,
- "[" represents a ■ piece that connects the top and has some ■ pieces that in the same connection component on the right,
- "]" represents a ■ piece that connects the top and has some ■ pieces that in the same connection component on the left,
- "][" represents a ■ piece that connects the top and has some ■ pieces that in the same connection component on both the left and right,
- "[]" represents a ■ piece that connects the top and has no other connections,
- "(" represents a ■ piece that does not connect the top and has some ■ pieces that in the same connection component on the right,
- ")" represents a ■ piece that does not connect the top and has some ■ pieces that in the same connection component on the left,
- ")(" represents a ■ piece that does not connect the top and has some ■ pieces that in the same connection component on both the left and right.

According to Theorem 3, each simplified border state has a parenthesis sequence representation. For example, the border state in Figure 5 is a simplified border state, we can represent this border state to the parenthesis sequence as "[","]","("," ",")", where "[" and "]" is the first two ■ pieces that connect the top, "("," ",")" is the last 3 pieces that do not connect the top.

To find the number of the simplified border states, we enumerate the starting positions and the ending positions of the ■ pieces that connect the top. Then the border state will be divided in to 3 parts: the left part, the middle part and the right part, as Figure 13 shows.

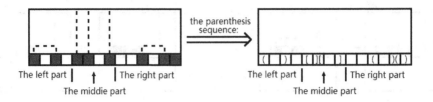

Fig. 13. The 3 parts of a border state

For the left and the right part, the parenthesis sequences have only 4 symbols " ","(",")",")(", and they obeys the rule as Figure 14 shows. Thus the number of the simplified border states with size x that off the top can be upper bounded as the number of the sequence with length x that can be accepted by the rules of Figure 14 shows.

	current state	the symbol of the state	$\square$	$[($	$)]$	$)[($	
The starting state →	1	$\square$	1	2	3	4	
	2	$[($	1	0	0	5	**The next state**
	3	$)]$	1	0	0	0	('0' means the
	4	$)[($	1	0	3	4	sqauence is
	5	$[($ $)[($	1	0	0	4	not accaptable)

Next symbol

Fig. 14. The rules of accepting the left part and the right part of the parenthesis sequences according to the conditions of simplified border states

It is easy to calculate that the number of the sequence with length x that can be accepted by the rules of Figure 14 shows can be upper bounded as $O(\lambda^x)$ with $\lambda = 2.7459...$, which is the real root of $\lambda^5 - 2\lambda^4 - 2\lambda^3 - 1 = 0$.

Now we consider the middle part. After replacing "[" to "(", "]" to ")", "][" to ")(", "[]" to "()", we find that the sequence with length $N > 3$ that satisfies the conditions of simplified border states are all obeys the rules of Figure 14 shows, except when $N = 1$, there is one more sequence "()", when $N = 2$, there is one more sequence "(",")" and when $N = 3$, there is one more sequence "(",")(",")". Thus the number of the simplified border states with size x that in the middle part can also be upper bounded as $O(\lambda^x)$.

Now we are ready to give an upper bound of the number of (k, N)-filled simplified border state. We enumerate the starting positions and the ending positions of the ■ pieces that connect the top. Let the starting position be i, ending position be j, then there are at most

$$\Sigma_{i,j=1}^{N}[O(\lambda^{i-1}) \cdot O(\lambda^{j-i+1}) \cdot O(\lambda^{N-j})]$$
$$= \Sigma_{i,j=1}^{N} O(\lambda^{N})$$
$$= O(poly(N) \cdot \lambda^{N})$$

(k, N)-filled simplified border states.

Now the number of (k, l)-filled simplified border states also can be upper bounded. Since the upper bound of the number of (k, N)-filled simplified border state is $O(poly(N) \cdot \lambda^{N})$, the upper bound of the number of (k, l)-filled simplified border states is no more than

$$O(poly(l) \cdot \lambda^{l}) \cdot O(poly(N - l) \cdot \lambda^{N-l}) = O(poly(N) \cdot \lambda^{N})$$

Thus for $1 \le i \le N, 1 \le j \le N$, we have $R_{i,j} = O(poly(N) \cdot \lambda^{N})$. So the time complexity calculating $C(E)$ is $R = \Sigma_{i,j=1}^{N} R_{i,j} = O(poly(N) \cdot \lambda^{N})$. with $\lambda = 2.7459...$, which reveals Theorem 1.

6   Conclusion

We propose a dynamic programming algorithm to compute the crossing probability on the square tessellation of a connection game with random move order, and analyze its time complexity. The time complexity contains an exponential term and a $poly(N)$ term. We get the exponential term by listing the rules of the border states that need to obey. We conjecture that the exponential term in the time complexity is optimal.

As we know, square tessellation is just one of the three regular tessellations. The other two regular tessellations are triangle tessellation and hex tessellation. We believe that similar dynamic programming algorithms can be designed to compute the crossing probabilities on triangle and hex tessellations, and the time complexities of these algorithms should also contain an exponential term and a $poly(N)$ term. We can also get the exponential terms by listing the rules of the border states on triangle and hex tessellations that need to obey.

In the future, we will try to find out the exact expressions of the $poly(N)$ terms in the time complexities, which is a more challenging task.

References

1. Browne, C.: Connection Games. A.K. Peters, Ltd. (2005)
2. Peres, Y., Schramm, O., Sheffield, S., Wilson, D.B.: Random-turn hex and other selection games. American Mathematical Monthly 114, 373–387 (2007)
3. Payne, S., Robeva, E.: Artificial intelligence for Bidding Hex, arXiv preprint arXiv:0812.3677 (2008)
4. Yang, Y., Zhou, S., Li, Y.: Square++: Making A Connection Game Win-Lose Complementary and Playing-Fair. Entertainment Computing 4(2), 105–113 (2013)
5. Tromp, J., Farneback, G.: Combinatorics of Go. In: Proceedings of the Fifth International Conference on Computer and Games, Turin, Italy (2006)

On Star-Cover and Path-Cover of a Tree[*]

Jie You[1], Qilong Feng[1], Jiong Guo[2], and Feng Shi[1]

[1] School of Information Science and Engineering, Central South University, China
[2] Universität des Saarlandes, Campus E 1.7, Saarbrücken 66123, Germany

Abstract. We study two restricted versions of the SUBFOREST ISOMORPHISM problem, the STAR-COVER OF A TREE (SCT) and the PATH-COVER OF A TREE (PCT) problems. Both problems are NP-complete. The problems are closely related to a number of well-studied problems, including the problems SUBGRAPH ISOMORPHISM, TREE EDITING, and GRAPH PACKING. We show that the problems SCT and PCT are fixed-parameter tractable. Thorough development of parameterized algorithms and kernelization algorithms for these problems are presented.

1 Introduction

In this paper, we study the complexity of the following SUBFOREST ISOMORPHISM problem: given a forest F and a tree T with the same number of vertices, decide if F is a subgraph of T.

The problem is closely related to a number of well-studied algorithmic graph problems. First of all, the problem is a restricted version of the famous NP-complete problem SUBGRAPH ISOMORPHISM. If the forest F is also a tree, then the SUBFOREST ISOMORPHISM problem becomes the TREE ISOMORPHISM problem, which can be solved in linear time [10]. However, if F is allowed to be a general forest, then the SUBFOREST ISOMORPHISM problem is NP-complete [9]. In fact, even when the forest F is a collection of stars or a collection of paths, the SUBFOREST ISOMORPHISM problem remains NP-complete [3].

Another related problem is the TREE EDITING problem [4], which has found applications in several diverse areas such as pattern recognition [7,13], computational biology [2,18], XML database [16], image analysis [12], and chemistry [15]. In particular, the well-known *tree-bisection-and-reconnection* (TBR) distance for two phylogenetic trees T_1 and T_2 is to find a minimum set E_1 of edges in T_1 so that the forest $T_1 \setminus E_1$ becomes a subforest of T_2 [1,14]. Thus, the SUBFOREST ISOMORPHISM problem can be regarded as an intermediate step that assembles the trees in $T_1 \setminus E_1$ to make the tree T_2.

The third problem related to the SUBFOREST ISOMORPHISM problem is the GRAPH PACKING problem, where we pack a collection of graph patterns into a given graph. The GRAPH PACKING problem includes the famous GRAPH MATCHING problem and has applications ranging from information theory to

---

[*] This work is supported by the National Natural Science Foundation of China under Grants (61103033, 61173051, 61232001).

J. Chen, J.E. Hopcroft, and J. Wang (Eds.): FAW 2014, LNCS 8497, pp. 298–308, 2014.

the design of efficient statistical experiments [17]. In fact, the SUBFOREST ISO-MORPHISM problem can be regarded as the task of packing the trees in the given forest F into the given tree T.

In the current paper, we are focused on two restricted versions of the SUB-FOREST ISOMORPHISM problem. We say that a graph G is a *star* if G is the complete bipartite graph $K_{1,h}$ for some integer $h \geq 0$. Our first version requires that all connected components in the given forest F be stars, while our second version requires that all connected components in the given forest F be paths. The formal formulations of the problems are given as follows.

STAR-COVER OF A TREE (SCT)
Input: A forest $F = (V_F, E_F)$ consisting of k disjoint stars
             and a tree $T = (V_T, E_T)$, with $|V_F| = |V_T|$;
Question: Is F a subgraph of T?

PATH-COVER OF A TREE (PCT)
Input: A forest $F = (V_F, E_F)$ consisting of k disjoint paths
             and a tree $T = (V_T, E_T)$, with $|V_F| = |V_T|$;
Question: Is F a subgraph of T?

Both problems have been studied in literature. Baumbach, Guo, and Ibrag-imov studied the problem SCT, and proved that the SCT problem is NP-complete, even when the tree in the input instance has a diameter bounded by a constant [3]. They also presented a parameterized algorithm that solves the SCT problem in time $O(|V_F|^{2h+2}h)$, where h is the number of *distinct* stars in the forest F. They asked, as an open problem, whether the SCT problem is fixed-parameter tractable, i.e., if the problem is solvable in time $f(h)|V_T|^{O(1)}$ for a function $f(h)$ that is independent of $|V_T|$.

Baumbach, Guo, and Ibragimov [3] also proved that the PCT problem is NP-complete. Another problem closely related to PCT, the MINIMUM CUT/PASTE DISTANCE BETWEEN TREES problem (MCPDT), was studied by Kirkpatrick *et al.* [11]. A tree is a *spider* if it contains at most one vertex of degree larger than 2. By a reduction from a known NP-hard problem, the MINIMUM COMMON INTEGER PARTITION (MCIP) problem [6], Kirkpatrick *et al.* [11] showed that even when the input trees are restricted to spiders, the MCPDT problem is already NP-hard. It is fairly straightforward to construct a polynomial-time reduction from the MCPDT problem on spiders to the PCT problem, which shows that even when the tree in the input instance is a spider, the PCT problem is already NP-complete.

In the current paper, we use the number k of connected components in the forest F as the parameter, and study the parameterized complexity of the SCT and PCT problems. For the SCT problem, we first present a polynomial-time reduction rule that leads directly to a simple parameterized algorithm of running time $O(3^k n^{O(1)})$ based on a branch-and-bound process. Then, by a more careful analysis on the structure of the tree T, we show how to improve the running time of the branch-and-bound algorithm to $O^*(2.42^k n(k + \log n))$. Finally, by taking advantage of the algorithm for the SCT problem proposed in [3], which runs in polynomial time when the number of distinct stars in the forest F is bounded by a constant, we show that the SCT problem can be solved in time $O((2 + \epsilon)^k n^{O(1)})$, for any constant $\epsilon > 0$. We also present a polynomial-time kernelization algorithm for the SCT problem that returns a kernel of size $O(k^3)$. For the PCT problem, we develop a parameterized algorithm that is based on branch-and-bound and dynamic programming and solves the PCT problem in time $O(4^k n^{O(1)})$.

All trees and forests in our discussion are unlabeled and unrooted.

2   Covering a Tree with Stars

We study the SCT problem in this section. A *type-$K_{1,i}$ star* is a graph that is isomorphisc to the complete bipartite graph $K_{1,i}$ with $i \geq 0$. A type-$K_{1,i}$ star in which the degree-i vertex is v will be written as S_v^i, where v is called the *center* of the star, and i is the *size* of the star. For a type-$K_{1,1}$ star, we arbitrarily pick a vertex of it as its center, and for a type-$K_{1,0}$ star, we let its unique vertex be the center. A star S_v^i is *larger* than another star S_w^j if $i > j$.

Let (F_0, T) be an instance of the SCT problem, where the number of vertices in the forest F_0 is equal to that in the tree T, and we need to decide if F_0, which is a collection of stars, is a subgraph of T. Suppose that the forest F_0 consists of k_0 stars. Then F_0 is a subgraph of T if and only if there are $k_0 - 1$ edges in T whose removal makes T become F_0. Therefore, instead of embedding the forest F_0 into the tree T, we consider how to remove $k_0 - 1$ edges in T to make T become F_0. Since the tree T will become a general forest during this edge removing process, we will consider a slightly more generalized version of the SCT problem, the STAR-COVER OF A FOREST problem, which is formally defined as follows: Given two forests F_0 and F^* with the same number of vertices, where F_0 is a collection of disjoint stars. The question is that whether F_0 is a subgraph of F^*. Note that the parameter $k = k_0 - k^*$ is equal to the number of edges in F^* minus the number of edges in F_0. Therefore, the parameter k is the number of edges we should remove from the forest F^* to make it become the forest F_0.

2.1   Pairing Stars in F_0 and F^*

Suppose that the forest F^* has a connected component that is a star. We show that in this case, if F_0 is a subgraph of F^*, then we can always directly map a star in F_0 to a subgraph of F^*. This is implemented by the algorithm PairStars given in Figure 1.

The correctness of the algorithm PairStars is given by the following lemma.

Lemma 1. *Suppose that the algorithm PairStars$(F_0, F^*; S_w^q)$ outputs two forests $\tilde{F}_0$ and $\tilde{F}^*$, which are subgraphs of F_0 and F^*, respectively. Then F_0 is a subgraph of F^* if and only if $\tilde{F}_0$ is a subgraph of $\tilde{F}^*$.*

Algorithm PairStars$(F_0, F^*; S_w^q)$
Input: (F_0, F^*) is an instance of SCF, and S_w^q is an isolated star in F^*
1.   let F_0^q be the set of stars in F_0 that are not larger than S_w^q;
2.   **if** $F_0^q = \emptyset$ **then** stop('F_0 is not a subgraph of F^*')
3.   **else** pick a largest star S_v^p in F_0^q;
3.1   arbitrarily remove $q - p$ edges in S_w^q to make a star S_w^p in F^*;
3.2   map S_v^p in F_0 to S_w^p in F^*, and remove them from F_0 and F^*,
respectively.

Fig. 1. An algorithm that directly maps a star in F_0 to a subgraph of F^*

2.2   Parameterized Algorithms for the SCF Problem

Algorithm PairStars suggests an effective pre-processing on instances of the SCF
problem: as long as the forest F^* contains stars, we can repeatedly call the
algorithm PairStars, which will identify a star in the forest F_0 and directly
determine its image in F^*, then remove the two stars from F_0 and F^*. This
process can continue until F^* contains no stars. Now if F^* is not empty, then
F^* must contain a simple path $P = [v_0, v_1, v_2, v_3]$ of length 3. Thus at least one
e of the three edges on the path P cannot be in the image of an isomorphism
from F_0 to a subgraph of F^*. Therefore, this edge e can be removed from F^*
while F_0 remains as a subgraph of the new forest F^*.

   This suggests a branch-and-search algorithm to solve the SCF problem: if the
forest F^* has a connected component that is a star, then we call the algorithm
PairStars to remove a star in both F_0 and F^*, otherwise, we can find a simple
path P of length 3 in F^*, on which we branch on the three edges of P. Since
each branch decreases the number of edges in F^* by 1, the running time of
this algorithm is bounded by $3^k n^{O(1)}$, where k is the parameter of the instance
(F_0, F^*), which is equal to the number of edges in F^* minus the number of edges
in F_0.

   In order to develop more efficient algorithms for the SCF problem, we apply
more careful analysis on the structures of the simple path P of length 3 in F^*.
To simplify our discussion, for an isomorphism σ from F_0 to a subgraph of F^*,
we will simply say that an edge e in F^* *is in* σ (resp. *is not in* σ) if the edge e
is (resp. is not) in the image of σ.

   We start with the following lemma.

Lemma 2. *Let* (F_0, F^*) *be an instance of SCF, and let* $e_1 = [x, y]$ *and* $e_2 = [y, z]$
be two distinct edges in F^* *that share a common end* y. *If* F_0 *is a subgraph of*
F^*, *then for every isomorphism* σ *from* F_0 *to a subgraph of* F^*, *either at least*
one of e_1 *and* e_2 *is not in* σ, *or no edge of the form* $[x, w]$ *or* $[w, z]$ *in* F^* *is in*
σ, *for* $w \neq y$.

Now let $P = [v_0, v_1, v_2, v_3]$ be a simple path of length 3 in the forest F^*, where
v_0 is a leaf. If the vertex v_2 has degree 2 in F^*, then we can have a more efficient
branch strategy on the edges in P, as shown by the following lemma.

Lemma 3. *Let (F_0, F^*) be an instance of SCF and let $P = [v_0, v_1, v_2, v_3]$ be a simple path in F^*, where v_0 is a leaf and v_2 has degree 2. If F_0 is a subgraph of F^*, then there is an isomorphism σ from F_0 to a subgraph of F^* such that at least one of the edges $[v_1, v_2]$ and $[v_2, v_3]$ is not in σ.*

The algorithm is given in Figure 2. We denote by $F^* \setminus E'$ the forest F^* with the edges in E' (but not the ends of these edges) removed.

Algorithm SCF-I(F_0, F^*)
Input: (F_0, F^*) is an instance of the SCF problem;
Question: Is F_0 a subgraph of F^*?

1. **if** $F^* = \emptyset$ **then** return YES;
2. **if** F^* contains an isolated star S_w^q
      then $(\tilde{F}_0, \tilde{F}^*) = \text{PairStars}(F_0, F^*; S_w^q)$;   return SCF-I$(\tilde{F}_0, \tilde{F}^*)$;
3. **if** F^* contains a simple path $P = [v_0, v_1, v_2, v_3]$ of length 3, where v_0 is a leaf **then**
                  if v_2 is of degree 2 in F^*
3.1   **then** branch: return SCF-I$(F_0, F^* \setminus \{[v_1, v_2]\})$;
                            return SCF-I$(F_0, F^* \setminus \{[v_2, v_3]\})$;
3.2   **else** branch: return SCF-I$(F_0, F^* \setminus \{[v_0, v_1]\})$;
                            return SCF-I$(F_0, F^* \setminus \{[v_1, v_2]\})$;
                            return SCF-I$(F_0, F^* \setminus E_{v_2}\})$.
    \\ E_{v_2} is the set of edges in F^* that are of the form $[v_2, w]$ with $w \neq v_1$.

Fig. 2. An $O(2.42^k n(k + \log n))$-time algorithm for the SCF problem

Note that if the forest F_0 is not a subgraph of F^*, then each of the branches of the algorithm SCF-I(F_0, F^*) will eventually reach an instance (F_0, F^*) in which the forest F^* contains a star S_w^q but F_0 has no star not larger than S_w^q. Therefore, the branch will stop in step 2 and return NO when the subroutine PairStars$(F_0, F^*; S_w^q)$ is executed.

Theorem 1. *The algorithm SCF-I(F_0, F^*) solves the problem SCF and has running time $O(2.42^k n(k + \log n))$, where n is the size of the instance (F_0, F^*), and k is the number of edges in F^* minus the number of edges in F_0.*

It is possible to develop an algorithm of running time $O(b^k n^{O(1)})$ to solve the problem SCF, where b is constant smaller than 2.42, if we are willing to pay in the polynomial component $n^{O(1)}$ in the complexity with a polynomial of higher degree . To achieve this, we will use a result of Baumbach, Guo, and Ibragimov [3].

It has been shown in [3] that the SCT problem (i.e., STAR-COVER OF A TREE) can be solved in time $O(n^{2c+3})$, where c is the number of stars of different types in F_0. In particular, if the size of the stars in F_0 is bounded by c, then certainly

F_0 has at most $c + 1$ stars of different types (note that the size of a star can be 0), so the SCT problem can be solved in time $O(n^{2c+5})$. This algorithm can be easily extended to solve the SCF problem (i.e., STAR-COVER OF A FOREST) in which the size of the stars in F_0 is bounded by c: for this, we can simply link the connected components in F^* by stars of size $c + 2$ to make it become a tree, as shown in Figure 3, and add a proper number of stars of size $c + 2$ to F_0. Therefore, The SCF problem for instances (F_0, F^*) in which the stars in F_0 have their size bounded by c can be solved in time $O(n^{2c+9})$. With this preparation,

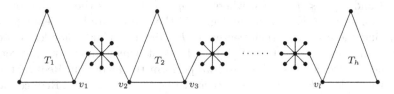

Fig. 3. Linking the connected components in the forest F^* to make it a tree

we are now ready to present our second algorithm for the SCF problem, which is given in Figure 4, where c is a constant to be decided later.

Algorithm SCF-II$(F_0, F^*; c)$
Input: (F_0, F^*) is an instance of the SCF problem;
Question: Is F_0 a subgraph of F^*?

1. **if** $F^* = \emptyset$ **then** return YES;
2. pick a vertex v_1 in F^* that has the largest degree in F^*;
3. **if** $\deg(v_1) \leq c$ **then** solve the instance (F_0, F^*) using the algorithm in [3];
4. **if** v_1 is the center of an isolated star $S_{v_1}^q$ in F^*
       then $(\tilde{F}_0, \tilde{F}^*) = $ PairStars$(F_0, F^*; S_{v_1}^q)$;   return SCF-II$(\tilde{F}_0, \tilde{F}^*)$;
5. **else** pick a simple path $P = [v_1, v_2, v_3]$ of length 2 in F^*;
5.1   branch: return SCF-II$(F_0, F^* \setminus \{[v_1, v_2]\})$;
               return SCF-II$(F_0, F^* \setminus \{[v_2, v_3]\})$;
               return SCF-II$(F_0, F^* \setminus E_{v_1})$.
    \\ E_{v_1} is the set of edges in F^* incident to v_1 but not including $[v_1, v_2]$.

Fig. 4. An $O((2 + \epsilon)^k n^{O(1)})$-time algorithm for the SCF problem

Lemma 4. *Fix a constant $c \geq 1$, the algorithm SCF-II$(F_0, F^*; c)$ solves the problem SCF in time $O(b^k n^{O(1)})$, where n is the size of the instance (F_0, F^*), k is the number of edges in F^* minus that in F_0, and b is the unique root larger than 2 of the polynomial $x^c - 2x^{c-1} - 1$.*

This concludes the following result in this section.

Theorem 2. *For any constant $\epsilon > 0$, the problem SCF can be solved in time $O((2 + \epsilon)^k n^{O(1)})$.*

Moreover, the results in this section are correct for SCT problem.

2.3   A Cubic Kernel for the SCF Problem

In this subsection we introduce a cubic kernel for SCF problem. Suppose that stars in F_0 are ordered (by the degree ascending) and there are two adjacent stars S_v^q and $S_w^p(q < p)$ in F_0, where $p - q > k + 2$, k is the number of edges in F^* minus that of F_0. Since the degree of those stars are larger enough, then we can find the images of their centers in F^* directly. Thus, some leaf nodes of the stars that are not less than S_w^p are redundant, which implies we can remove them and their images. In this case, we perform the algorithm ReduceStars to reduce the scale of F_0 and F^*. The correctness of the algorithm ReduceStars is given by the following lemmas.

To simplify the discussion below, if an edge e has an end point of v, we simply say that e is an edge of v. For an isomorphism σ from F_0 to F^*, we say that an edge e of F^* is in σ if the edge e is in the image of σ.

Algorithm ReduceStars($F_0, F^*; S_w^p, q$)

Input: (F_0, F^*) is an instance of SCF, S_w^p is a star in F_0 and q is an integer;

1.   let F_0^p be the set of stars in F_0 that are not less than S_w^p;

2.   let V_s be the set of nodes in F^*, where the degree of each node in it is not less than p ;

3.   **if** $|F_0^p| \neq |V_s|$ **then** stop('F_0 is not a subgraph of F^{*}')

4.   **else**

4.1   arbitrarily remove $p - (q + k + 2)$ leaf nodes for each star in F_0^p;

4.2   arbitrarily remove $p - (q + k + 2)$ leaf nodes from $N(u)$ for each $u \in V_s$;

Fig. 5. An reduction rule of the instance (F_0, F^*)

Lemma 5. *Given an instance (F_0, F^*) of SCF problem, where F_0 is ordered by the degree ascending. Let S_v^q and S_w^p be two adjacent stars of F_0, where $p - q > k + 2$. Let F_0^p be the set of stars that are not less than S_w^p. Thus, there is a set of nodes V_s in F^* where the degree of each node in it is not less than p and $|V_s| = |F_0^p|$. Moreover, for each node in V_s, there are more than $p - q - (k + 2)$ neighbours of it are leaf nodes.*

Lemma 6. *Suppose that the algorithm ReduceStars($F_0, F^*; S_w^p, q$) returns an instance $(\tilde{F}_0, \tilde{F}^*)$, which are subgraphs of F_0 and F^*, respectively. Then F_0 is a subgraph of F^* if and only if $\tilde{F}_0$ is a subgraph of $\tilde{F}^*$.*

If the degree of the smallest star in F_0 is larger than $k+2$. By using the algorithm ReduceStars($F_0, F^*, S_w^p, 0$), we can reduce the instance (F_0, F^*), where S_w^p is the smallest star in F_0.

Algorithm SCF-kernel($F_0; F^*$)
Input: (F_0, F^*) is an instance of the SCF problem;
Output: An instance $(\tilde{F}_0, \tilde{F}^*)$;
1. Order the stars of F_0 by the degree ascending;
2. **if** the degree of the smallest star S_w^p in F_0 is larger than $k + 2$ **then**
    $(\tilde{F}_0, \tilde{F}^*) = $ KernelStars($F_0, F^*, S_w^p, 0$); return $SCF - kernel(\tilde{F}_0; \tilde{F}^*)$;
3. **While** there are two adjacent stars S_v^q and S_w^p that $p - q > k + 2$ **do**
    $(F_0, F^*) = $ KernelStars(F_0, F^*, S_w^p, q);
4. Return (F_0, F^*)

Fig. 6. A kernelization algorithm for SCF problem

Theorem 3. *The algorithm SCF-kernel(F_0, F^*) returns an instance $(\tilde{F}_0, \tilde{F}^*)$, then the scale of the new instance is yielded by $O(k^3)$, where k is the difference between the number of edges of F_0 and that of F^*.*

3   An $O(4^k)$ Algorithm for Path-Cover of a Tree

We study the PCT problem in this section. A *crossing* is a vertex in F^* that has degree larger than two. Let p be a simple path, $len(p)$ indicates the number of nodes in p. Let (F_0, T) be an instance of PCT problem, where F_0 consists of k_0 paths and the number of vertices in F_0 is equal to that of T. The question is to ask whether F_0 is a subgraph of T. Moreover, F_0 is a subgraph of T if and only if T can be converted into F_0 via a removal of edges. Similar with SCF problem, the discussion on PCT problem focuses at a general version, which is PCF problem, PATH-COVER OF A FOREST: Given two forests F_0 and F^* with the same number of vertices, where F_0 is a collection of paths. The parameter of PCF problem is $k = k_0 - k^*$, where k_0 and k^* are the number of edges in F_0 and F^*, respectively.

If there is a crossing in F^*, thus we use algorithm BranchPaths to deal with, which is given in Figure 7.

Let $(F_0, \tilde{F}^*)$ be the new instance returned by algorithm BranchPaths, such that F^* is a collection of paths. Then we have:

Path Reduction. Let $\tilde{p}$ be a path of $\tilde{F}^*$ and p be a path in F_0. If $len(\tilde{p}) = len(p)$, then remove $\tilde{p}$ and p from $\tilde{F}^*$ and F_0, respectively.

Lemma 7. *Suppose that Path Reduction outputs an instance $(\ddot{F}_0, \hat{F}^*)$, which are subgraphs of F_0 and $\tilde{F}^*$, respectively. Then F_0 is a subgraph of $\tilde{F}^*$ if and only if $\ddot{F}_0$ is a subgraph of $\hat{F}^*$.*

Algorithm BranchPaths(F_0, F^*)
Input: (F_0, F^*) is an instance of PCF;
1. let v be a crossing of F^*;
2. **if** v is not exit **then** return(F_0, F^*);
3. **for** all distinct combines $\{e_1, e_2\}$ of edges of v **do**
3.1   return BranchPaths($F_0, F^*/(E_v/\{e_1, e_2\})$);
\\ E_v is the set of edges of v

Fig. 7. A branching algorithm for PCF problem

By the Lemma 7, there is no path in $\ddot{F}_0$ that has the same length with any path in $\hat{F}^*$. Thus, each path in $\hat{F}^*$ is a image of more than one paths in $\ddot{F}_0$. Next we will show a dynamic programming algorithm on this instance, where one path is mapped to another one, then they are mapped from top to down continuously.

The Algorithm. Let $(\ddot{p}_1, \ldots, \ddot{p}_l)$ be $\ddot{F}_0$ and $(\hat{p}_1, \ldots, \hat{p}_m)$ be $\hat{F}^*$. Let $B_j^i \in S^i$ be a sub-set of $\ddot{F}_0$, where S^i is a collection of sub-sets that consist of i paths of $\ddot{F}_0$ and B_j^i is in S^i for $j \in [1, |S^i|]$. Let $T[B_j^i, x]$ be a mapping from B_j^i to $\{\hat{p}_1, \hat{p}_2, \ldots, \hat{p}_x\}$. $T[B_j^i, x]$ stores a value h that records the minimum edges needed to be removed. To motivate the algorithm, we initialize $T[\{\ddot{p}\}, 1]$ for each $\ddot{p} \in \ddot{F}_0$. If $len(\ddot{p}) < len(\hat{p}_1)$, then $T[\{\ddot{p}\}, 1] = 1$, otherwise, $T[\{\ddot{p}\}, 1] = \infty$. The dynamic programming equation is given in Equation 1, where $\hat{p}_{x+1}$ is a path of $\hat{F}^*$, and $\ddot{p}$ is a path of $\ddot{F}_0$.

$$T[B_j^{i+1}, x+1] = Min \begin{cases} T[B_j^{i+1}, x] + \hat{p}_{x+1}; \\ \min_{\ddot{p} \in B_j^{i+1}, B_k^i = B_j^{i+1}/\ddot{p}} \{T[B_k^i, x+1] + \ddot{p}\} \end{cases} \qquad (1)$$

Let f be an increment of $T[B_j^i, x]$ while a path of $\ddot{F}_0$ or $\hat{F}^*$ is added into $T[B_j^i, x]$. Since a path $\ddot{p}$ of $\ddot{F}_0$ cannot be mapped to two distinct paths in $\hat{F}^*$, then f is ∞ if $len(B_k^i \bigcup \ddot{p}) > \sum_{x=1}^{x=x} len(\hat{p}_x)$. Moreover, $\hat{p}_x$ is longer than $\ddot{p}$, if $\hat{p}_x$ has the image of $\ddot{p}$, thus in this case $f = 1$. Lemma 7 shows that $\hat{p}_x$ is equal to $\ddot{p}$, then $f = 0$. The minute detail of f is given in Equation 2, where $\hat{p}$ is a path of $\hat{F}^*$, and $\ddot{p}$ is a path of $\ddot{F}_0$.

$$\begin{cases} f(B_k^i, x, \ddot{p}) = \begin{cases} 1, & len(B_k^i \bigcup \ddot{p}) < \sum_{x=1}^{x=x} len(\hat{p}_x); \\ 0, & len(B_k^i \bigcup \ddot{p}) = \sum_{x=1}^{x=x} len(\hat{p}_x); \\ \infty, & \text{Other.} \end{cases} \\ f(B_j^i, x, \hat{p}) = \begin{cases} 0, & len(B_j^i) = \sum_{x=1}^{x=x} len(\hat{p}_x); \\ \infty, & \text{Other.} \end{cases} \end{cases} \qquad (2)$$

Algorithm PCF-Alg($F_0; F^*$)
Input: (F_0, F^*) is an instance of PCF problem;

1.   **for** each F_0, F^* returned by BranchPaths(F_0, F^*) **do**;
2.   call path reduction process;
3.   initialize $T[\{\ddot{p}\}, 1]$ for $\ddot{p} \in F_0$;
4.   **for** x range from $[2, |F^*|]$ **do**
         for i range from $[1, |F_0|]$ **do**
          for B_j^i in S^i **do**
           for $\ddot{p}$ in F_0/B_j^i **do**
            $T[B_j^i \bigcup \ddot{p}, x] = \min(T[B_j^i, x] + \ddot{p}, T[B_j^i \bigcup \ddot{p}, x])$;
            $T[B_j^i, x + 1] = T[B_j^i, x] + \hat{p}_{x+1}$;
5.   **if** $T[F_0, |F^*|] \neq \infty$ **then** return YES; **else** return NO;

Fig. 8. The full algorithm for PCF problem

The full algorithm of PCF problem is given in Figure 8, and the correctness of it is given by the following theorem.

Theorem 4. *The algorithm PCT-Alg(F_0, F^*) is running in $O(4^k n^{O(1)})$, where k is the difference between the number of connected components of F_0 and that of F^*.*

4   Conclusion

Subgraph isomorphism is one of the oldest problems in the world and the calculation of the problem is hard. We are interested in whether the problem is FPT by using the number of the connected components in F_0 as the parameter. Moreover, the problem is NP-hard even the forest consists of stars or paths. In this paper, we discuss on the two patterns, stars and paths, and prove they are all FPT. We give an $O(k^3)$ kernel and an $O((2 + \epsilon)^k n^{O(1)})$ algorithm for STAR-COVER OF A FOREST problem, an $O(4^k n^{O(1)})$ algorithm for PATH-COVER OF A FOREST problem. The further job for us is to work on a general model of isomorphism problem.

References

1. Amir, A., Keselman, D.: Maximum agreement subtree in a set of evolutionary trees: metrics and efficient algorithms. SIAM J. Comput. 26(6), 1656–1669 (1997)
2. Anderson, S.: Graphical representation of molecules and substructure-search queries in MACCS. Journal of Molecular Graphics 2, 8–90 (1984)
3. Baumbach, J., Guo, J., Ibragimov, R.: Covering Tree with Stars. In: Du, D.-Z., Zhang, G. (eds.) COCOON 2013. LNCS, vol. 7936, pp. 373–384. Springer, Heidelberg (2013)
4. Bille, P.: A survey on tree edit distance and reltaed problems. Theoret. Comput. Sci. 337, 217–239 (2005)

5. Chen, J., Kanj, I., Jia, W.: Vertex cover: further observations and further improvements. J. Algorithms 41(2), 280–301 (2001)
6. Chen, X., Liu, L., Liu, Z.: On the minimum common integer partition problem. ACM Trans. Algorithm 5(12), 1–18 (2008)
7. Conte, D., Foggia, P., Sansone, C., Vento, M.: Thirty years of graph matching in pattern recognition. International Journal of Pattern Recognition 18(3), 265–298 (2004)
8. Cormen, T., Leiserson, C., Rivest, R., Stein, C.: Introduction to Algorithms. McGraw-Hill, Boston (2001)
9. Garey, M.R., Johnson, D.S.: Computers and Intractability, A Guide to the Theory of NP-Completeness. W.H. Freeman, San Francisco (1979)
10. Hopcroft, J.E., Wong, J.K.: Linear time algorithm for isomorphism of planar graphs. In: Proc. 6th Annual ACM Symp. Theory of Computing (STOC 1974), pp. 172–184 (1974)
11. Kirkpatrick, B., Reshef, Y., Finucane, H., Jiang, H., Zhu, B., Karp, R.: Comparing pedigree graphs. Journal of Computational Biology 19(9), 998–1014 (2012)
12. Klein, P., Tirthapura, S., Sharvit, D., Kimia, B.: A tree-edit-distance algorithm for comparing simple, closed shapes. In: Proc. 11th Annual ACM-SIAM Symp. Discrete Algorithms (SODA 2000), pp. 696–704 (2000)
13. Pelillo, M., Siddiqi, K., Zucker, S.W.: Matching hierarchical structures using association graphs. IEEE Trans. Pattern Analysis and Machine Intelligence 21(11), 1105–1119 (1999)
14. Shi, F., Wang, J., Chen, J., Feng, Q., Guo, J.: Algorithms for parameterized maximum agreement forest problem on multiple trees. Theoret. Comput. Sci. (in press)
15. Stobaugh, R.E.: Chemical substructure searching. Journal of Chemical Information and Computer Sciences 25, 271–275 (1985)
16. Yang, H., Lee, L., Hsu, W.: Finding hot query patterns over an XQuery stream. The International Journal of Very large Data Bases 13(4), 318–332 (2004)
17. Yuster, R.: Combinatorial and computational aspects of graph packing and graph decomposition. Computer Science Review 1(1), 12–26 (2007)
18. Zhang, K., Shasha, D.: Simple fast algorithms for the editing distance between trees and related problems. SIAM J. Comput. 18, 1245–1262 (1989)

On the Complexity of Constrained Sequences Alignment Problems

Yong Zhang[1,2,*], Joseph Wun-Tat Chan[3], Francis Y.L. Chin[2,**],
Hing-Fung Ting[2,***], Deshi Ye[4,†], Feng Zhang[5,1], and Jianyu Shi[2,6,‡]

[1] College of Mathematics and Computer Science, Hebei University, China
[2] Department of Computer Science, The University of Hong Kong, Hong Kong
{yzhang,chin,hfting}@cs.hku.hk
[3] College of International Education, Hong Kong Baptist University, Hong Kong
cswtchan@gmail.com
[4] College of Computer Science, Zhejiang University, China
yedeshi@zju.edu.cn
[5] School of Mathematical Sciences, Universiti Sains Malaysia, Malaysia
amyfzhang@gmail.com
[6] School of Life Science, Northwestern Polytechnical University
jianyushi@nwpu.edu.cn

Abstract. We consider the problem of aligning a set of sequences subject to a given constrained sequence, which has applications in computational biology. In this paper we show that sequence alignment for two sequences A and B with a given distance function and a constrained sequence of k identical characters (say character c) can be solved in $O(\min\{kn^2, (t-k)n^2\})$ time, where n is the length of A and B, and t is the minimum number of occurrences of character c in A and B. We also prove that the problem of constrained center-star sequence alignment (CCSA) is NP-hard even over the binary alphabet. Furthermore, for some distance function, we show that no polynomial-time algorithm can approximate the CCSA within any constant ratio.

1 Introduction

Sequence alignment for DNA or protein sequence is one of the fundamental problems in computational biology or bioinformatics. It is primarily used to discover similarity in a set of biological sequences. By analyzing the alignment

* Research supported by NSFC 11171086 and Natural Science Foundation of Hebei Province A2013201218.
** Research supported by HK RGC grant HKU-711709E and Shenzhen basic research project (NO. JCYJ20120618143038947).
*** Research supported by HK RGC grant HKU-716412E.
† Research supported by NSFC 11071215.
‡ Research supported by Fundamental Research Foundation of Northwestern Polytechnical University in China (Grant No. JC201164) and China Postdoctoral Science Foundation (Grant No. 2012M521803).

J. Chen, J.E. Hopcroft, and J. Wang (Eds.): FAW 2014, LNCS 8497, pp. 309–319, 2014.

and the similarity of the biological sequences, one can infer similar functions or structures of these sequences.

Given two sequences, S_1 and S_2, the optimal sequence alignment with the minimum alignment score can be solved in $O(n^2)$ time, where n is the length of the longer sequence. When there are $m \geq 3$ sequences, the optimal sequence alignment problem which minimizes the sum of all-pairs scores becomes NP-hard [2]. There are a number of heuristics which approximate the optimal alignment, some with guaranteed worst case approximation ratio, and some with good performance in practice, e.g., BLAST[1], Clustal W[7]. One of the approximation algorithms in [5], based on the center-star alignment, can approximate the optimal alignment within a factor of $2 - 2/m$ in $O(mn^2)$ time, where m is the number of sequences.

The work by Tang et al. [8] is among the first to incorporate constraints into the sequence alignment problem to take into account biologists' information. Formally, the *constrained* sequence alignment problem has a constrained sequence as an additional input, and it is required that every character in the constrained sequence must appear in an entire column of the alignment and in the same order as the constrained sequence. In Tang et al.'s algorithm [8], both the time and space complexity are $O(kn^4)$, where k is the length of the constrained sequence. Chin et al. [3] improved the time and space complexities of the algorithm by Tang et al. from $O(kn^4)$ to $O(kn^2)$ for finding the optimal constrained sequence alignment of two sequences. In [3], an approximation algorithm, similar to the one for multiple sequences alignment, which also uses the optimal center-star alignment to approximate the optimal constrained alignment for $m \geq 3$ sequence, has an approximation ratio of no more than $2 - 2/m$. However, the proposed algorithm for finding the optimal constrained center-star sequence alignment (CCSA) for m sequences takes $O(Cmn^2)$ time, where C is the total number of occurrences of the constrained sequence in the m sequences. As C can be exponential, the time complexity for CCSA can be exponential.

In this paper we give an $O((t - k)n^2)$ algorithm to solve the constrained sequence alignment problem for two sequences when the constrained sequence consists of k characters and all the k characters are the same (say character c), and t is the minimum number of occurrences of character c in the two sequences. Combined with the previous result, this problem can be solved in $O(\min\{kn^2, (t - k)n^2\})$ time. Then, we show that the CCSA problem for multiple sequences is NP-complete even when the size of the alphabet set is two. We further show the inapproximability result of this CCSA problem for some distance function.

2   Preliminaries

Let Σ be the set of alphabets. For s, a sequence of n characters over Σ, $s[x..y]$ denotes the substring of s from the x-th character to the y-th character of s, where $1 \leq x < y \leq n$. In particular, let $s[x]$ denote the x-th character of s.

We define the *Pair-wise Sequence Alignment* (PSA) of two sequences s and t as two sequences s' and t' such that s' and t' have the same length n' and removing all space characters "−" from s' and t' gives s and t respectively. Let function α define the alignment positions, i.e., $\alpha(s, i)$ denotes the position of the character in s' corresponds to character $s[i]$. So $s'[\alpha(s, i)] = s[i]$. For a given distance function $\delta(x, y)$ which measures the mutation distance between two characters, where $x, y \in \Sigma \cup \{-\}$, the pair-wise score of two length n' sequences s' and t' is defined as $\sum_{1 \leq x \leq n'} \delta(s'[x], t'[x])$.

In the multiple sequence alignment (MSA) problem, we are given m sequences $S = \{s_1, s_2, ..., s_m\}$ with maximum length n. A MSA is an alignment matrix A, with m rows and $n'(\geq n)$ columns of characters over $\Sigma \cup \{-\}$, such that removing space characters from the i-th row of A gives s_i for $1 \leq i \leq m$. The sum-of-pairs (SP) score of an MSA A is defined to be the sum of the pair-wise scores of all pairs of the sequences, i.e.,

$$score(A) = \sum_{1 \leq i < j \leq m} \sum_{1 \leq p \leq n'} \delta(A_{i,p}, A_{j,p})$$

where $A_{i,p}$ and $A_{j,p}$ are the characters at the i- and j-th row and the p-th column of A, respectively. It was shown in [2,9] that finding an alignment matrix with the minimum sum-of-pair alignment score is NP-Hard.

In the constrained multiple sequence alignment problem (CMSA), we are given, in addition to the inputs of the MSA problem, a constrained sequence $C = c_1 c_2 ... c_k$, where C is a common subsequence of all $s_i \in \{s_1, s_2, ..., s_m\}$. The solution of a CMSA problem is a constrained alignment matrix A which is an alignment matrix such that each character in C appears in an entire column of A and also in the same order, i.e. there exists a list of integers $\{g_1, g_2, ..., g_k\}$ where $1 \leq g_1 < ... < g_k \leq n'$ and for all $1 \leq i \leq m$ and for all $1 \leq j \leq k$, we have $A_{i,g_j} = c_j$.

Let A be a CMSA matrix for $S = \{s_1, s_2, ..., s_m\}$ and the constrained sequence C. Define $score(A)$ to be the score of the CMSA matrix A. Let $A_{S,C}^*$ be the optimal CMSA matrix and $A_{S,C}$ be the CMSA matrix derived by an approximation algorithm. The approximation algorithm is said to have an approximation ratio r if and only if for any S and C,

$$\frac{score(A_{S,C})}{score(A_{S,C}^*)} \leq r.$$

In this paper we consider the problem of *constrained center-star sequence alignment (CCSA)*. This problem is similar to the CMSA problem since the goal is also to find an optimal alignment matrix with constrained sequence. However, CCSA has a different way of calculating the score of an alignment matrix. In CMSA, we minimize the sum of all pair-wise scores of the alignment matrix. In CCSA, we minimize, for $s \in S$, the sum of the pair-wise scores of s with every sequence in $S - \{s\}$. Precisely, the score of an alignment matrix A in CCSA is defined to be

$$\min_{1 \leq i \leq m} \sum_{1 \leq j \leq m, j \neq i} \sum_{1 \leq p \leq n'} \delta(A_{i,p}, A_{j,p}).$$

For an optimal alignment matrix A, we call a particular sequence s_ℓ the *center sequence* if the score of A is

$$\sum_{1 \le j \le m, j \ne \ell} \sum_{1 \le p \le n'} \delta(A_{\ell,p}, A_{j,p}).$$

3   A Special Case of the Constrained Pair-Wise Sequence Alignment Problem

In this section we give an algorithm to solved a special case of the constrained pair-wise sequence alignment (CPSA) problem. Given two sequences $A = a_1 a_2 \ldots a_n$ and $B = b_1 b_2 \ldots b_n$, a constrained sequence $C = c_1 c_2 \ldots c_k$, and a distance function δ, the CPSA problem is to find the minimum-score PSA, A' and B', such that A' and B' have the same length $n' \ge n$ and $A'[g_i] = B'[g_i] = c_i$ for some integers $1 \le g_1 < g_2 < \ldots < g_k \le n'$. The special case we consider is when $c_1 = c_2 = \ldots = c_k$, i.e., the k characters are all the same. Let that character be c and denote $C = c^k$.

First, we review the algorithm for finding the CPSA presented in [3]. Let $S(i, j, p)$ denote the optimal CPSA score for sequence $A[1..i]$ and $B[1..j]$ with the constrained sequence c^p and the distance function δ. The function $S(i, j, p)$ can be defined recursively as follows.

$$S(i, j, p) = \min \begin{cases} S(i-1, j-1, p-1) & \text{if } a_i = b_j = c \\ S(i-1, j-1, p) + \delta(a_i, b_j) & \text{if } i, j > 0 \\ S(i-1, j, p) + \delta(a_i, -) & \text{if } i > 0 \\ S(i, j-1, p) + \delta(-, b_j) & \text{if } j > 0 \end{cases}$$

For the boundary cases, $S(0, 0, 0) = 0$ and $S(0, 0, p) = \infty$ for $p > 0$.

This algorithm computes all the entries $S(i, j, p)$ in $O(kn^2)$ time because $i, j \le n$ and $p \le k$ and each entry can be computed in constant time. Without loss of generality, assume that the number of occurrences of character c in B is t and it is no more than the number of occurrences of c in A. Consider the case when k approaches t, the above algorithm takes $O(tn^2)$ time. In the following we give a more efficient algorithm when $k \to t$. In particular, the algorithm takes $O(n^2)$ time when $k = t$ and $O((t - k)n^2)$ time when $k \to t$.

Now we show how to find $S(m, n, k)$ in $O((t - k)n^2)$ time. Let $T(i, j, q)$ denote the optimal PSA score for sequences $A[1..i]$ and $B[1..j]$ and the distance function δ such that in the PSA, A' and B', there are exactly q occurrences of character c in B' matched with a character (in A') other than c. Precisely, there are exactly q integers $g_1, g_2, \ldots, g_q$ with $B'[g_i] = c$ and $A'[g_i] \ne c$. The function $T(i, j, q)$ can be defined recursively as follows.

$$T(i, j, q) = \min \begin{cases} T(i-1, j-1, q) + \delta(a_i, b_i) & \text{if } i, j > 0 \\ T(i-1, j, q) + \delta(a_i, -) & \text{if } i > 0 \\ T(i, j-1, q-1) + \delta(-, c) & \text{if } b_j = c \\ T(i, j-1, q) + \delta(-, b_j) & \text{if } b_j \ne c \end{cases}$$

In the recursive step, there are three choices for the optimal PSA to proceed. (Note that the third and fourth cases are mutually exclusive.) The optimal PSA could either (i) match a_i with b_j (ii) match a_i with space, or (iii) match b_j with space. Obviously, the optimal PSA should yield the minimum score among these three choices. For (i), the PSA would include the distance between a_i and b_j and proceed with shortened inputs $A[1..i-1]$ and $B[1..j-1]$, where exactly q occurrences of character c in $B[1..j-1]$ will match with a character other than c. For (ii), the PSA would include the distance between a_i and the space character "-". Since only a_i is matched but not b_j, the PSA proceeds with $A[1..i-1]$ and $B[1..j]$ and it still requires exactly q occurrences of character c in $B[1..j]$ matched with a character other than c. For (iii), the PSA would include the distance between b_j and the space character "-" and proceed with $A[1..i]$ and $B[1..j-1]$. However, it differs for the case where $b_j = c$ and $b_j \neq c$. If $b_j = c$, we have this character c matched with a character "-", which is not c. Thus the subsequent PSA requires only $q-1$ occurrences of character c in $B[1..j-1]$ matched with a character other than c. Otherwise, it still requires exactly q occurrences of character c in $B[1..j-1]$ matched with a character other than c.

Remind that in this section our target is to find the CPSA with constrained sequence c^k. In fact, it is equivalent find the minimum-score PSA among the PSA with exactly i occurrences of character c in B matched with a character other than c for $i = 0$ to $t - k$. Therefore, the required CPSA is the minimum-score PSA among those correspond to $T(m, n, i)$ for $0 \leq i \leq t - k$. In our algorithm we can compute the entries $T(i, j, q)$ for all $i, j \leq n$, $q \leq t - k$. Since each entry can be computed in constant time, the algorithm can find CPSA in $O((t-k)n^2)$ time.

Combining the algorithm of [4] with time complexity $O(kn^2)$ and our new algorithm with time complexity $O((t-k)n^2)$ if $k < t$ and $O(n^2)$ if $k = t$, we can compute the required CPSA efficiently as shown in the following theorem.

Theorem 1. *The CPSA problem for two sequences with length n, the constrained sequence of c^k, and a given distance function can be solved in $O(\min\{k, t-k\}n^2)$ if $k < t$ and in $O(n^2)$ if $k = t$ where t is the minimum number of occurrences of character c in these two sequences.*

4   NP-Hardness of CCSA

In this section we prove that CCSA is NP-Hard over binary alphabet, i.e., the size of alphabet set is 2. To prove this, we give a polynomial time reduction for the NP-hard problem *Maximum Independent Set (MIS)* [6] to CCSA. Consider the decision version of MIS as follows.

Given a graph $G = (V, E)$ and an integer k, is there a subset $V' \subseteq V$ of vertices for $|V'| = k$ such that each edge in E is incident on at most one vertex in V'?

For any instance of the decision version of MIS which includes a graph $G = (V, E)$ and an integer k, we give a polynomial-time transformation to an instance of CCSA which includes the alphabet set Σ, a set of sequences S over Σ, a constrained sequence C, and a distance function δ. We define the transformation as Π, i.e., $\Pi(G, k) = (\Sigma, S, C)$. Let $V = \{v_1, v_2, ..., v_n\}$ and $E = \{e_1, ...e_m\}$. The transformation $\Pi(G, k)$, constructed according to G and k, is shown as follows. We have $\Sigma = \{a, b\}$ and $S = \{t_1, ..., t_m, s_1, ..., s_m\}$, where $t_1 = ... = t_m = (aba)^n$ and each s_i is an encoding corresponding to an edge e_i. Suppose $e_i = (v_p, v_q)$ with $p < q$, then $s_i = (bab)^{p-1}(baa)(bab)^{q-p-1}(aab)(bab)^{n-q}$. The constrained sequence $c = b^k$. The distance function δ between any two characters in $\Sigma \cup \{\text{-}\}$ of the constructed CCSA instance is defined as follows.

	a	b	$-$
a	0	1	2
b	1	0	2
$-$	2	2	0

Consider $s_i = (bab)^{p-1}(baa)(bab)^{q-p-1}(aab)(bab)^{n-q}$ and $t_j = (aba)^n$ for $1 \leq i, j \leq m$ and $e_i = (v_p, v_q)$ where $p < q$. Two possible alignments of s_i and t_j are

$$s_i' = \text{-}(bab)^{p-1}(baa)(bab)^{q-p-1}(aab)(bab)^{n-q}$$
$$t_j' = (aba)^n\text{-} \tag{1}$$

and

$$s_i' = (bab)^{p-1}(baa)(bab)^{q-p-1}(aab)(bab)^{n-q}\text{-}$$
$$t_j' = \text{-}(aba)^n. \tag{2}$$

It can be seen that both alignments yield a score of $n + 3$. The following lemma shows that except the above two alignments all other alignments of s_i and t_i yield a score greater than $n + 3$.

Lemma 1. *For any $1 \leq i, j \leq m$, the alignments of s_i and t_j according to Alignments (1) and (2) yield a score of $n + 3$. All other alignments of s_i and t_j yield a score greater than $n + 3$.*

Proof. Suppose edge $e_i = (v_p, v_q)$, then $s_i = (bab)^{p-1}(baa)(bab)^{q-p-1}(aab)(bab)^{n-q}$. We also have $t_j = (aba)^n$.

We consider all possible alignments of s_i and t_j and divide them into 7 cases. These cases are characterized by how the sub-string "baa" in s_i "overlaps" with the p-th "aba" in t_j. The positions of the characters of sub-string "baa" in s_i (as well as the p-th "aba" in t_j) are $3p - 2, 3p - 1$ and $3p$, respectively. The following cases consider all possible relationships between the values of $\alpha(s, 3p - 2), \alpha(s, 3p - 1), \alpha(s, 3p), \alpha(t, 3p - 2), \alpha(t, 3p - 1)$. and $\alpha(t, 3p)$.

Case (1): $\alpha(t, 3p - 2) > \alpha(s, 3p)$. In this case the corresponding sub-string "baa" in s_i appears before the p-th "aba" in t_j in the alignment. We further consider the sub-case $\alpha(s, 3p) < \alpha(t, 3p - 2) \leq \alpha(s, 3p + 1)$. The other sub-case $\alpha(s, 3p + 1) < \alpha(t, 3p - 2)$ can be proved similarly. The alignment can be seen as breaking each of s_i and t_j into two parts and aligning corresponding parts of s_i and t_j as follows.

$$\frac{s_i}{t_j}\begin{Vmatrix}(bab)^{p-1}\mathbf{baa}\\(aba)^{p-1}\end{Vmatrix}\begin{matrix}(bab)^{q-p-1}(aab)(bab)^{n-q}\\\mathbf{aba}(aba)^{n-p}\end{matrix}$$

The first part of s_i, denote by s_i^1 has $p+1$ "a" and $2p-1$ "b", the first part of t_j, denote by t_j^1 has $2p-2$ "a" and $p-1$ "b". Even if all "a" in s_i^1 and all "b" in t_j^1 are matched, there are still $p-3$ "a" and p "b" unmatched. Therefore, the score of the first part of this alignment is at least $p+3$ because if all unmatched $p-3$ "a" align with $p-3$ "b" there are still 3 "b" that must align with space characters. For the same reason, the score of second part of this alignment is at least $n-p+4$. Hence, the total score of this pairwise alignment is at least $n+7$.

Case (2): $\alpha(s, 3p-1) < \alpha(t, 3p-2) \le \alpha(s, 3p)$. The alignment can be seen as breaking each of s_i and t_j into two parts and aligning corresponding parts of s_i and t_j as follows.

$$\frac{s_i}{t_j}\begin{Vmatrix}(bab)^{p-1}\mathbf{ba}\\(aba)^{p-1}\end{Vmatrix}\begin{matrix}a(bab)^{q-p-1}(aab)(bab)^{n-q}\\\mathbf{aba}(aba)^{n-p}\end{matrix}$$

Similarly to the counting in Case (1), we have the score of the first part of the alignment at least $p+2$ and the score of the second part of the alignment at least $n-p+2$. Hence, the total score of the alignment is at least $n+4$.

Case (3): $\alpha(s, 3p-2) < \alpha(t, 3p-2) \le \alpha(s, 3p-1)$. The alignment can be seen as breaking each of s_i and t_j into two parts and aligning corresponding parts of s_i and t_j as follows.

$$\frac{s_i}{t_j}\begin{Vmatrix}(bab)^{p-1}\mathbf{b}\\(aba)^{p-1}\end{Vmatrix}\begin{matrix}\mathbf{aa}(bab)^{q-p-1}(aab)(bab)^{n-q}\\\mathbf{aba}(aba)^{n-p}\end{matrix}$$

In the first part, the minimal alignment score is $p+1$, and this happens only when $s_i^{1\prime} = (bab)^{p-1}b$ and $t_j^{1\prime} = \text{-}(aba)^{p-1}$. Otherwise, the score is greater than $p+1$. For the second part, consider the sub-string of "aa", the $(n-p-1)$ sub-strings of "bab", and the sub-string of "aab", which form s_i^2. The score of each of "aa" and the $(n-p-1)$ "bab" is at least 1. Moreover, there must be at least one space character to be inserted to s_i^2 for the alignment with t_j^2. So the minimum score is $n-p+2$ and it happens only when $s_i^{2\prime} = aa(bab)^{q-p-1}(aab)(bab)^{n-q}$- and $t_j^{2\prime} = (aba)^{n-p+1}$. Thus the minimum score for aligning s_i and t_j in this case is $n+3$ and it happens only when $s_i' = (bab)^{p-1}(baa)(bab)^{q-p-1}(aab)(bab)^{n-q}$- and $t_j' = \text{-}(aba)^n$. Otherwise, the score is greater than $n+3$.

Case (4): $\alpha(t, 3p-2) \le \alpha(s, 3p-2)$ and $\alpha(t, 3p) \ge \alpha(s, 3p)$. We further consider the sub-case $\alpha(s, 3p-3) < \alpha(t, 3p-2) \le \alpha(s, 3p-2)$ and $\alpha(s, 3p) \le \alpha(t, 3p) < \alpha(s, 3p+1)$. The other sub-cases can be proved similarly. The alignment can be seen as breaking each of s_i and t_j into three parts and aligning corresponding parts of s_i and t_j as follows.

$$\frac{s_i}{t_j}\begin{Vmatrix}(bab)^{p-1}\\(aba)^{p-1}\end{Vmatrix}\begin{matrix}\mathbf{baa}\\\mathbf{aba}\end{matrix}\begin{Vmatrix}(bab)^{q-p-1}(aab)(bab)^{n-q}\\(aba)^{n-p}\end{Vmatrix}$$

The alignment score for the first part is at least $p+3$, for the second part is at least 2, and for the third part is at least $n-p+2$. Thus the alignment score of s_i and t_j is at least $n+7$.

Case (5): $\alpha(t, 3p-2) \leq \alpha(s, 3p-2)$ and $\alpha(s, 3p-1) \leq \alpha(t, 3p) \leq \alpha(s, 3p)$. The alignment can be seen as breaking each of s_i and t_j into two parts and aligning corresponding parts of s_i and t_j as follows.

$$\frac{s_i \| \quad (bab)^{p-1}\mathbf{ba} | \mathbf{a}(bab)^{q-p-1}(aab)(bab)^{n-q}}{t_j \| (aba)^{p-1}\mathbf{aba} | (aba)^{n-p}}$$

Similarly to that of Case (3), in the first part, the minimal alignment score is $p+1$, and this happens only when $s_i^{1'} = $ -$(bab)^{p-1}ba$ and $t_j^{1'} = (aba)^{p-1}$. Otherwise, the score is greater than $p+1$. For the second part, the minimum alignment score is $n-p+2$ and it happens only when $s_i^{2'} = a(bab)^{q-p-1}(aab)(bab)^{n-q}$ and $t_j^{2'} = (aba)^{n-p}-$. Thus the minimum score for aligning s_i and t_j in this case is $n+3$ and it happens only when $s_i' = $ -$(bab)^{p-1}(baa)(bab)^{q-p-1}(aab)(bab)^{n-q}$ and $t_j' = (aba)^n-$. Otherwise, the score is greater than $n+3$.

Case (6): $\alpha(t, 3p-2) \leq \alpha(s, 3p-2)$ and $\alpha(s, 3p-2) \leq \alpha(t, 3p) \leq \alpha(s, 3p-1)$. The alignment can be seen as breaking each of s_i and t_j into two parts and aligning corresponding parts of s_i and t_j as follows.

$$\frac{s_i \| \quad (bab)^{p-1}\mathbf{b} | \mathbf{aa}(bab)^{q-p-1}(aab)(bab)^{n-q}}{t_j \| (aba)^{p-1}\mathbf{aba} | (aba)^{n-p}}$$

Similar to that in Case (2), we can see that the total score of this alignment is at least $n+4$.

Case (7): $\alpha(t, 3p) < \alpha(s, 3p-2)$. The alignment can be seen as breaking each of s_i and t_j into two parts and aligning corresponding parts of s_i and t_j as follows.

$$\frac{s_i \| \quad (bab)^{p-1} | \mathbf{baa}(bab)^{q-p-1}(aab)(bab)^{n-q}}{t_j \| (aba)^{p-1}\mathbf{aba} | (aba)^{n-p}}$$

Similar to that in Case (1), we can see that the total score of this alignment is at least $n+7$. $\quad\square$

Corollary 1. *If the score for the constrained center-star sequence alignment is at most $m(n+3)$, the center sequence must be one of t_j for $1 \leq j \leq n$.*

Proof. By Lemma 1, the alignment score between s_i and t_j for any i and j is at least $n+3$. It is obvious that the alignment score between s_i and s_j for any $i \neq j$ is more than 0. Thus, if s_i for some i is the center sequence, the total alignment score must be greater than $m(n+3)$. Therefore, we can assume that the center sequence must be one of t_j for $1 \leq j \leq n$. $\quad\square$

Lemma 2. *There is an independent set for G of size k if and only if there is a constrained center sequence alignment for the transformed instance $\Pi(G, k)$ of CCSA with score $m(n+3)$.*

Proof. First, if G has an independent set Φ of size k, we give a constrained center sequence alignment with score $m(n+3)$. The alignment has t_1 (or any one of t_j for $1 \leq j \leq m$) as the center sequence. We have $t_j' = $ -$(aba)^n-$ for $1 \leq j \leq n$. For a sequence s_i corresponding to edge $e_i = (v_p, v_q)$ with $p < q$, we have s_i' as follows.

- If $v_p \in \Phi$, $s_i' = --s_i = --(bab)^{p-1}(baa)(bab)^{q-p-1}(aab)(bab)^{n-q}$,
- If $v_p \notin \Phi$, $s_i' = s_i-- = (bab)^{p-1}(baa)(bab)^{q-p-1}(aab)(bab)^{n-q}--$,

Recall that the constrained sequence is b^k. We prove that the constrain is satisfied in the alignment. Consider a vertex $v_x \in \Phi$. We can see that $t_j'[3x] = b$ for $1 \leq j \leq m$. For an edge $e_i = (v_x, v_y)$ incident to v_x and if $x < y$, then $s_i' == --(bab)^{x-1}(baa)(bab)^{y-x-1}(aab)(bab)^{n-y}$ and $s_i'[3x] = b$. If $y < x$, since $v_y \notin \Phi$, $s_i' == (bab)^{y-1}(baa)(bab)^{x-y-1}(aab)(bab)^{n-x}--$ and we also have $s_i'[3x] = b$. For an edge e_i not incident to v_x, we can also verify that $s_i'[3x] = b$. Therefore, the constrain is satisfied.

For the alignment score, by Lemma 1, we can see that the score of the alignment t_1' and s_i' for each $1 \leq i \leq m$ is $n+3$ and the score of the alignment t_1' and t_j' for each $1 \leq j \leq m$ is 0. Thus the total score is $m(n+3)$.

Second, if there is a constrained center sequence alignment with score $m(n+3)$, we prove that there is an independent set Φ for G of size k. We prove by contradiction. Assume that the maximum independent set of G is of size $k' < k$.

By Corollary 1, we can assume that t_1 is the center sequence. By Lemma 1, we can further assume that the pair-wise alignment score between t_1 and each of s_i for $1 \leq i \leq n$ is exactly $n+3$ and the alignment follows either Alignment (1) or (2). As the constrain b^k is satisfied, we assume the x-th "b" in b^k appear in column ℓ_x of the alignment matrix, in which the whole column should consist of "b" only. Suppose that column ℓ_x intersects with the $h(x)$-th "aba" of t_1, and $h(x)$-th "bab"/"baa"/"aab" of s_i for $1 \leq i \leq n$. We define a subset of vertex Φ that includes all $v_{h(x)}$ for $1 \leq x \leq k$.

We prove that Φ is an independent set of G. For any two vertices $v_{h(x)}$ and $v_{h(y)}$ in Φ for $x < y$, we claim that edge $e_i = (v_{h(x)}, v_{h(y)})$ does not exist. If not, there is a sequence $s_i = (bab)^{h(x)-1}(baa)(bab)^{h(y)-h(x)-1}(aab)(bab)^{n-h(y)}$. It can be verified that to follow Alignment (1) or (2), there is no way to have both the "b" in "baa" and the "b" in "aab" of s_i appear in column ℓ_x and column ℓ_y of the alignment matrix, respectively. Thus s_i, as well as e_i, does not exist. Therefore, Φ is an independent set of G and it is of size $k > k'$, which is a contradiction. □

By Lemma 2, we can reduce the NP-hard problem of MIS to CCSA with binary alphabet, and hence we have the following theorem.

Theorem 2. *CCSA (Constrained Center-Star Sequence Alignment) with binary alphabet is NP-Hard.*

5 Inapproximability of CCSA

In this section we show that, unless $P = NP$, no polynomial-time algorithm can solve the CCSA problem of arbitrary distance function within any constant approximation ratio $r > 0$. If there is such algorithm, we show that the MIS problem can be solved in polynomial time. In particular, we will focus on the CCSA problems where the distance function does not satisfy the triangle inequality.

We show a new reduction from MIS to CCSA which is similar to that in Section 4. The new transformation $\Pi'(G, k) = (\Sigma, S, c, \delta)$ is defined as follows. The alphabet set consists of one more character than before, i.e., $\Sigma = \{a, b, c\}$. The set of sequence $S = \{t_1, ..., t_m, s_1, ..., s_m\}$, where $t_1 = ... = t_m = c(aba)^n c$ and for each edge $e_i = (v_p, v_q)$ with $p < q$ $s_i = c(bab)^{p-1}(baa)(bab)^{q-p-1}(aab)(bab)^{n-q}c$. The constrained sequence is still $C = b^k$. The new distance function δ between any two characters in $\Sigma \cup \{-\}$ is defined as follows.

	a	b	c	$-$
a	0	1	2	B
b	1	0	2	B
c	2	2	0	0
$-$	B	B	0	0

where $B = r \cdot m(n + 3) + 1$. Note that the distance function may not satisfy the triangle inequality.

Similar to Lemma 2, we can show the new transformation Π' also yields a polynomial-time reduction from MIS to CCSA.

Lemma 3. *There is an independent set for G of size k if and only if there is a constrained center sequence alignment for the transformed instance $\Pi'(G, k)$ of CCSA with score $m(n + 3)$.*

In fact, for the transformed instance $\Pi'(G, k)$, one can obtain the optimal alignment if the guaranteed score is at most $(r \cdot m(n + 3))$.

Lemma 4. *If there is a constrained center sequence alignment for the transformed instance $\Pi'(G, k)$ of CCSA with score at most $(r \cdot m(n + 3))$, then this alignment has score exactly $m(n + 3)$.*

Proof. Consider the case that when there is a center constrained sequence alignment of $\Pi'(G, k)$ with score at most $(r \cdot m(n + 3))$. Since the distance between space character and a or b is greater than $r \cdot m(n + 3)$, the alignment between s_i and t_1 (ignoring the "c") for any i must follow Alignment (1) or (2). Hence, the score of each pair-wise alignment is exactly $n + 3$, and thus the total alignment score is $m(n + 3)$.   □

As a result, if there is an algorithm that gives a solution of $\Pi'(G, k)$ with approximation ratio r, then we can determine if an independent set for G of size k exists or not. If the algorithm gives an alignment of score at most $(r \cdot m(n + 3))$, then by Lemma 4 the optimal alignment has score $m(n + 3)$ and then by Lemma 3, there is an independent set for G of size k. If the algorithm gives an alignment of score greater than $(r \cdot m(n + 3))$, then the optimal alignment has score greater than $m(n + 3)$ and then by Lemma 3, there is no independent set for G of size k. Therefore, we have the following theorem for the inapproximability of CCSA.

Theorem 3. *Unless $P = NP$, there is no polynomial-time algorithm for CCSA with constant approximation ratio.*

6 Conclusion

We have studied the CPSA problem when the constrained sequence is a string of k identical characters and gave an $O(n^2)$ algorithm when $k = t$, where t is the minimum number of occurrences of that character in these two sequences, i.e, the largest number of that character in the constrained sequence. However, for some $k < t$ and problem instances, this CPSA problem might take $O(n^3)$ time. It is not sure whether this CPSA problem can be solved in strictly less than $O(n^3)$ time for all k. Since the solution for the center-star alignment problem can provide a good approximation for the MSA problem, the solution for the CCSA problem can also give a good approximation for the CMSA problem. Unfortunately, the CCSA problem has been proved to be NP-complete and equally difficult to find an approximate solution for the CCSA problem for some distance function. Thus, it remains open whether there exists a good approximation algorithm for the CMSA problem.

References

1. Altschul, S.F., Gish, W., Miller, W., Myers, E.W., Lipman, D.J.: Basic local alignment search tool. Journal of Molecular Biology 215(3), 403–410
2. Bonizzoni, P., Vedova, G.D.: The complexity of multiple sequence alignment with sp-score that is a metric. Theor. Comput. Sci. 259(1-2), 63–79 (2001)
3. Chin, F.Y.L., Ho, N.L., Lam, T.W., Wong, P.W.H.: Efficient constrained multiple sequence alignment with performance guarantee. Journal of Bioinformatics and Computational Biology 3(1), 1–18 (2005)
4. Chin, F.Y.L., Santis, A.D., Ferrara, A.L., Ho, N.L., Kim, S.K.: A simple algorithm for the constrained sequence problems. Information Processing Letters 90, 175–179 (2004)
5. Gusfield, D.: Efficient methods for multiple sequence alignment with guaranteed error bounds. Bulletin of Methematical Biology 55, 141–154 (1993)
6. Garey, M., Johnson, D.: Computers and Intractability: A guide to the theory of NP-completeness. W. H. Freeman and Company, San Francisco (1979)
7. Larkin, M.A., Blackshields, G., Brown, N.P., Chenna, R., McGettigan, P.A., McWilliam, H., Valentin, F., Wallace, I.M., Wilm, A., Lopez, R., Thompson, J.D., Gibson, T.J., Higgins, D.G.: ClustalW and ClustalX version 2. Bioinformatics 23(21), 2947–2948 (2007)
8. Tang, C.Y., Lu, C.L., Chang, M.D.-T., Tsai, Y.-T., Sun, Y.-J., Chao, K.-M., Chang, J.-M., Chiou, Y.-H., Wu, C.-M., Chang, H.-T., Chou, W.-I.: Constrained multiple sequence alignment tool development and its application to rnase family alignment. Journal of Bioinformatics and Computational Biology 1(2), 267–287 (2003)
9. Wang, L., Jiang, T.: On the complexity of multiple sequence alignment. Journal of Computational Biology 1(4), 337–348 (1994)

On the Advice Complexity
of One-Dimensional Online Bin Packing*

Xiaofan Zhao[1] and Hong Shen[2,3,**]

[1] School of Computer and Information Technology
Beijing Jiaotong University, Beijing, China
[2] School of Information Science and Technology
Sun Yat-sen University, China
[3] School of Computer Science
University of Adelaide, Australia
11112083@bjtu.edu.cn, hongsh01@gmail.com

Abstract. In this paper, we study the problem of the online bin packing with advice. Assume that there is an oracle with infinite computation power which can provide specific information with regard to each incoming item of the online bin packing problem. With this information, we want to pack the list L of items, one at a time, into a minimum number of bins. The bin capacity is 1 and all items have size no larger than 1. The total size of packed items in each bin cannot exceed the bin capacity. Inspired by Boyar et al's work [1] of competitive ratio $\frac{4}{3}$ [1] with two advice bits per item., we show that if the oracle provides three bits of advice per item, applying a different item classification scheme from Boyar et al's we can obtain an online algorithm with competitive ratio $\frac{5}{4}OPT + 2$ to pack list L.

Keywords: bin packing, online algorithm, computation with advice, competitive ratio.

1 Introduction

The classical bin packing problem has been studied extensively for more than thirty years. Hence this problem has numerous applications in the real word, such as loading trucks and shipments [2], memory allocation [3]. The online version of this problem can be defined as follow: Given a list L of items $l_1, l_2, \ldots$ arriving in an online manner, each of which has size $0 < s(l_i) \leq 1$, and an infinite number of unit-capacity bins of side length 1, we are asked to irrevocably pack the current item into a bin without any information on the next items. The total size of packed items in each bin cannot exceed the bin capacity. The goal is to use as few bins as possible.

---

* This work is supported by National Science Foundation of China under its General Projects funding #61170232, State Key Laboratory of Rail Traffic Control and Safety independent research grant RS2012K011, and Sun Yat-sen University 985 Project funding.
** Corresponding author.

J. Chen, J.E. Hopcroft, and J. Wang (Eds.): FAW 2014, LNCS 8497, pp. 320–329, 2014.

There is a rich literature on the classical online bin packing problem. As the first algorithm, Johnson [4] showed that the Next Fit algorithm has performance ratio 2 for one-dimensional online bin packing; Lee and Lee [5] proposed $Harmonic_M$ algorithm which is M-space bounded with an asymptotic performance ratio no less than 1.69103 when M goes to infinity. After that, many other Harmonic class algorithms have been devised. The asymptotic competitive ratio was improved from 1.63597 [5] to 1.61217 [6] and stopped at 1.58889 by Seiden [7] till now, whose algorithm named Super Harmonic. For the lower bound, Yao [8] first proved that no online bin packing algorithm can be better than $\frac{3}{2}$. New lower bounds then recurred continually from 1.53635 [9] to 1.54014 [10]. Very recently, for certain classes of bin packing algorithms, Balogh et al. gave a new lower bound of 1.54037 [11].

The offline version of this problem is NP-hard, so does its online variant [12]. Since it is impossible in general to produce the best solution when computation occurs online, we consider approximation algorithms. The standard measure of performance (output quality) of an online approximation algorithm is called asymptotic competitive ratio, which is the ratio of the online algorithm's performance versus that of the optimal offline algorithm on any given input when the problem size goes to infinity. For the bin packing problem, denote by $OPT(L)$ the performance of an optimal offline bin packing algorithm on a sequence of items L, by $A(L)$ the performance of an online approximation algorithm A, where performance refers to the number of used bins. The *asymptotic competitive ratio* of an online approximation algorithm A is defined to be

$$R_A^\infty = \lim_{n\to\infty} \sup \sup_L \left\{ \frac{A(L)}{OPT(L)} | OPT(L) = n \right\}.$$

Let $\mathcal{O}$ be a set of all on-line bin packing algorithms. The optimal *asymptotic competitive ratio* is defined as

$$R_{OPT}^\infty = \inf_{A\in\mathcal{O}} R_A^\infty.$$

The online algorithm packs items only depending on the previous incoming items, while the offline algorithm knows the information of all items in list L. This is obviously unfair. In a real application it is impossible to know nothing about the items needed to be packed. A piece of information about the incoming item may be given upon its arrival. So Dobrev et al. [13] first proposed a new way of characterizing the complexity of online problems with advice. Assume that there is an oracle with infinite computing power which can quickly have a look at the whole list L. The oracle provides some information accompanied with each arriving item. Combining these information and the information of the arrived items, the asymptotic competitive ratio of the online algorithm for bin packing can be improved.

Actually, there is a trade-off between competitive ratio and advice complexity. If we are given as much advice as possible, the optimal packing can be obtained. The cost is that we have to use more bits of advice and the advice complexity is higher. If the oracle provides only limited advice, the advice complexity could be decreased, while the asymptotic competitive ratio of the algorithm could be worse since we do not have enough auxiliary information to help us to do the packing.

In recent years, online algorithm with advice has been applied to many classical problems [14, 15, 1, 16–18]. Because of its great application value, there has been an increasing interest in studying the online problem with advice [19–21].

In this paper, we only concentrate on the online bin packing with advice. Boyar et al. [1] first consider this problem. They design an algorithm with $2n + o(n)$ bits of advice and achieve a competitive ratio of $\frac{4}{3} + \varepsilon$. The idea is to classify items into different classes depending on the final optimal packing and pack different classes of items using distinct algorithms. After that, Renault et al. [22] give an online algorithm with advice that is $(1 + \varepsilon)$-competitive and use $O(\frac{1}{\varepsilon} \log \frac{1}{\varepsilon})$ bits of advice per request, for $0 < \varepsilon < \frac{1}{2}$.

Based on the idea of paper [1], we design an algorithm with competitive ratio $\frac{5}{4}$ by providing three bits of advice per item. We divide the interval $(0, 1]$ into more sub intervals than that in paper [1]. According to these sub intervals, we classify the items by their sizes. The three bits of advice show the different packing strategy of each incoming items. Our method is simpler and easier to understand. Furthermore, we reduce the advice bits per item to two and adjust our algorithm to keep the competitive ratio $\frac{5}{4}$.

The remaining part is organized as follows: We first give the definition of online bin packing with advice and theories previous proved in other paper in section 2. In section 3, we modify the classification of items in paper [1] and design a roughly $\frac{5}{4}$-competitive ratio algorithm with three bits of advice per item. Section 4 concludes this paper and gives the future researching direction.

2   Preliminaries

In this part, we give the definition of online bin packing problem with advice in paper [1]. There are also two important results.

Definition 1. *[1] In the online bin packing problem with advice, the input is a sequence of items $\sigma = < x_1, ..., x_n >$, revealed to the algorithm in an online manner $(0 < x_i \leq 1)$. The goal is to pack these items in the minimum number of bins of unit size. At time step t, an online algorithm should pack item x_t into a bin. The decision of the algorithm to select the target bin is a function of $\phi, x_1, ..., x_{t-1}$, where ϕ is the content of the advice tape. An algorithm A with advice tape ϕ is c-competitive with advice complexity $s(n)$ if there exists a constant c_0 such that, for every n and for each input sequence σ of length at most n, there exists some ϕ such that $A(\sigma) \leq cOPT(\sigma) + c_0$, and at most the first $s(n)$ bits of the tape have been accessed by the algorithm. An algorithm A is optimal if $A(\sigma) = OPT(\sigma)$.*

Corollary 1. *[1] Consider the bin packing problem for packing sequences of length n. To achieve a competitive ratio of $\frac{5}{4} - \delta$, in which δ is a small, but fixed positive number, an online algorithm needs to receive $\Omega(n)$ bits of advice.*

3   A $\frac{5}{4}$-Competitive Algorithm with Three Bits of Advice per Item

Boyar's idea for optimal algorithm with linear advice is that to classify items not only by their sizes, but also by the packing patterns of the final solution. Motivated by this idea, we differentiate the used bins from each other depending on the items that each bin has contained. The difference between our algorithm and Boyar's algorithm is that the item classification of our algorithm is finer and the bin types are defined by a distinct sized item set. Different packing strategies are adopted for different classes of items. All of these can be achieved through the given advice string of each item. Since we have three bits of advice per item, there are eight types of advice strings in total. Table 1 shows the meaning of each type of advice string in our algorithm.

The detail meaning of each advice accompanied with each item is list in table 1. The first column "AS" gives the content of advice string. There are at most eight advice strings as we have three bits for each item. The second column "BT" gives the bin type which the item with certain advice string should be packed into. The last column shows the different meaning of each advice string represents. Clearly, different advice strings point out different packing patterns and different types of bin the item should be packed.

First, we give the classification of items in our algorithm. Roughly, items are classified into large, small and tiny. Items with size larger than or equal to $\frac{1}{5}$ belong to large items. Items with size smaller than $\frac{1}{5}$ are small. We also call small items with size smaller than $\frac{1}{20}$ tiny. Denote t_i to be the type of large items, the large items are classified into finer categories according to harmonic sequence as follows:

$t_i = (\frac{1}{i+1}, \frac{1}{i}]$, where $1 \leq i \leq 4$.

Furthermore, t_1 is divided into four subclasses:

$t_1^1 = (\frac{4}{5}, 1]$, $t_1^2 = (\frac{2}{3}, \frac{4}{5}]$, $t_1^3 = (1 - \triangle, \frac{2}{3}]$, $t_1^4 = (\frac{1}{2}, 1 - \triangle]$,

and t_2 is divided into two subclasses as follow:

$t_2^1 = (\frac{1}{3}, \triangle]$, $t_2^2 = (\triangle, \frac{1}{2}]$,

For the sake of simplicity, assume that $\triangle = \frac{2}{5}$.

Finally, we select five types to form three groups each of which consists of items from two different types. They are type t_4 and type t_1^3 make up group 1, type t_3 and type t_1^3 make up group 2, type t_2^1 and type t_1^4 make up group 3. These types of items will be packed two in one bin, both of which belong to the same group. The details will be discussed in subsection 3.2.

We give the following two definitions to determine the classification of the final packed bins.

Definition 2. *Given a set S of items, if the total size of items in S belongs to the interval $[\frac{1}{20}, \frac{1}{5}]$, we call this set S of items is a critical set.*

This sequence can be used to differentiate the bins in final packing. Furthermore, items contained in these bins are also classified. Apparently, all items in set S are small items.

Table 1. The meanings of advice strings with three bits per item

AS	BT	Meaning
000	critical	item should be packed into a critical bin
010		large item of group 1 and its matching companion has not come
011		large item of group 1 and its matching companion has come
100		large item of group 2 and its matching companion has not come
101	non-critical	large item of group 2 and its matching companion has come
110		large item of group 3 and its matching companion has not come
111		large item of group 3 and its matching companion has come
001		large item or tiny item which is packed by using harmonic

Definition 3. *Consider that our algorithm has packed all the items of the sequence L. Denote all the used bins to be $b_1, b_2, \ldots, b_i, \ldots$. If the set of items packed in b_i contain one or more critical sets, b_i is called a critical bin, otherwise b_i is called a non-critical bin.*

3.1 Outline of Our Algorithm

The main idea of our algorithm is as follow: For an incoming item l_i with advice string M, we first judge which types of bin l_i should be placed into. This depends on M. Then according to the size of l_i, we pack it into a corresponding bin by using the following algorithms. If l_i should be packed into a critical bin, we use algorithm 1 to pack it. If l_i should be packed into a non-critical bin, we use algorithm 2 to pack it. Algorithm 3 is the subroutines of algorithm 2.

The outline of algorithm online bin packing with advice is as follow: For an incoming item l_i with advice string M, if $M =$ "000", use critical algorithm to pack it; else, use non-critical algorithm to pack it.

Algorithm 1. Critical Algorithm for Critical Bin

---

1. If l_i is large, we pack it in the remaining space using Best Fit strategy. Keep the total size of large items do not exceed $\frac{19}{20}$. Updates the level of the bin to reflect the actual size of l_i.
2. If l_i is small, we pack it in the reserved space of bins which contain large items by using First Fit strategy. Update the occupancy of the current bin to reflect the total size of the packed items. Open a new bin for l_i if necessary.

3.2 Upper Bound and Its Analysis

Theorem 1. *Given each item three bits of advice, there exists an online algorithm A for bin packing problem with cost at most $\frac{5}{4}OPT(\sigma) + 2$.*

Proof. Consider that at the end of the packing process, each item of list L has been packed into either a critical bin or a non-critical bin. Our analysis is consist of two parts. One part is to analyze the least occupancy of all critical bins and

---

Algorithm 2. Non-Critical Algorithm for Non-Critical Bin

---

1. If M = "001" and l_i is large, pack it into the bin which hold items with same type of l_i by using harmonic algorithm; Open a new bin if necessary.
2. Else M = "001" and l_i is tiny, pack it into non-critical bin which do not hold large items with M = "001" by using First Fit strategy. Open a new bin if necessary.
3. Else $M \in (010, 011, 100, 101, 110, 111)$ and l_i is large, pack it into a bin by using matching algorithm.

Algorithm 3. Matching Algorithm in Non-Critical Bin

---

1. If M = "010", open a new bin and pack l_i into this bin;
2. If M = "011", find a bin which hold an item with advice string M = "010" and pack l_i into this bin. Close this bin.
3. If M = "100", open a new bin and pack l_i into this bin;
4. If M = "101", find a bin which hold an item with advice string M = "100" and pack l_i into this bin. Close this bin.
5. If M = "110", open a new bin and pack l_i into this bin;
6. If M = "111", find a bin which hold an item with advice string M = "110" and pack l_i into this bin. Close this bin.

---

the other part is to analyze the least occupancy of all non-critical bins. Then the upper bound of our algorithm can be obtained by the minimum occupancy of the two values.

First, we consider the packing for critical bins. From the definition, we know that each critical bin has one or more critical sets. Thus, we reserve free space for items in the critical set. Since the total size of items in one critical set belongs to interval $[\frac{1}{20}, \frac{1}{5}]$, at most four critical sets could co-exist with a large item. We reserve space for items which have a total size of at least $\frac{1}{20}$ and at most $\frac{1}{5}$ for one critical set. The expanding rate of reserved space of each critical set in each open bin should keep balance. Obviously, the least occupancy appears when the bin is only filled with four critical sets in each bin.

Lemma 1. *Except for the last bin, the utilization of each critical bin is at least $\frac{4}{5}$ according to our algorithm.*

Proof. Define bins that only contain small and tiny items to be red bins, otherwise blue bins. Two scenarios may happen when the packing of list L has been done.

Case 1: In the final packing, there is at least one red bin. In this case, the reserved space of each blue bin has all been filled up with small items and the blue bin cannot hold any more small items. Thus, we open new red bin to contain small items. Obviously, all red bins have an occupancy of at least $\frac{4}{5}$ except the last one. Now we consider the least occupancy of the blue bins. Once small items are packed into bin, the occupancy of the reserved space changes into the actually total amount of all small items plus the large items. Large items are packed by using Best Fit strategy until the total size of all large items in this bin exceed $\frac{19}{20}$ or no sufficient free space can hold the incoming small item. The later is the same as the red bins and more easier to analyze. The occupancy rate is obviously at least $\frac{4}{5}$ for each blue bin. For the former, if the total sizes of packed large items exceed $\frac{19}{20}$, we can move the last packed item to other bin by using Best Fit. Since the number of small items is more than that of large items, there is no need to open a new bin when free reserved space still exists. The left space caused by removing a large item can be fill up by a more suitable large item or some small items. The analysis of the later case is still the same as red bin. The former can be considered as the previous situation and processed alike until the free space is all filled up by small items. This repeated process lead to an utilization of at least $\frac{4}{5}$ in each bin.

Case 2: There is no red bin in the final packing. Assume that there are bins with an utilization less than $\frac{4}{5}$. We will prove that there is a contradiction. Let b_i be the first bin that has occupied space less than $\frac{4}{5}$. If b_i is the last bin, all bins except the last one has an occupied space no less than $\frac{4}{5}$. Thus Lemma 1 is true. Now, we consider that b_i is not the last one. Since all bins contain one or more critical set, we conclude the next bin b_{i+1} should contain at least one critical set of items with total size at most $\frac{1}{5}$. According to our packing strategy, using First Fit, small items of critical set in b_{i+1} should be packed into b_i if and only if b_i has a free space of at least $\frac{1}{5}$. We consecutively use First Fit to pack small items from the bins after b_i into b_i until it has an utilization at least $\frac{4}{5}$. At this time, if b_{i+1} becomes the first bin that has occupancy less than $\frac{4}{5}$, we can do the same thing like that with bin b_i. Repeat this until that we can get a bin b_j $(j > i)$ occupied less than $\frac{4}{5}$ and the bins after b_j have no critical sets. This is a contradiction since our assumption is that all bins are critical bins and have at least one critical set of items. If b_{i+1} has occupied space no less than $\frac{4}{5}$ after supplying the gaps in b_i, it is impossible for other bins b_k $(k > i + 1)$ to become a bin with less than $\frac{4}{5}$ occupied depending on First Fit. In this scenario, all bins include b_i are at least $\frac{4}{5}$ full.

Consequently, in both two cases, all critical bins except the last one have an occupancy at least $\frac{4}{5}$.

Now we consider the occupancy of each non-critical bin.

Since each non-critical bin does not contain critical set, the items packed in these bins are big items or very small items (tiny items). The total sizes of sub sequence of these items are either larger than $\frac{1}{5}$ or smaller than $\frac{1}{20}$. The packing of these two types of items are considered separately in our packing strategy.

We pack large items by using two different methods which called matching algorithm and harmonic algorithm. There are three groups of large items which are packed by matching algorithm. They are items of group 1 composed by type t_4 and type t_1^3, group 2 composed by type t_3 and type t_1^3, group 3 composed by type t_2^1 and type t_1^4. There are three types of items which are packed by harmonic algorithm. They are type t_4, type t_2^2 and type t_1^1. Obviously, items belong to the same type may be packed by distinct algorithm. The decision on which algorithm to choose depends on the bits of advice that each item has.

The matching algorithm packs exactly two large items together in one bin and is realized by the help of advice. We call these two large items packed in one bin the matching companion. Our advice tape is consist of three bits of advice per item. There are three groups of big items need to be packed. We use the first two bits of advice to identify the three groups. Let "01" denote group 1, "10" denote group 2 and "11" denote group 3. The last one bit of advice is used to judge if the matching companion of current item has already arrived or not. Let "1" denote the matching item has already come and "0" denote the matching item has not come yet. If the matching companion of the incoming item has already come, we pack the incoming item into one of the previous open bins by Best Fit. If not comes yet, we open a new bin to accommodate the current item. For example, the advice bit of an item is "011" means that the incoming item is big item and will be packed by matching algorithm. It belongs to group 1 and the matching companion has already come. We then pack this item into one of the previous open bins which also have advice "01" in the first two bits by using Best Fit.

Lemma 2. *The occupancy of each non-critical bin which is filled by matching algorithm is at least $\frac{4}{5}$.*

Proof. For group 1, one item has size no less than $\frac{1}{5}$ and the other item has size no less than $\frac{3}{5}$. They are packed together in one bin. Obviously, the total occupied area of the bin is no less than $\frac{4}{5}$. Similarly, the total occupied area of bins which belong to group 2 and group 3 is no less than $\frac{17}{20}$ and $\frac{5}{6}$ respectively. The occupancy are all larger than $\frac{4}{5}$.

Lemma 3. *The occupancy of each non-critical bin which is filled up by harmonic algorithm is at least $\frac{4}{5}$.*

Proof. From the proof of matching algorithm, the first two bits of advice "00" has not been used. We sign the advice bits "001" to items which packed by using harmonic algorithm. These items are type t_4, type t_2^2 and type t_1^1. Using harmonic algorithm, we pack four type t_4 items, two type t_2^2 items and one type t_1^1 item respectively in each non-critical bin and the occupancy is at least $\frac{4}{5}$.

Now we consider the tiny items packed into non-critical bins. These items also have an advice bits of "001". If all tiny items are packed alone in one or more bins, the critical sets may exist in these bins which contradicts our assumption. So we just pack these items in the bins which have already contained big items. That means these tiny items are scatted packed into each opened non-critical bins by using First Fit. The limitation is that the total size of tiny items in each bin can not exceed $\frac{1}{20}$, otherwise we open a new bin to accommodate tiny items only. Noticing that there is at most one bin which only contains tiny items. Since each non-critical bin all have an occupancy larger than $\frac{4}{5}$ by Lemma 2 and Lemma 3, the following Lemma 4 is true:

Lemma 4. *For the packing of non-critical bins, except the last bin, the utilization of each bin is at least $\frac{4}{5}$ according to our packing strategy.*

The three bits of advice are used as in the proof of Lemma 2 and Lemma 3. There is still one type of advice bit "000" which is not mentioned. We use "000" to denote items which should be packed into critical bins. Thus all items can be identified by their advice and packed by different strategies. Let the amount of items packed into critical bins and non-critical bins is c_1 and c_2, respectively. The cost of an optimal packing of these sequence σ is $OPT(\sigma) = c_1 + c_2$. By using Lemma 1 and Lemma 4, the cost of our algorithm A on sequence σ is $A(\sigma) \le \frac{5}{4}c_1 + 1 + \frac{5}{4}c_2 + 1 = \frac{5}{4}OPT(\sigma) + 2$.

4 Conclusion

In this paper, we presented a modified algorithm for online bin packing problem with advice that improves the known competitive ratio $\frac{4}{3}$ [1] to $\frac{5}{4}$. Through integration, we successfully refined our algorithm to maintain the $\frac{5}{4}$-competitive ratio but only use three bits of advice per item. Our improvment is achieved by applying a refined scheme to partition interval $(0, 1]$ by further dissecting the large items into more different types so as to improve the packing ratio. In constrast, Boyar's method partitions the interval $(0, 1]$ and packs items with different advice by using different algorithms.

Next we will extend our method to multidimensional bin packing problem with advice. We will also explore how to extend this bin packing with advice algorithm to other bin packing with relaxations problems.

References

1. Boyar, J., Kamali, S., Larsen, K.S., López-Ortiz, A.: Online bin packing with advice. CoRR, abs/1212.4016 (2012)
2. Xing, W.: A bin packing problem with over-sized items. Operations Research Letters 30(2), 83–88 (2002)
3. Garey, M.R., Graham, R.L., Ullman, J.D.: Worst-case analysis of memory allocation algorithms. In: STOC, pp. 143–150 (1972)
4. Johnson, D.S.: Fast algorithms for bin packing. J. Comput. Syst. Sci. 8, 272–314 (1974)
5. Lee, C.C., Lee, D.T.: A simple on-line bin-packing algorithm. J. ACM 32(3), 562–572 (1985)
6. Ramanan, P., Brown, D.J., Lee, C.C., Lee, D.T.: On-line bin packing in linear time. J. Algorithms 10(3), 305–326 (1989)
7. Seiden, S.S.: On the online bin packing problem. J. ACM 49(5), 640–671 (2002)
8. Yao, A.C.-C.: New algorithms for bin packing. J. ACM 27, 207–227 (1980)
9. Liang, F.M.: A lower bound for on-line bin packing. Inf. Process. Lett. 10(2), 76–79 (1980)
10. van Vliet, A.: An improved lower bound for on-line bin packing algorithms. Inf. Process. Lett. 43, 277–284 (1992)
11. Balogh, J., Békési, J., Galambos, G.: New lower bounds for certain classes of bin packing algorithms. Theor. Comput. Sci. 440-441, 1–13 (2012)
12. Garey, M.R., Johnson, D.S.: Computers and Intractability; A Guide to the Theory of NP-Completeness. W. H. Freeman & Co., New York (1990)

13. Dobrev, S., Královič, R., Pardubská, D.: How much information about the future is needed? In: Geffert, V., Karhumäki, J., Bertoni, A., Preneel, B., Návrat, P., Bieliková, M. (eds.) SOFSEM 2008. LNCS, vol. 4910, pp. 247–258. Springer, Heidelberg (2008)

14. Böckenhauer, H.-J., Komm, D., Královič, R., Královič, R.: On the advice complexity of the k-server problem. In: Aceto, L., Henzinger, M., Sgall, J. (eds.) ICALP 2011, Part I. LNCS, vol. 6755, pp. 207–218. Springer, Heidelberg (2011)

15. Böckenhauer, H.-J., Komm, D., Královič, R., Rossmanith, P.: On the advice complexity of the knapsack problem. In: Fernández-Baca, D. (ed.) LATIN 2012. LNCS, vol. 7256, pp. 61–72. Springer, Heidelberg (2012)

16. Dobrev, S., Královič, R., Markou, E.: Online graph exploration with advice. In: Even, G., Halldórsson, M.M. (eds.) SIROCCO 2012. LNCS, vol. 7355, pp. 267–278. Springer, Heidelberg (2012)

17. Forišek, M., Keller, L., Steinová, M.: Advice complexity of online coloring for paths. In: Dediu, A.-H., Martín-Vide, C. (eds.) LATA 2012. LNCS, vol. 7183, pp. 228–239. Springer, Heidelberg (2012)

18. Renault, M.P., Rosén, A.: On online algorithms with advice for the k-server problem. In: Solis-Oba, R., Persiano, G. (eds.) WAOA 2011. LNCS, vol. 7164, pp. 198–210. Springer, Heidelberg (2012)

19. Böckenhauer, H.-J., Komm, D., Královič, R., Královič, R., Mömke, T.: On the advice complexity of online problems. In: Dong, Y., Du, D.-Z., Ibarra, O. (eds.) ISAAC 2009. LNCS, vol. 5878, pp. 331–340. Springer, Heidelberg (2009)

20. Emek, Y., Fraigniaud, P., Korman, A., Rosén, A.: Online computation with advice. In: Albers, S., Marchetti-Spaccamela, A., Matias, Y., Nikoletseas, S., Thomas, W. (eds.) ICALP 2009, Part I. LNCS, vol. 5555, pp. 427–438. Springer, Heidelberg (2009)

21. Hromkovič, J., Královič, R., Královič, R.: Information complexity of online problems. In: Hliněný, P., Kučera, A. (eds.) MFCS 2010. LNCS, vol. 6281, pp. 24–36. Springer, Heidelberg (2010)

22. Renault, M.P., Rosn, A., van Stee, R.: Online algorithms with advice for bin packing and scheduling problems. CoRR, vol. abs/1311.7589 (2013)

Dynamic Matchings
in Left Weighted Convex Bipartite Graphs

Quan Zu, Miaomiao Zhang, and Bin Yu

School of Software Engineering, Tongji University, Shanghai, China
{7quanzu,miaomiao,0yubin}@tongji.edu.cn

Abstract. We consider the problem which is dynamically maintaining
a maximum weight matching in a left weighted convex bipartite graph
$G = (V, E)$, $V = X \cup Y$, in which each $x \in X$ has an associated weight,
and neighbors of each $x \in X$ form an interval in the ordered Y set. The
maintenance includes update operations (vertices and edges insertions
and deletions) and query operations (inquiries of a vertex matching information). We reduce this problem to the corresponding unweighted
problem and design an algorithm that maintains the update operations
in $O(\log^3 |V|)$ amortized time per update. In addition, we develop a
data structure to obtain the matching status of a vertex (whether it is
matched) in constant worst-case time, and find the pair of a matched
vertex (with which it is matched) in worst-case $O(k)$ time, where k is
not greater than the cardinality of the maximum weight matching.

Keywords: Dynamic Matching, Weighted Convex Bipartite Graph,
Matroid, BST, Implicit Representation.

1 Introduction

Let $G = (V, E), V = X \cup Y$, be a bipartite graph in which no two vertices in
X (or Y) are adjacent. A bipartite graph is a convex bipartite graph (CBG) if
neighbors of each $x \in X$ form an interval in Y, provided Y is totally ordered.
A left weighted convex bipartite graph (LWCBG) is a CBG in which each left
vertex $x \in X$ has an associated weight and the weight of each edge is equal to
that of its left endpoint in X.

A subset M of E is a *matching* if no two edges are adjacent in M. A *maximum
cardinality matching* (MCM) is a matching that contains the maximum number
of edges. The weight of a matching is the sum of the weights of its edges. A
maximum weight matching (MWM) is a matching that contains the maximum
weight. The MWM problem in an LWCBG naturally corresponds to the classical
maximum scheduling problem, in which the input is a collection of unit-time jobs,
each with an associated release time, deadline and weight, and the objective is
to schedule a maximum weight subset of the jobs.

The MCM problem in a CBG has been extensively researched. Glover [7]
proposed an algorithm to produce the greedy matching by iterating each $y \in Y$
according to the total order and matching y to the unmatched x of the earliest

J. Chen, J.E. Hopcroft, and J. Wang (Eds.): FAW 2014, LNCS 8497, pp. 330–342, 2014.
© Springer International Publishing Switzerland 2014

deadline in the set $\{x \in X : (x, y) \in E\}$. The implementation of *Glover's algorithm* runs in $O(|V| \log |V|)$ time with a priority queue data structure. Later, Lipski and Preparata [9], as suggested by Hopcroft, transformed this problem to the off-line minimum problem [1] which can be solved in nearly linear time by using the Union-Find data structure [15]. The running time of the off-line minimum problem and thus the matching problem were reduced to $O(|X| + |Y|)$ time [5]. Finally, Steiner and Yeomans [14] found an $O(|X|)$ time algorithm by separating the graph into distinct components.

The MWM problem in a CBG has serval variants in terms of the definition of the edge weight. In a right weighted CBG (each $y \in Y$ associated with a weight), Katriel [8] obtained an $O(|E| + |V| \log |V|)$ time algorithm to find an MWM. To compute an MWM in an LWCBG, by exploring the matroid framework, Lipski and Preparata [9] proposed an $O(|X|^2 + |X||Y|)$ time solution, and Plaxton [10] developed an algorithm running in $O(|X| + k \log^2 k)$ amortized time ($k \leq \min\{|X|, |Y|\}$) using a hierarchical data structure. Recently, Plaxton [11] pointed out the MWM problem in a general weighted CBG (each $v \in V$ associated with a weight) can be solved in $O(|V| \log^2 |V|)$ time, and improved the time bound of the problem in an LWCBG to $O(|V| \log |V|)$ time.

In this paper, we consider the problem of dynamically maintaining a maximum weight matching in an LWCBG. The maintenance includes update operations of insertions and deletions of vertices and edges, and query operations of finding whether a vertex is matched and which a matched vertex is paired with. We observe that this problem of an LWCBG can be reduced to the corresponding problem of an unweighted CBG.

Brodal et. al. [3] pointed out that the dynamic matching in a CBG must be hard in the sense that a single insertion or deletion of a vertex may change all the previous matchings. However, they also observed that the number of element change is only one in the vertex sets corresponding to the matchings. From these observations, they developed an algorithm to solve the dynamic MCM problem in a CBG, based on the binary computation tree data structure devised by Dekel and Sahni [4] for parallel computing. The algorithm maintains the MCM in $O(\log^2 |V|)$ amortized time per update, reports the status of vertex (matched or unmatched) in constant worst-case time, and reports the mate of a matched vertex in $O(\min\{k \log^2 |X| + \log |X|, |X| \log |X|\})$ worst-case time, where $k < \min\{|X|, |Y|\}$, or $O(\sqrt{|X|} \log^2 |X|)$ amortized time.

The matching instance of an LWCBG forms a matroid [9][10] in which an *independent* set is the left vertex set participating in a matching. In section 3, it is presented that in terms of the weight there is an *optimal* independent set corresponding to an MWM. We design an algorithm to maintain the optimal independent set in the update operations. The maintenance includes finding the vertex of the minimum (maximum) weight in the *replaceable* (*compensable*) set (i.e., the set of left vertices covered by the alternating path). The replaceable and compensable sets can be computed by the modification of an unweighted dynamic algorithm, e.g., Brodal et. al.'s algorithm. By this means, we reduce the weighted dynamic matching problem to the corresponding unweighted problem and the update operations can be maintained in $O(\log^3 |V|)$ amortized time per update. Furthermore, we find out that the MWM M computed from the independent set can be partitioned into components such that in each component the right vertices covered by M form an interval. From this observation, in section 4, we develop a data structure to obtain the matching status of a vertex (whether it is matched) in constant worst-case time, and find the pair of a matched vertex (with which it is matched) in worst-case $O(k)$ time, where k is not greater than the cardinality of the maximum weight matching. Section 5 concludes our work by offering some remarks.

2   Preliminaries

In this section, we specify the terminologies of a left weighted convex bipartite graph, which will be used throughout the remainder of the paper. We also define the sequence of the update and query operations supported in our algorithm.

2.1   Terminologies and Definitions

In an LWCBG $G = (V, E)$, $V = X \cup Y$, we use X and Y to denote the left and right vertex set, respectively. Y is totally ordered and each $y \in Y$ is represented by a distinct integer. Each $x \in X$ has a start point $\mathtt{BEG}[x]$ and an end point $\mathtt{END}[x]$, and for all $y \in [\mathtt{BEG}[x], \mathtt{END}[x]], (x, y) \in E$. We assume that there is no isolated x since it is easy to be verified and removed in a preprocessing. Each x is also associated with a distinct integer index $\mathtt{ID}[x]$ and a nonnegative weight $w(x)$. We define a total order over the set of X. We say x has higher priority than x' if $w(x) > w(x')$ or $w(x) = w(x')$ && $\mathtt{ID}[x] > \mathtt{ID}[x']$, which called *weight-id* total order. We primarily use this order in this paper and adopt this convention of the comparison for x unless stated otherwise.

We use M to denote an MWM of G and use Z to denote the set of left vertices that are covered by M, i.e., $Z = \{x \in X : \exists y, (x, y) \in M\}$. We say Z corresponds to M. Z is also identified as *independent* in the matroid framework. It is the fact that any instance of matching in an LWCBG forms a matroid $\mathcal{M} = (S, \mathcal{I})$, where the ground set S is X and $\mathcal{I}$ is the set of the independent subset Z' of X, i.e., there is a matching M' corresponding to Z'. We use $w(M)$ and $w(Z)$ to denote the weight of M and that of Z, respectively. By definition of LWCBG, $w(M) = w(Z)$. It is straightforward that the independent set of an MWM has the maximum weight among all the independent sets.

An MWM of an LWCBG is also an MCM if every edge has a nonnegative weight, which is proved in [11][13]. So in our work, we maintain the MWM in an LWCBG with the maximum cardinality.

Here, we define the maintenance of an MWM in an LWCBG which consists of a sequence of update operations and query operations. The update operations include:

- $\mathtt{Insert}(x)$: a vertex $x \notin X$ with $(\mathtt{ID}[x], \mathtt{BEG}[x], \mathtt{END}[x], w(x))$ is inserted in X, i.e., $X = X \cup x$; for all $y \in [\mathtt{BEG}[x], \mathtt{END}[x]], E = E \cup (x, y)$;

- `Delete(x)`: a vertex $x \in X$ is deleted from X, i.e., $X = X - x$; all edges adjacent with x are deleted;
- `Insert(y)`: a vertex $y \notin Y$ is inserted in Y, i.e., $Y = Y \cup y$; for all x such that $\text{BEG}[x] < y < \text{END}[x]$, $E = E \cup (x, y)$;
- `Delete(y)`: a vertex $y \in Y$, which is not equal to $\text{BEG}[x]$ or $\text{END}[x]$ of some x, is deleted from Y, i.e., $Y = Y - y$; all edges adjacent with y are deleted;

The arbitrary edge updates may break the convex property and the update operations only support the restricted edge updates. The updates of edges are implicitly operated by the updates of vertices. Inserting (x, y) is implemented by deleting x and then inserting x with the updated start or end point. Deleting an edge is similar. The query operations include:

- `Status Query`: obtain the status of whether an x or y is matched or not;
- `Pair Query`: given a matched x or y, find its mate in the MWM.

3 Dynamic Updates in Left Weighted CBG

In this section, we present our method of dynamically maintaining the MWM in an LWCBG under update operations, and the implementation of this method based on Brodal et al.'s algorithm.

3.1 Optimal Independent Set

In the update operations, we maintain the MWM M and the corresponding independent set Z that can be obtained by the following *greedy algorithm*: (i) sort the vertices in X according to the decreasing weight-id ordering, (ii) starting from $M = \emptyset$ and $Z = \emptyset$, scan X in this order; for any $x \in X$, if $Z + x$ is independent, i.e., M can be augmented by adding x, x is matched and M and Z are updated, or otherwise x is unmatched. We say the set Z selected by the greedy algorithm is the *optimal* independent set, OIS in short, and the following lemma can be implied from *Gale-optimal*([6], see also [9]).

Lemma 1. *(Gale-optimal): If $Z = \{z_1, \ldots, z_k\}$, $z_1 \geq \ldots \geq z_k$, is the optimal independent set, for any other independent set $Z' = \{z'_1, \ldots, z'_l\}$, $z'_1 \geq \ldots \geq z'_l$, it satisfies the condition $l \leq k$, $z_1 \geq z'_1, \ldots, z_l \geq z'_l$.*

To improve the reading, we let independent sets be sorted according to the decreasing weight-id order in the following of this section, unless stated otherwise. The below lemma follows directly from lemma 1.

Lemma 2. *Let $Z[i]$ denote the ith vertex in Z. There is no any other independent set Z' in which, $\exists j \in [1, |Z'|]$, $Z'[j] > Z[j]$.*

The optimal independent set corresponds to the MWM, and we maintain the invariant of the OIS in our dynamic update operations. Suppose at some state, we have the optimal independent set Z. In the insertion operations, when a new left vertex x is inserted, if Z can be augmented, i.e., $Z + x$ is independent, $Z = Z + x$ and Z is optimal in $X + x$. Otherwise, we may first locate the minimum vertex r in the *replaceable* set $R(x) \subseteq Z$, i.e., there is an alternating path from x to each $x' \in R(x)$, or $R(x) = \{x' \in Z : Z - x' + x \in \mathcal{I} \text{ and } S = X + x\}$ from matroid perspective. If $R(x) \neq \emptyset$ and $r < x$, the vertex r is replaced by x in Z, and the resulting independent set is optimal; if $R(x) = \emptyset$ or $r > x$, nothing changes. We have the following theorem for the dynamically insertion.

Theorem 1. *Let $r \in R(x)$ be the minimum one. If $Z' = Z + x$ is independent, Z' is the optimal independent set in $X + x$; otherwise, if $x > r$, $Z'' = Z + x - r$ is the optimal independent set in $X + x$, and if $x < r$, Z is still the optimal one.*

Proof. We first consider $Z' = Z + x$ is independent. Suppose Z' is not optimal and let $\hat{Z}$ be the optimal one, $\hat{Z} \neq Z'$. If $x \notin \hat{Z}$, it is impossible since then $\hat{Z}$ is an independent set in X with $|\hat{Z}| > |Z|$. If $x \in \hat{Z}$, we show that there is a contradiction. Let x be the ith element in Z', i.e., $x = Z'[i]$. Let $x = \hat{Z}[j]$. Let $Z[i,j]$ denote the interval in Z from i to j. If $i > j$, $Z'[i] = \hat{Z}[j] > \hat{Z}[i]$, a contradiction, since by lemma 2, $\hat{Z}[i] \geq Z'[i]$ for $\hat{Z}$ is optimal. Then we have $Z'[1, i-1] = \hat{Z}[1, i-1]$, since, by Lemma 1, $\hat{z}_1 \geq z'_1, \ldots, \hat{z}_{i-1} \geq z'_{i-1}$ for $\hat{Z}$ is optimal in $X + x$, and $z'_1 \geq \hat{z}_1, \ldots, z'_{i-1} \geq \hat{z}_{i-1}$ since $Z' - x = Z$ is optimal in X and $\hat{Z}[1, i-1]$ is an independent set in X. If $i < j$, i.e., $Z'[i] \neq \hat{Z}[i]$, then, by Lemma 2, $\hat{Z}[i] > Z'[i] = x > Z'[i+1] = Z[i]$. But since $\hat{Z}[1, i]$ is an independent set in X, by Lemma 2, $Z[i] \geq \hat{Z}[i]$, a contradiction. Thus $i = j$ and $Z[i] = \hat{Z}[i] = x$. Similarly as above, $Z'[i+1, |Z'|] = \hat{Z}[i+1, |Z'|]$, implied from, by Lemma 1, $\hat{z}_{i+1} \geq z'_{i+1}, \ldots, \hat{z}_{|\hat{Z}|} \geq z'_{|Z'|}$ since $\hat{Z}$ is optimal in $X + x$, and $z'_{i+1} \geq \hat{z}_{i+1}, \ldots, z'_{|Z'|} \geq \hat{z}_{|\hat{Z}|}$ since $Z' - x = Z$ is optimal in X and $\hat{Z} - x$ is an independent set in X. Hence, it is implied that $Z' = \hat{Z}$, a contradiction with our assumption that $\hat{Z}$ is optimal and $\hat{Z} \neq Z'$.

Then we consider $Z' = Z + x$ is not independent.

- $R(x) = \emptyset$ or $r > x$. We prove Z is optimal in $X + x$. Suppose Z is not and let $\hat{Z}$ be the optimal one in $X + x$, $\hat{Z} \neq Z$. If $x \notin \hat{Z}$, it is straightforward that $\hat{Z}$ is also optimal in X, a contradiction. If $x \in \hat{Z}$, we will show that it is also a contradiction. Let $x = \hat{Z}[i]$. We have $\hat{Z}[1, i-1] = Z[1, i-1]$, implied from, by Lemma 1, $\hat{z}_1 \geq z_1, \ldots, \hat{z}_{i-1} \geq z_{i-1}$ since $\hat{Z}$ is optimal in $X + x$, and $z_1 \geq \hat{z}_1, \ldots, z_{i-1} \geq \hat{z}_{i-1}$ since $\hat{Z}[1, i-1] \in \mathcal{I}$ in X in which Z is optimal . By Lemma 2 and $x \notin Z$, $x = \hat{Z}[i] > Z[i] = \max\{x' : x' \in Z[i, |Z|]\}$. From the matroid augmentation axiom (if $I_1 \in \mathcal{I}$, $I_2 \in \mathcal{I}$ and $|I_1| < |I_2|$, then $\exists e \in I_2 - I_1$ such that $I_1 \cup e \in \mathcal{I}$), since $Z \in \mathcal{I}$ in X and $|\hat{Z}[1, i]| < |Z|$, $\hat{Z}[1, i]$ can be augmented to an independent set $\hat{Z}'$ of size $|Z|$ with the elements in $Z - \hat{Z}[1, i] = Z[i, |Z|]$. $\hat{Z}' = Z - r + x$ where $r \in Z[i, |Z|]$, which is a contradiction since $R(x) \neq \emptyset$ and $r \in R(x)$ with $x > r$ ($x = \hat{Z}[i] > Z[i] \geq r$).
- $R(x) \neq \emptyset$ and $r < x$. We prove $Z'' = Z + x - r$ is optimal in $X + x$. Suppose Z'' is not and let $\hat{Z}$ be the optimal one in $X + x$. Let $x = Z''[i]$. Let $r = Z[j]$. There must be $i \leq j$ since otherwise, $r = Z[j] > Z[i] = Z''[i-1] > Z''[i] = x$, which is a contradiction.

- Prove $Z[1, i-1] = Z''[1, i-1] = \hat{Z}[1, i-1]$. Since there is only one element difference of Z and Z'' and $i \leq j$, $Z[1, i-1] = Z''[1, i-1]$. Suppose x is the kth element in $\hat{Z}$. If $k \geq i$, by Lemma 1, same as the proof in $R(x) = \emptyset$ case, $Z''[1, i-1] = \hat{Z}[1, i-1]$. If $k < i$, $\hat{Z}[k] = Z''[i] < Z''[k]$, which is a contradiction by Lemma 1.

- Prove $x = Z''[i] = \hat{Z}[i]$. By Lemma 2, we have $\hat{Z}[i] \geq Z''[i]$. If $\hat{Z}[i] > Z''[i]$, $\hat{Z}[i] \neq x$ and $\hat{Z}[i] \in X$, implying $\hat{Z}[1, i]$ is an independent set in X. Since Z is optimal in X, $Z[i] \geq \hat{Z}[i]$. If $i = j$, $Z''[i] = x > r = Z[i] \geq \hat{Z}[i]$, which is a contradiction by Lemma 2. If $i < j$, $Z[i] = Z''[i+1] < Z''[i]$, i.e., $\hat{Z}[i] < Z''[i]$, also a contradiction. Thus, $Z''[i] = \hat{Z}[i]$.

- Prove $Z[i, j-1] = Z''[i+1, j] = \hat{Z}[i+1, j]$. If $i = j$, it is trivially true. If $i < j$, it is straightforward that $Z[i, j-1] = Z''[i+1, j]$. By Lemma 1, $\hat{Z}[k] \geq Z''[k]$ for $k \in [i+1, j]$. If $\hat{Z}[k] = Z''[k]$ for all $k \in [i+1, j]$, then $Z''[i+1, j] = \hat{Z}[i+1, j]$. Otherwise, let k be the first integer in $[i+1, j]$ such that $\hat{Z}[k] > Z''[k]$. Similarly, $\hat{Z}' = \hat{Z}[1, i-1] \cup \hat{Z}[i+1, k]$ is an independent set in X with $\hat{Z}'[k] > Z[k]$, a contradiction by Lemma 2.

- Prove $r = Z[j] \notin \hat{Z}[j+1, |\hat{Z}|]$. Suppose $r \in \hat{Z}[j+1, |\hat{Z}|]$, which implies $\hat{Z}[1, j] \cup r \in \mathcal{I}$ in $X + x$, by matroid subset axiom (if $I \in \mathcal{I}$ and $I' \subseteq I$, $I' \in \mathcal{I}$). If $j + 1 = |\hat{Z}|$, then $\hat{Z}[j+1] = r$. We have $Z - \hat{Z} = Z[j+1]$ and $\hat{Z} - Z = x$, which implies $Z[j+1] \in R(x)$ and $Z[j+1] < r$, a contradiction. If $j + 1 < |\hat{Z}|$, i.e., $|\hat{Z}[1, j] \cup r| < |Z''|$, by matroid augmentation axiom, $\hat{Z}[1, j] \cup r$ can be augmented to an independent set $\hat{Z}'$ of size $|Z''|$ with the elements in $Z'' - \{\hat{Z}[1, j] \cup r\} = Z''[j+1, |Z''|]$, which implies $\exists k \in [j+1, |Z''|]$ such that $Z[k] = Z''[k] \notin \hat{Z}'$. Then, same as above, $Z'' - \hat{Z}' = Z''[k]$ and $\hat{Z}' - Z'' = r$, which implies $Z[k] \in R(x)$ and $Z[k] < Z[j] = r$, a contradiction. Hence, $\hat{Z}[1, j] \cup r \notin \mathcal{I}$ in $X + x$ and $r = Z[j] \notin \hat{Z}[j+1, |\hat{Z}|]$.

- Prove $Z[j+1, |Z|] = Z''[j+1, |Z|] = \hat{Z}[j+1, |Z|]$. It is straightforward that $Z[j+1, |Z|] = Z''[j+1, |Z|]$. Let's first prove $\hat{Z}[j+1] = Z''[j+1]$. By Lemma 2, $\hat{Z}[j+1] \geq Z''[j+1]$. Suppose $\hat{Z}[j+1] > Z''[j+1]$. By matroid augmentation axiom, since $Z[1, j] \in \mathcal{I}$ in $X + x$, $Z[1, j]$ can be augmented by one element from $\hat{Z}[1, j+1] - Z[1, j] = \{x, \hat{Z}[j+1]\}$. In the above, we have proved that $Z[1, j] \cup x = \hat{Z}[1, j] \cup r \notin \mathcal{I}$, which implies $\bar{Z} = Z[1, j] \cup \hat{Z}[j+1]$ is independent in $X + x$. Then we have $\bar{Z}[j+1] = \hat{Z}[j+1] > Z''[j+1] = Z[j+1]$, which is a contradiction by Lemma 2 since $\bar{Z}$ is also an independent set in X in which Z is optimal. Hence, $\hat{Z}[j+1] = Z''[j+1]$. Similarly, $Z''[j+2, |Z|] = \hat{Z}[j+2, |Z|]$.

From above, it implies $Z'' = \hat{Z}$, a contradiction. That concludes our proof.

□

In the x deletion operation, we first locate the *compensable set* $C(x) \subseteq X - Z$, i.e., $C(x) = \{x' \in X - Z : Z - x + x' \in \mathcal{I} \text{ and } S = X - x\}$. If $C(x) = \emptyset$, only x is deleted and $Z - x$ is the OIS in $X - x$. Otherwise, selecting the maximum vertex $c \in C(x)$, the vertex x is replaced by c in Z, and the resulting independent set is optimal. Due to space limitation, we omit the proof of the following theorem for x deletion and the descriptions of the other update operations, which can be maintained similarly.

Theorem 2. *Let $c \in C(x)$ be the maximum one. If $C(x) \neq \emptyset$, $Z' = Z - x + c$ is the optimal independent set in $X - x$; otherwise, $Z'' = Z - x$ is the optimal independent set in $X - x$.*

3.2 Dynamic Updates

The theorems in the last subsection reduce the problem of maintaining dynamic update for an LWCBG to that of a CBG. We implement the update operations for an LWCBG based on some modifications of Brodal et al.'s data structure developed for a CBG. In the x insertion operation, they maintains the *greedy matching* by replacing the vertex that has the largest end point in the replaceable set. In our algorithm, we use their structure to locate the replaceable set $R(x)$, select in it the minimum vertex r according to the weight-id order, and replace r with x. For the other updates, the modifications are similar. The computation for the replaceable set can be implemented in $O(\log^2 |V|)$ worst-case time via augmenting their structure with augmented BST data structures. Then the maintenance of the OIS takes $O(\log^3 |V|)$ amortized time per update. Due to space limitation, please refer to the details in the Appendix.

4 Dynamic Queries in Left Weighted CBG

In an LWCBG, right vertices are not associated with weights. Given the OIS, the selection of the set of the matched y will not change the weight of the matching. Thus, the dynamic pair query for an LWCBG can be maintained directly by using Brodal et al.'s algorithm in $O(\min\{k \log^2 |X| + \log |X|, |X| \log |X|\})$ worst-case time, where $k < \min\{|X|, |Y|\}$, or $O(\sqrt{|X|} \log^2 |X|)$ amortized time. In this section, we present an alternative approach to maintain the dynamic query operations in worst-case $O(k)$ time, where k is not greater than the cardinality of the MWM, based on the data structure we develop.

4.1 Disjoint Matching Interval

The dynamical update operations maintain the optimal independent set Z from which the corresponding matching M is computed according to Glover's greedy matching rule. Let Z and W be the left and right vertex set covered by M, respectively. We observe that M can be partitioned into components such that in each component the right vertices covered by M form an interval. We denote the vertex set covered by each component as the *Disjoint Matching Interval* (DMI in short). Formally, each DMI D_i contains a left vertex set $DX_i \subseteq Z$ and a right vertex set $DY_i \subseteq W$, where DY_i is an interval in Y. $|DX_i| = |DY_i|$ and $\forall x \in DX_i, \exists y \in DY_i, (x, y) \in M$, i.e., there is a bijection from DX_i to DY_i. The DMI structure may make some improvement of efficiency in the pair query operations. We compute DMIs from Z by the following static algorithm.

First, the vertices in Y is sorted in increasing order of $\text{ID}[y]$. For each $y \in Y$, we define $A(y) = \{x \in Z : \text{BEG}[x] = y\}$, and iterate each $x \in Z$ to initialize these sets. Then we scan Y to find the first y such that $A(y) \neq \emptyset$, and create a new DMI D_i by adding $A(y)$ into DX_i and y into DY_i. The DMI D_i is constructed by keeping iterating y with adding y into DY_i and adding $A(y)$ into DX_i if $A(y)$ is not empty, until the vertex y' such that $A(y') = \emptyset$ and $|DX_i| = |DY_i|$ before adding y'. We then proceed to find the next y such that $A(y) \neq \emptyset$ to create the next DMI, and so on, until all $x \in Z$ are added in DMIs.

Each DMI contains a DX set and a DY set. All DMIs constitute the set of DMI $\mathcal{D}$. The sequence of the operations maintained in $\mathcal{D}$ is as following:

- $minx(\textbf{DMI } D_i)$: Return the minimum x in DX_i.
- $maxx(\textbf{DMI } D_i)$: Return the maximum x in DX_i.
- $miny(\textbf{DMI } D_i)$: Return the minimum y in DY_i.
- $maxy(\textbf{DMI } D_i)$: Return the maximum y in DY_i.
- $find(\textbf{Z } x)$: Return the DMI D_i containing x, provided x is matched.
- $find(\textbf{W } y)$: Return the DMI D_i containing y, provided y is matched.
- $create(\textbf{Z } x, \textbf{W } y)$: Create a new DMI D_i in $\mathcal{D}$ with $DX_i = \{x\}$ and $DY_i = \{y\}$, provided x and y are not in any existing DMIs.
- $del(\textbf{DMI } D_i)$: Delete D_i from $\mathcal{D}$.
- $concatenate(\textbf{DMI } D_i, \textbf{DMI } D_j)$: Update D_i with $DX_i = DX_i || DX_j$ and $DY_i = DY_i || DY_j$, provided DY_i and DY_j are consecutive in Y. Delete D_j from $\mathcal{D}$.
- $split(\textbf{DMI } D_i, \textbf{W } y)$: Delete y and separate D_i into D_i and D_j, provided $y \in DY_i$. Let y is the kth element in DY_i. Update D_i with $DX_i = DX_i[1, k-1]$ and $DY_i = DY_i[1, k-1]$, and create D_j in $\mathcal{D}$ with $DX_j = DX_i[k+1, |DX_i|]$ and $DY_j = DY_i[k+1, |DY_i|]$. If y is the last in DY_i, D_j will not be created.

We now discuss the data structures to implement $\mathcal{D}$ and DMIs. First, for $\mathcal{D}$, we maintain an augmented red-black tree with a node for each DMI D_i; the key of each node is $miny(D_i)$. To represent a DMI D_i, we maintain two 2-3 trees [1], named the left tree and the right tree, while the left for DX_i and the right for DY_i. Each leaf in the left tree and the right tree corresponds to each $x \in DX_i$ and $y \in DY_i$, respectively. For each node v in the trees, we attach an integer "$size$" field to count the leaves in the subtree rooted at v.

The $minx, maxx, miny, maxy, find, create, del, concatenate$ operations can run in $O(\log |Y|)$ worst-case time by standard techniques. The $split$ operation needs to find the index of y, which is computed by calculating the sum of the size of all the left brothers of the nodes on the path from the root to y. The operation proceeds with the use of the standard split technique of 2-3 trees, achieving implementation in $O(\log |Y|)$ worst-case time.

It is the fact that Z and W can change only by one element per update [2][3]. By using the Union-Find structure [5], a vertex's matching status (matched or unmatched) can be obtained in constant worst-case time. To find the matching pair of a matched vertex, we first locate the DMI D_i containing the vertex. If the matching of D_i have not been computed or have been updated since the last pair query, it is computed by using the linear time CBG matching algorithm of Steiner and Yoemans [14] on the subgraph induced by $DX_i \cup DY_i$ in $O(|DX_i|)$ worst-case time, where $|DX_i| \leq |M|$.

4.2 Dynamic Pair Queries

Proceeding with the vertex insertion or deletion in Z or W, the DMI structure is maintained by the dynamic algorithm, presented in this subsection, in worst-case $O(\log k)$ time, where $k \leq |Z|$, by using implicit representation. Due to the space limitation, here we only discuss the case of $Z = Z - x$, and the other cases can be implemented similarly.

To maintain a dictionary for Y, we use a red-black tree TY with a node for each $y \in Y$. We can locate the predecessor ($pred(y)$) and successor ($succ(y)$) of y in $O(\log |Y|)$ worst-case time.

Delete x. Suppose x is in the DMI D_i. The deletion of x may split D_i into two DMIs. In D_i, define $a(j) = |A(y_j)|$ (recalling $A(y) = \{x \in DX : \text{BEG}[x] = y\}$), where y_j is the $j^{th} y$ in DY_i. Also define

$$Diff(j) = \sum_{k \in [miny(D_i), j]} a(k) - |\{y \in DY_i : miny(D_i) \leq y \leq y_j\}| \quad (1)$$

We maintain the invariant that $Diff(j) \geq 0$ for $j \in [miny(D_i), maxy(D_i)]$ and $Diff_{maxy(D_i)} = 0$. When deleting x, $\text{BEG}[x] = y_h$, $Diff(j)$ decrease one for all $j \in [h, k-1]$, where k is the split point, i.e., y_k is the first y with $k \geq h$ and $Diff(k) = 0$ before the deletion. If $k = maxy(D_i)$, D_i only shrinks one element from the last. The algorithm is depicted by the DMI operations as following:

1: $D_i = find(\text{BEG}[x])$
2: $j = findSplitPoint(D_i, x)$
3: $split(D_i, j)$

The $findSplitPoint$ operation may iterate all elements in DY_i and run in linear worst-case time. To meet the logarithmic time bound, we augment our data structures to implicitly represent the information. In the dictionary tree TY, we augment each node y with an integer "$numx$" field that is equal to $|A(y)|$, i.e., $a(j)$ if y is the jth in D_i. In the 2-3 tree for DY_i in the DMI D_i, we augment an integer "$diff$" field to each node (including leaves). Define $d(u, v) = \sum_{w \in P[u,v)} w.diff$, where v is an ancestor of u and $P[u, v)$ is the path from u to v excluding v. Also define $d(u) = d(u, root)$, where $root$ is the root node of the tree. $Diff(j) = d(j)$, i.e., $Diff(j)$ is represented by the path from the leaf to the root. And to each node u (excluding leaf) we augment an integer "min" field that is equal to $\min\{d(k, u) : k \in leaves(u)\}$, where $leaves(u)$ denotes the set of leaves contained in the subtree rooted at u. Thus, $\min\{Diff(k) : k \in leaves(u)\} = u.min + d(u)$.

There are two stages in the deletion of x with $\text{BEG}[x] = y_h$. First find the minimum k such that $k \geq h$ and $Diff(k) = 0$. Then split D_i at k, and decrease $Diff(j)$ for $j \in [h, k)$ by adjusting the associated $diff$ and min fields. In the first stage, we traverse the tree along the path $P(root, y_h]$ top-down while computing $d(u)$ for each node u on the path. Then from the leaf y_h we traverse back along the path $P[y_h, root)$ bottom-up to check each node u on the path, until there is a right brother v of u with $v.min + d(v) = 0$, where $d(v) = v.diff + d(u.parent)$. We select the first such right brother v if there is more than one and v is the root of the subtree containing the split point. Next we locate the first leaf $y_k \in leaves(v)$ such that $Diff(k) = 0$. We traverse the path from v to the leaf by selecting the first child v' with $v'.min + d(v') = 0$ in each node. The whole search operation traverses the path between the root and the leaf at most three times, which is in $O(\log |Y|)$ time.

In the second stage, after the split, D_i contains the leaves $[min_y(D_i), y_{k-1}]$. Let γ be the least common ancestor of the leaves y_h and y_{k-1} in the tree of D_i (γ can be located in logarithmic time). For each node u on the path from γ to y_h excluding γ, if u has right brothers, decrease $diff$ in each right brother node. In the end decrease $diff$ in y_h and decrease $numx$ of y_h in TY. Then we update the min field for each node u on the path $P[y_h, root)$ bottom-up by setting $u.min = \min\{v.min + v.diff\}$, where v is a child of u. The second stage traverses the paths top-down and then bottom-up in $O(\log |Y|)$ time.

Theorem 3. *The dynamic algorithm of Section 3 maintains the maximum weight matching in $O(\log^3 |V|)$ amortized time per update, and the dynamic algorithm of Section 4 obtains the matching status of a vertex in constant worst-case time and finds the pair of a matched vertex in worst-case $O(k)$ time, where k is not greater than the cardinality of a maximum weight matching.*

5   Concluding Remarks

Song et al. [12] pointed out that the Independent Domination Set problem is tractable for some tree convex bipartite graphs and developed a polynomial algorithm. We will investigate whether the technique of the present paper can be used to obtain the dynamic version of the problem.

References

1. Aho, A.V., Hopcroft, J.E., Ullman, J.D.: The Design and Analysis of Computer Algorithms. Addison-Wesley (1974)
2. Berge, C.: Two theorems in graph theory. Proceedings of the National Academy of Sciences of the United States of America 43(9), 842–844 (1957)
3. Brodal, G.S., Georgiadis, L., Hansen, K.A., Katriel, I.: Dynamic matchings in convex bipartite graphs. In: Kučera, L., Kučera, A. (eds.) MFCS 2007. LNCS, vol. 4708, pp. 406–417. Springer, Heidelberg (2007)

4. Dekel, E., Sahni, S.: A parallel matching algorithm for convex bipartite graphs and applications to scheduling. Journal of Parallel and Distributed Computing 1(2), 185–205 (1984)
5. Gabow, H.N., Tarjan, R.E.: A linear-time algorithm for a special case of disjoint set union. J. Comput. Syst. Sci. 30(2), 209–221 (1985)
6. Gale, D.: Optimal assignments in an ordered set: An application of matroid theory. Journal of Combinatorial Theory 4(2), 176–180 (1968)
7. Glover, F.: Maximum matching in a convex bipartite graph. Naval Research Logistics Quarterly 14(3), 313–316 (1967)
8. Katriel, I.: Matchings in node-weighted convex bipartite graphs. INFORMS Journal on Computing 20(2), 205–211 (2008)
9. Lipski Jr., W., Preparata, F.P.: Efficient algorithms for finding maximum matchings in convex bipartite graphs and related problems. Acta Inf. 15, 329–346 (1981)
10. Plaxton, C.G.: Fast scheduling of weighted unit jobs with release times and deadlines. In: Aceto, L., Damgård, I., Goldberg, L.A., Halldórsson, M.M., Ingólfsdóttir, A., Walukiewicz, I. (eds.) ICALP 2008, Part I. LNCS, vol. 5125, pp. 222–233. Springer, Heidelberg (2008)
11. Plaxton, C.G.: Vertex-weighted matching in two-directional orthogonal ray graphs. In: Cai, L., Cheng, S.-W., Lam, T.-W. (eds.) ISAAC 2013. LNCS, vol. 8283, pp. 524–534. Springer, Heidelberg (2013)
12. Song, Y., Liu, T., Xu, K.: Independent domination on tree convex bipartite graphs. In: Snoeyink, J., Lu, P., Su, K., Wang, L. (eds.) AAIM 2012 and FAW 2012. LNCS, vol. 7285, pp. 129–138. Springer, Heidelberg (2012)
13. Spencer, T.H., Mayr, E.W.: Node weighted matching. In: Paredaens, J. (ed.) ICALP 1984. LNCS, vol. 172, pp. 454–464. Springer, Heidelberg (1984)
14. Steiner, G., Yeomans, J.S.: A linear time algorithm for maximum matchings in convex, bipartite graphs. Computers and Mathematics with Applications 31(12), 91–96 (1996)
15. Tarjan, R.E.: Efficiency of a good but not linear set union algorithm. J. ACM 22(2), 215–225 (1975)

6 Appendix

In this section, we review the update operations part of Brodal et al.'s algorithm and then present our algorithm for dynamic updates in an LWCBG.

6.1 Brodal et al.'s Dynamic Updates in CBG

Brodal et al.'s algorithm [3] is based on the binary computation tree data structure developed by Dekel and Sahni [4]. Let $S = \{s_1, \ldots, s_k\} = \{\text{BEG}[x] : \forall x \in X\}$. Let $s_{k+1} = \max(y) + 1$. Thus Y is partitioned into k intervals $[s_i, s_{i+1} - 1]$, $i \leq k$. In the tree, the leaf s_i corresponds to the right vertex set $Y(s_i) = [s_i \ldots s_{i+1} - 1]$ and the left vertex set $X(s_i) = \{x \in X : \text{BEG}[x] = s_i\}$. In an internal node P, $X(P)$ and $Y(P)$ are respectively the union of the corresponding sets in its children. The root node corresponds to the universal X and Y sets.

Let $s(P)$ and $t(P)$ be the smallest and largest y in $Y(P)$, respectively. Each node P associates with several auxiliary subsets of $X(P)$, which are $M(P)$, $T(P)$ and $I(P)$. $M(P)$ covers Glover's matching of the subgraph induced by $X(P)$ and $Y(P)$, $T(P) = \{x \in X(P) \backslash M(P) : \text{END}[x] > t(P)\}$, and $I(P)$ contains the remainder of $X(P)$. P also associates with an *equal-start structure*, which is a subgraph G' induced by $X'(P)$ and $Y(P)$. $X'(P)$ is $M(P)$ if P is a leaf; if P is internal $X'(P)$ is $T(L) \cup M(R)$, where L and R are respectively P's left and right child. It is safe to assume that for every $x \in X'(P)$ $\text{BEG}[x] = s(R)$ by Theorem 2.1 in [4]. Thus in the *equal-start structure* of P, the matched x set $M'(P)$ can be computed efficiently, as well as its two other auxiliary sets $T'(P)$ and $I'(P)$. Then $M(P) = M(L) \cup M'(P)$, $T(P) = T(R) \cup T'(R)$ and $I(P) = I(L) \cup I(R) \cup I'(P)$.

To maintain the dynamic updates in the *equal-start structure* of P, $Y(P)$ is translated into $[1 \ldots m]$ and $X'(P)$ is translated accordingly. Then each $y \in Y(P)$ is an integer $i \in [1, m]$ and each $x \in X'(P)$ is represented as a pair $(s(x), t(x))$. For each i, define a_i as the number of matched $x \in X'(P)$ with $t(x) \le i$. Suppose x' is to be inserted. Let j be the smallest integer such that $j \ge t(x')$ and $a_j = j$. If such a j does not exist, x' is added into $M'(P)$. Otherwise, the replaceable set can be proved to be $\{x \in X'(P) : t(x) \le j\}$, in which the vertex with the largest $\text{END}[x]$ is replaced by x' in $M(P)$ and moved into $T(P)$ or $I(P)$.

6.2   Dynamic Updates in LWCBG

In the process of inserting x, it is inserted in each node on the path from the leaf corresponding to x to the root of the tree. In a node P, if $M(P)$ can be augmented, x is added in it; otherwise, some $x' \in M(P)$ is replaced by x.

If P is a leaf, only the computation in the *equal-start structure* of P is involved. Let j be the smallest integer such that $j \ge t(x)$ and $a_j = j$; if such a j does not exist, let $j = m + 1$. If $j = m + 1$, $M(P)$ is augmented by adding x. If $j = m$ and $\text{End}[x'] > t(P)$, where x' has the largest $\text{End}[x]$ in $M(P)$, x' is replaced by x in $M(P)$ and moved into $T(P)$. Otherwise, the minimum $x' \in R(x)$ is replaced by x if $x' < x$, where $R(x) = \{x \in M(P) : t(x) \le j\}$.

If P is an internal node, the computation may involve both P and its children. If $x \in I(L)$ or $x \in I(R)$, it implies the insertion of x fails and the process terminates. Otherwise, suppose x was previously inserted into the left child L. If $M(L)$ was augmented, $M(P)$ is augmented by adding x. If $I(L)$ was augmented by x', x' is replaced by x in $M(P)$. Otherwise, $T(L)$ was augmented by x', x' is added into $X'(P)$, and the *equal-start structure* of P is updated. Let j be defined similarly as in a leaf. Let j' be the largest integer such that $j' < s(x)$ and $a_{j'} = j'$; if such a j' does not exist, let $j' = 0$. If j is equal to $m + 1$ or m, the situations are similar as that in a leaf. Otherwise, $j < m$ and the minimum $x' \in R(x)$ will be replaced. $R(x)$ may contain some subset $R'(x)$ of $M'(P)$ and some subset $R''(x)$ of $M(L)$. Let $R'(x) = \{x \in M'(P) : j' < t(x) \le j\}$. Let $\Delta = T(L) \cap R'(x)$. If Δ is empty, $R(x) = R'(x)$. If not, $R'(x) = \{x \in M'(P) : t(x) \le j\}$ and $R''(x)$ is computed by traversing the path from P to the very left leaf in its subtree until the node with empty Δ. The computation in each node on the path is similar as that in P. By traversing back to P, the minimum $r \in R'(x) \cup R''(x)$ is selected with updating the sets in the nodes on the path containing r. Then r is replaced by x in $M(P)$ and moved into $I(P)$. On the other hand, suppose x was previously inserted into the right child R. If the replaced x' in R is in $M(P)$, then it is still replaced by x in $M(P)$. Otherwise, x is added into $X'(P)$ and the computation is similar as that of the left child.

The computation for the replaceable set, including the Δ computation and the minimum selection, can be implemented by augmenting Brodal et al.'s structure with augmented BSTs. It runs in $O(\log^2 |V|)$ worst-case time since path traversals are involved, which is different from their algorithm. Then the dynamic update operations are maintained in $O(\log^3 |V|)$ amortized time per update.

Author Index